BEYOND THE
STARS

Our Origins and the Search for
Life in The Universe

D1369636

BEYOND THE
STARS

Our Origins and the Search for
Life in The Universe

Paolo Saraceno

Institute for Astrophysics and Space Planetology INAF, Italy

Translated by

David Goodstein

California Institute of Technology, USA

McConnell Library Radford University

WITHDRAWN

 **World Scientific**

NEW JERSEY · LONDON · SINGAPORE · BEIJING · SHANGHAI · HONG KONG · TAIPEI · CHENNAI

Published by

World Scientific Publishing Co. Pte. Ltd.

5 Toh Tuck Link, Singapore 596224

USA office: 27 Warren Street, Suite 401-402, Hackensack, NJ 07601

UK office: 57 Shelton Street, Covent Garden, London WC2H 9HE

British Library Cataloguing-in-Publication Data
A catalogue record for this book is available from the British Library.

Cover image courtesy of Heikenwaelder Hugo © 1998.

BEYOND THE STARS
Our Origins and the Search for Life in the Universe

Copyright © 2012 by World Scientific Publishing Co. Pte. Ltd.

All rights reserved. This book, or parts thereof, may not be reproduced in any form or by any means, electronic or mechanical, including photocopying, recording or any information storage and retrieval system now known or to be invented, without written permission from the Publisher.

For photocopying of material in this volume, please pay a copying fee through the Copyright Clearance Center, Inc., 222 Rosewood Drive, Danvers, MA 01923, USA. In this case permission to photocopy is not required from the publisher.

ISBN 978-981-4295-53-6
ISBN 978-981-4295-54-3 (pbk)

Typeset by Stallion Press
Email: enquiries@stallionpress.com

Printed in Singapore by Mainland Press Pte Ltd.

Acknowledgements

This book has been possible thanks to the numerous contributions I had from colleagues and friends. Without their suggestions, the book would probably not have been possible.

For the English version, I have to thank Prof. Peter Clegg of the Queen Mary College of London for the several suggestions he gave in the revision of the first part of the book and Erina Pizzi, for the fantastic work she provided for the figures. I also thank Howard Smith of the Harvard-Smithsonian Center for Astrophysics for his suggestion and Mario Cavagnaro for his careful reading of the final manuscript.

I also thank the colleagues expert in the different fields discussed in the book, they made possible the first version of the text among them I have to remember Prof Patrizia Dall'Olio of the M. L. King Secondary School, Masate Milan, Prof Alberto Franceschini of the University of Padua, Prof Paolo De Bernardis of the University of Rome, Prof Bruno Carli of the Istituto di Fisica applicata of CNR Florence, Dott. Giovanni Valsecchi of the Institute of Planetology and Space Physics of Rome and Prof Olga Rickards of the Institute of Molecular Anthropology of the University of Rome.

I am also indebted to the many friends who, with their enthusiasm, comments and discussions push me to continue this long work, among them I remember Anna Orcese, Renato Orfei, Luigi Spinoglio and Michele Pestalozzi.

Paolo Saraceno
Institute for Astrophysics
and Space Planetology
INAF, Italy

Preface

About a year ago, a close friend of mine, Professor Franco Scaramuzzi, gave me as a present a book, *Il Caso Terra* by Paolo Saraceno. I put the book aside for future reading but soon had forgotten all about it until a few months later. I came across the book again, and began reading it in earnest.

I quickly realized this was a genuinely important book, about the Universe and had been written with lay readers in mind. It laid out clearly the mechanisms that underlay all of the many features it dealt with. After reading book halfway, I wrote to Franco to say the book deserved to be translated into English. Franco quickly put me in touch with Paolo, who had begun his career in physics as Franco's student.

I didn't know Paolo, but I wrote to him saying his book was important and deserved to be translated into English. Paolo wrote back suggesting that I do the translation! This suggestion came as a considerable shock to me, given my weak grasp of the Italian language, but I took the challenge and translated the first chapter. I found the writing so lucid I was able to translate it easily and translate it I did. I spent about an hour a day at it for the next six months, and now you will be able to read the result.

In the meantime, Paolo and I had become good friends. Paolo has now a retired as an astrophysicist at the (first) University of Rome and he, Franco and I, together with our wives, had spent happy evenings together talking about all sorts of things at Paolo's home in Rome.

Any awkwardness or difficulty in understanding the English text is my fault alone. I did my best to be true to Paolo's elegant Italian, and this, sometimes, caused problems. But I am hopeful I had succeeded in rendering into English this beautiful book Paolo had written, for everyone to enjoy.

David Goodstein
Rome, 2009

Contents

Part 1

Origins

What is there beyond everything?
Have you ever stayed out late, just to look at the stars?
Long enough to make your head spin
Not because you're bent back,
But because you looked so far away

The more the night is dark, the further you can see in the heavenly sky
Have you ever thought what's behind the stars?
Other stars of course, but what's behind them?

Jostein Gaarder

Allesia, 11 years old, a primary school student in Perugia, continues in such a way:

There will be some angels,
Or some meteors
From the smallest galaxy
There will be Jesus, there will be paradise
There will be very small stars

There will be our desires and
Our dear animals who are dead,
There will be my Grandfather and many other people
There will be my past and my future.

Our Origins

Chapter 1

1.1 The Ancient Questions

On a moonless night we might find ourselves stretched out on a deserted beach or high up on a mountain, far away from the lights that have diminished our view of the sky. If the sky is clear and there are no obstructions on the horizon, the spectacle we see is truly extraordinary: surrounded by myriads of stars we are immersed in an enchanted world. Above us the great luminous stripe of the Milky Way sweeps across the sky from one side to the other. We recognize, depending on the season, the principal constellations: the great W of Cassiopeia, the sword of Orion high in the winter sky, the Scorpion low in the summer sky, the Swan and the Pleiades — constellations with ancient Greek and Arabic names, invented by people who, in contrast to us while gazing at the sky, were fascinated by such beauty, and created stories and myths to explain the origins of stars. Wherever we look, there are luminous points: the brightest few are planets; the others, the majority, are stars. Farthest away, not quite visible to the naked eye, are galaxies which, like ours (the Milky Way), consist of hundreds of billions of stars. Then if we have a telescope, and point it to an apparently empty zone of the sky, we discover new luminous points, stars still fainter and farther away. Everywhere we look we see a great show that is only an infinitesimal part of our universe.

At such moments, it is difficult not to ask ourselves, "What is the origin of all that we see, of the stars, the planets, and of our own life?" It is difficult not to ask ourselves whether we have our roots in the universe, or if there might exist other forms of life around stars. These are questions that humans have always asked, and for which they have offered answers drawn more from the realm of myth and fantasy than from science. Today, however, for the first time in human history, one can try to give some answers to the problems of our origins, thanks to the extraordinary progress made during the last century in the fields of physics, chemistry, astronomy, geology and biology. The techniques used are those of science. Perhaps they are cold, but the answers have the same fascination as did the ancient myths.

The conviction that we could answer these questions is so strong that, in the 1990s, American scientists decided to give the name "Origins" to a series of investigations, space missions and large instruments on the Earth devoted to answering two simple questions:

— Where do we come from? What is the origin of the universe, of the stars, planets, of the elements that constitute our world and, finally, of life? How did life begin on the Earth?
— Are we alone in the universe? Are there observations that will enable us to determine if forms of life similar to ours exist on distant planets?

The choice of the name "Origins" and of these two themes is an attempt to convince the greater American public of the importance of continuing this research and, above all, that it is worth the investment of the enormous financial resources required. Consider, for example, the Hubble Space Telescope (Figure 1.1), an orbiting observatory built by the United States with contributions from Europe, some of whose images we shall use in this book. The cost of the instrument and of managing the mission was more than 4 billion Euros, an amount similar to the estimated cost of the bridge over the Straits of Messina, just for this single space mission, although admittedly one of the costliest.

Aside from political ends, there is, among the scientists who proposed the Origins program, the conviction that something extraordinary had happened. Science had begun responding to ancient questions like those of the origins of the universe and of man, and to understand the incredible sequence of events that had made possible the emergence of an extraordinary planet like ours on which life has been able to reproduce and to evolve.

The awareness of what was going on, brought the past Director of the Origins Program of NASA to write : "... *When the answers to these questions are known, our civilizations will evolve and we will have new visions of who we are and what our future might be. We have already learned enough to appreciate that the universe is enormous and ancient, but life — tiny and transient — is its precious jewel.*"

In 2004, the European Space Agency (ESA) began a program similar to that of NASA, called Cosmic Visions (Section 12.1). The questions posed by European researchers are similar to those of the Origins program, and the need to find answers has by now involved scientists of all nations.

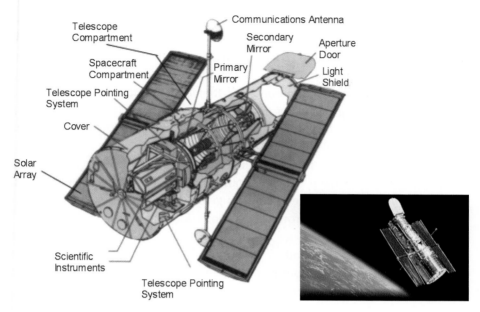

Figure 1.1: *The Hubble Space Telescope (HST) from which several of the images reported in this book were taken. On the bottom right is a picture of HST orbiting around the Earth, on the left is a schematic picture of the satellite. Despite the "relatively small mirror" (2.4 m), the quality of the optics and the fact that it operates outside the atmosphere made HST the most sensitive optical instrument ever built. Launched in 1990, HST, after the last maintenances of 2008 and 2009, should function until 2004, with no other intervention. Then HST will fall to Earth like all satellites inside a geosynchronous orbit (Section 10.1). (© STScI & NASA.)*

1.2   You Can't Answer Everything

By now, many results are known. We know, for example that the universe was born 13.7 billion years ago in a great explosion, the Big Bang, and that since then it has continued to expand and cool proportionately.[1] We know, therefore, that everything we can observe has a beginning in time and that at this beginning the universe was very hot and confined in a microscopic space.

One could ask what was there before the Big Bang, but this is one of those questions for which science has no answer, at least not today. Galileo taught us that science must be based on facts, on measurements, and what cannot be measured does not belong in the kingdom of science.[2] Since what occurred before the Big Bang is not observable, science cannot say anything about it. One can form theories and hypotheses, but they will exist only in the world of

---

1 *This is easy to understand: when a system expands, the energy it contains is distributed in an ever larger volume so it is diluted and diluting the energy means cooling or lowering the temperature.*

2 *In effect, the need of a greater rigor in science and to base it on measurement (also to separate science from magic) had been felt by Leonardo da Vinci a century before Galileo. In the "Trattato della Pittura", he described science like this: "No human investigation can be called science unless it has mathematical demonstrations; and if you say that science that begins and ends in the mind is true, this is not conceded and is negated for many reasons, first among them that in mental discourse there is no measurement, without which nothing can be known with certainty".*

fantasy. Today, to ask what existed before the Big Bang is like trying to reconstruct the path someone has taken in the desert from the footprints he has left in the sand: it is possible to go back in time, but not forever — only to the last sandstorm. Traces of what happened before the last sandstorm no longer exist, having been erased by the wind and lost forever. The sandstorm of the universe was the Big Bang. This does not mean that nothing existed before the Big Bang, only that science has nothing to say about it, at least with the techniques available today.

We must therefore admit that science cannot answer every question. Its domain is limited to what is observable and measurable, to measurements that must be repeatable by anyone yet always giving the same result. What is not observable or not deducible from observations does not belong to the world of science. That is the price science has to pay in order to construct its concepts on solid and irrefutable grounds.

The philosopher Karl Popper proposed a test to determine if a given argument can be qualified as science. According to this criterion, for every argument pertaining to science it must be possible to conceive a measurement that could falsify the proposed thesis, such that a negative result of the measurement would mean the thesis is mistaken.

The principle cannot be stated in a positive way by saying "to search for a measurement whose result will prove the thesis" because you cannot exclude *a priori* that there will be errors in the measurement or in the analysis, or that future measurements might lead to a different thesis. As an example, imagine a fisherman scientist who uses a net with a spacing of 5 cm. If he does not take into account that smaller fishes will escape through the holes (whose dimensions are the limits of the instrument), when he examines the catch he will conclude that there are no fish in nature smaller than 5 cm.

History has provided many examples of such errors: the Greeks believed that small flies and worms were born of the Earth by the transformation of inanimate matter into small animals, a theory that remained in force until 1800 when the microscopic eggs from which the flies were born were discovered (Section 5.1). Even the conviction that the Earth was at the center of the universe was the result of measurements, which were the most accurate that could be made with the available instruments (Section 3.1). In more recent times many measurements have led to conclusions that have later been found to be only an approximation of the reality. (An example is classical mechanics, which is an approximation to the theory of relativity and to quantum theory.) This will certainly recur in the future.

If we accept Popper's criterion, we find that there are some subjects that will never be part of science because they are not overturned when a predicted result turns out to be false. One of these is astrology. No astrologer, when faced with a horoscope that does not work, will admit that the theory on which it is based is wrong, as would a physicist when faced with a measurement that contradicts a theory. The universal law of gravitation would be disproven if a single planet were found not to be obeying it. Therefore he who believes in a horoscope does so, following his personal conviction; this is acceptable and possibly helps him feel better, but he does not have the certainty of a scientific result. He must always remember that it has never been proven that people born in the same period are destined to similar lives (a fundamental assumption on which astrology is based). If it were to be proven, astrology would be a science....

Other subjects, such as the study of the universe before the Big Bang, do not pass Popper's test because no one has devised observations capable of testing them; if such observations become possible some day, these topics will rightfully join the field of science (Section 2.11).

That does not mean that questions which science cannot falsify by measurement are meaningless, or that those who pose such questions are not motivated by sensible and fundamental aims. In such cases, one does not discuss the validity of the question, but rather the possibility that science can provide an answer. For those interested in going further into this argument, we suggest a book by the physicist Bridgman,[3] *The Logic of Modern Physics* written in 1927, in which is listed a series of questions that science could not answer, such as: Why does a positive charge attract a negative charge? Why does nature have laws? Can time have a beginning and an end? To these questions we might add (at least for now) the one we spoke a while ago: what was there before the Big Bang?

Nothing prevents metaphysics or religion from seeking answers to these questions, but their answers will be neither absolute nor unique. For this reason they will not be part of science. Only when we have this concept clearly in mind can we venture into the slippery domain of the origin of the universe, of stars, of life and of the existence of life far from Earth, an area in which it is easy to fall into fantasy.

1.3   The Importance of Doubt

Whilst the solidity of science is based on measurement, the progress in science is based on intuition, on the verification of hypotheses, on the generalization of the results, and on the capacity to construct theories and devise new measurements. We are speaking of very important mental tasks that show the human genius, that no computer can accomplish, not even the most powerful ones.

As long as an idea is not confirmed by a measurement, every doubt about it is not only legitimate but fundamental, because only doubt reduces the possibility of error and permits to weigh all possible cases. The scientist accepts nothing by faith and this distinguishes science from fundamentalism, be it political or religious, which has some preconceived idea as the departure point for its doctrine.

This attitude often makes scientists very contentious (because they doubt more about the theories of others). Democritus, for example, held the corpuscular theory of light against the opinion of the other Greek philosophers who saw light as a searchlight exploring its surroundings (Section A 1.1). There was the historic dispute between Hooke and Huygens on one side on the undulatory nature of light against the corpuscular model of Newton, a dispute that continued until the nineteenth century when a measurement settled the matter (almost...). In more recent times, there was controversy between those who believed in a steady state universe, like Fred Hoyle, and those who believed in the Big Bang, like George Gamow (Section 2.1). Still more recently there was the polemic between Montagnier and Gallo on the pathogenic agent that causes AIDS.

These are useful disputes that help the progress of science, leading to further measurements that resolve the issue. Such disputes are not always understood by non-scientists who want science to provide clear answers to everything and often prefer the illusory certainty of those who supply answers based on criteria that have nothing to do with the scientific method.

The doubt that accompanies every new theory is both the strength and the weakness of science: strength because it permits us to evaluate all possible solutions and new experiments that can prove what they claim; weakness because when science encounters fundamentalist positions (religious, political, or a combination of both) it comes out loser, at least for the moment. The fundamentalists, unlike the scientists, have no doubts, believe through faith, and offer a message that is clear and easy; and they have success because to believe is easier than to doubt.

Science, based on the comparison of ideas, is one of the highest manifestations of democracy. No one is put into advantage by its procedures: even if a wrong result is imposed by "political" reasons, it will be soon overturned by the scientific method: the past century has shown that science develops better in the countries where there are higher levels of democracy and that, where science has developed, it has brought welfare and wealth. The richest countries, in fact, are not those with the most natural resources, but rather those that have good universities and research institutes, and where ideas are circulating freely. Instead we see that where ideologies, with their certainties, succeed in imposing themselves on scientific thought, the final result has never been a contribution to humanity. They oppose an idea, not because someone has disproved it, but because it is inopportune, in conflict with the political or religious ideas of the moment. When people opposing to an idea have power, it often happens that they eliminate those who sustain it together with the idea itself. This was done in the past by the inquisition against those who sustained the ideas of Copernicus.

1.4   Are Science and Religion Compatible?

Sometimes it is claimed that science and religion are incompatible. That vision is not shared by, and is contradicted by, numerous scientists (among them Giordano Bruno, Galileo, Newton and Teilhard de Chardin). They did not see any contradiction between their religious sentiments and the work they were doing, even though, they had to pay a heavy price to maintain their ideas.

Science and religion address very different topics, and so they should never be in conflict. Religion is concerned with the transcendent, which is not observable by scientists, and gives rules of behavior (ethics) that cannot be in conflict with the results of measurements. Science, instead, is not concerned with ethics (can it be immoral that the Earth rotates about the Sun, or that man descended from the apes?) but only what is measurable. The utilization of scientific results can be a question of ethics, as it is with all human activities, but this cannot justify to contradict the result of a measurement.

St Francis, one of the most important interpreters of Christianity, would not have had any difficulty accepting Copernicanism or the modern theory of evolution. He recognized (just like Galileo, Newton and many other scientists) the work of God in all that is created and he accepted and loved it, calling the Sun, the Moon and every living he encountered brothers and sisters.

Michael Heller, a priest and mathematician, winner of the 2008 Templeton Prize for his study about the origin of the universe, when interviewed by the magazine *Physics World* on a possible conflict between being a scientist and a believer, said that he did not see any conflict because it is obvious that science reads the mind of God. This is a phrase that underlines the absurdity of the conflict that in the past and still today stands between science and faith.

The conflict between science and religion often occurs when religion abandons the transcendent, for reasons more political than spiritual, to impose on science ideas that conflict with observation. In this way religion becomes ideology, loses its force, and like all ideologies fears the freedom of thought. In such cases, religion is the first to be discredited because it is difficult to believe in anything transcendent which cannot be proved if, out of your belief, something that is in contradiction with the measurement is derived.

It was a risk clear to Galileo, who tried to convince the church to separate scientific problems from questions of faith so as to avoid the risk that faith might be contradicted by a measurement.[3]

---

3 Galileo, in his Works VII (541), writes: *"Tell the theologians that by wanting to make into matters of faith that Earth is standing and the Sun moves, they are in danger of being condemned of heresy when it will be definitively demonstrated that the Sun is standing and the Earth moves."* (Galileo was not completely certain of the Copernican theory because the measurements, possible at the time, could not show why the relative position of the fixed stars remained fixed if the Earth was moving (Section 3.1).

History is abundant with cases in which science and religion have been used for political purposes; a curious example is reported by Bertrand Russell in his book "The Wisdom of the West" in which he tells of an assassination perpetrated by the Pythagoreans to prevent the spreading of the discovery of irrational numbers for fear that it would menace their power.[4] In the fifth century AD, the Pythagoreans reigned at Taranto, legitimizing their power with the knowledge of numbers and geometry, raised to a religion. They discovered the irrational numbers (in the Pythagorean theorem a right triangle with two sides equal to 1 has a hypotenuse equal to the square root of 2 which is an irrational number) and decided to keep it hidden in order not to lose their power. Because of an internal struggle, one of them decided to reveal this discovery. He crossed the country telling everybody that the numbers were not perfect as the Pythagoreans were saying. They hunted him down and killed him.

The same fear (of losing power because of an objective fact) brought Giordano Bruno to the stake and Galileo to face abjuration and an early death. Today, the same reasons induce certain movements to oppose the idea of evolution of species, showing the fragility of human nature which cannot accept the world as it is.

In conclusion, the enemy of science is fundamentalism in all its forms. It fears knowledge and hides facts because it knows that an informed person can understand problems and make the right choices for his own welfare, choices that are not always in the interest of those in power. The methods of science may appear uncertain or unpleasing, but we must recognize that there is no better way of arriving at knowledge. Not accepting this method puts us at the mercy of those who will use our fears for their own interests.

1.5 Life in the Universe

Why take up the study of the origin of the world around us and the search for life in the universe, areas in which it is easy to fall into fantasy, having just invoked the scientific method and Galilean rigor? The reason is simple: the progress of space technology, in particular that of infrared astronomy, has rendered possible to devise space missions to search for forms of life similar to our own, based on the chemistry of carbon[5] (Section 5.9), on planets outside our Solar System. It is thought that this search could be carried out on a few hundred nearby stars that should have planetary systems similar to ours. It would be a sufficiently large sample to give statistical significance to the

---

4 *The irrational numbers are those that cannot be represented as a ratio of two integer; they are constituted of an infinite sequence of decimals; an example is the square root of 2 = 1.4142136.... From the practical viewpoint of the Greeks such numbers are nonsensical (thus their name) because they thought that all numbers had to represent things or fractions of things. They did not believe in numbers that were not reducible to one of these two categories.*

5 *These compounds form organic chemistry which is at the basis of all life.*

answer. In other words, it is possible today to search for the existence of life on a planet outside the Solar System, or at least it is possible to conceive an observation that would answer this question.

But how does one observe life? In risking losing readers who are in a hurry and would be satisfied by a simple answer, we anticipate what we will say in the last chapters in honor of Galileo: *the presence of life on a planet can be revealed by the presence of oxygen in its atmosphere.*

Oxygen oxidizes, that is, combines chemically with everything it finds, so that if it is not continually produced, it will quickly disappear in the atomic form. Huge quantities of oxygen in gaseous state, such as those that one finds in the atmosphere of the Earth, can only exist in the presence of a process like photosynthesis that produces it continuously. Since photosynthesis indicates the existence of living organisms, the presence of oxygen in the gaseous state would indicate the presence of life, perhaps primordial, but a life capable of evolving into something more complex. Without photosynthesis the oxygen would quickly disappear from the atmosphere of our Earth. As we will see, we are speaking of a delicate equilibrium that can easily be destroyed by humans.

This book is divided into three parts. In the first we will talk of the "Origins", of the roots our existence has in the cosmos. We will recount what is known about the origin of the universe, of the stars, of the planets, of the atoms of which we are composed and of life. We will see how the stars and the planets were born (Chapter 3), we will discover that we exist only because, the atoms of which we are made were and are formed inside stars (Chapter 4), atoms that did not exist when the universe was born. We will see how these rare atoms increased their abundance with the passage of time, increasing their concentration on a planet like the Earth, by means of a process that led to the birth of the Solar System, then through the vegetable species that absorbed them with their roots and leaves, and finally in the animals that fed on the vegetables.

We will only marginally consider the question of the origin of life, about which there is still much to be investigated. Just consider that no chemist, even in a controlled environment like that of a laboratory, has yet succeeded in bringing to life any of the primordial organism (bacteria) like those which inhabited our planet since the beginning. This is a process that nature managed with great ease: life was born as soon as the planet cooled (Section 6.3) in an atmosphere without oxygen (because as we have seen, it was life that created oxygen and not vice versa) and saturated with ammonia, methane and carbon dioxide, in extremely hostile conditions, even impossible for the existence of most of the bacteria existing today and for all species of animals and vegetables we know.

In the second part of the book, we will study in detail "the case of the Earth". We will tell the story of our planet from its origins up to today (Chapter 6) and we will tell finally of the role played by extinctions in the evolution of life and the risks that we run for its future, because of the degradation of the environment that we are causing (Chapter 7). We will discuss what is needed for a planet to be habitable, and to be able to develop biological species similar to those we know (Chapter 8). We will discover that the greenhouse effect is not always bad, and that without volcanoes and earthquakes dry land would not exist (Chapter 9) and that the Earth would have been a frozen world, with a tenuous atmosphere, like that of Mars. We will tell finally of the importance of the position of the Earth in our galaxy and our Solar System, and of the risks represented by the fall of meteorites (Chapter 10).

The third part of the book will be dedicated to the attempts of the last decades to get in touch with extraterrestrial intelligence (Chapter 11) and we will talk about how, in years to come, searches will be undertaken for "inhabitable" planets on which our species might live (Chapter 12). The search techniques will take advantage of the measurement of oxygen in their atmospheres, an extremely difficult measurement that requires very sensitive and sophisticated instrumentation in order not to be dazzled by the light of the nearby star.[6] This measurement is considered possible today, and is in the programs of America's National Aeronautics and Space Administration (NASA) and of the European Space Agency (ESA); such missions are likely to be undertaken around 2025.

---

6 *Given their large distance from us, the planet and the star appear very close to one another. One of the principal difficulties is in separating the weak light from the planet from the intense light of the star.*

The Beginning
of Everything

<u>Chapter 2</u>

2.1   The Big Bang

The question of the beginning of the universe is an old one, addressed by every
culture and religion since the dawn of time. It is a question every philosophy
poses, and is perhaps the very reason for the existence of philosophy. The
interpretation that philosophers have given to the origin of the world has oscil-
lated between two extremes: an immanent vision, in which the principle of the
universe is found in the universe itself — and for this reason it has always
existed and will always exist — and a creationist one, according to which the
world is the creation of a demiurge who is outside the universe and creates it;
the world has therefore a beginning and an end.

   Greek philosophy was permeated by the first view and rejected the possibility that the world had a beginning and an end. It maintained that what is eternal has a "superior perfection" to that which is created. Because the universe is
the place of the maximum conceivable laws, to which even the gods were
subject (even the gods cannot avoid their "destiny"), the Greek philosophers
deduced that it must be perfect and therefore eternal.

   In this universe, which was above everything, they recognized an immanent
and eternal physical principle from which everything else was derived. For
Thales, the first of the philosophers of the school of Miletus, the principle was
water; everything began with water and ended with water, because he observed
that everything which nourishes plants and animals is wet. In a certain sense,
he was a precursor to the idea of the necessity of water for life (Section 6.3).

For Heraclitus, the principle was fire, varying only in its form: fire produces gases, which precipitate in water, leaving behind a residue when the water evaporates — the solids. Gas, liquid and solid therefore represent the different forms of fire. For Anaximenes, the principle was air: in the processes of rarefaction and condensation he saw the manifestation of the synthesis of opposites. Analogous theories were proposed by all the philosophers: for Empedocles the principles were the four elements, for Democritus the atoms, for Pythagoras the mathematical order. The Greek philosophers studied the world like real scientists and were struck by physical, chemical and biological phenomena and, from them, they derived their first principles. They had a concrete perception of the world that Aristotle summed up with the phrase "nothing can arise out of nothing." For the Greeks, *if the universe did exist then it had always existed* and could never end — if you cannot go from nothing to something, the opposite is also true: *it is impossible to go from something that exists to nothing*. If the universe has always existed, it will always exist; time has no beginning and no end. The Big Bang, as the origin of time, is not in accordance with Greek thought.

Saint Augustine and the Christian theologians had a different idea. For them God exists outside of space and time, which exist only because He created them. The question of what God was doing before the creation of the world was nonsensical for Augustine, because time had not "yet" been created, and there was no "before." This reasoning did not please everyone, and every time Saint Augustine arrived at the end of his lessons there was someone asking? "What was God doing before the creation?", unable to accept the idea that there had been an instant without a "before." The legend says that Saint Augustine, exasperated by these repeated questions showing that people had not understood his thought, once responded that God was busy creating an inferno big enough to contain all those who were asking such stupid questions.

What did not please the disciples of Saint Augustine, and is still displeasing today, is the nothingness implicit in the idea that time has a beginning. We have seen, however, that there is a big difference between the vision of Saint Augustine and that of science. For Saint Augustine, the "before" did not exist because time had not been created; for science (Section 1.2), the question about what existed "before" the Big Bang makes no sense because currently there is no way of making a measurement to observe it (although we will see in Section 2.11 that such an observation may be devised one day). The example of the tracks in the sand (Section 1.2) reflects the state of things.

Even if the Big Bang theory does not negate the existence of a universe "before," saying only that it is not observable, the creationist aspects present in this theory are strong. To these we have to add the absurdity of an "initial singularity." If the universe is expanding in time, then going backward in time it will occupy an ever-smaller volume until it becomes infinitesimally tiny, a difficult concept to accept. We therefore understand why talented scientists sought to demonstrate that the theory was a mistake. The very notion of an origin, an instant from which time itself began, was too creationist for the enlightened spirit of many people. The idea of an Aristotelian universe, which has always existed and will always exist, seemed more plausible.

In 1928, a few years after the publications which showed that the universe was expanding (Section 2.3), the first theories of Thomas Gold and Herman Bondi appeared, explaining the existing observations with a model of a "stationary universe," in which time had no beginning and the universe was not concentrated in a point. Later, to explain the observed expansion without admitting the existence of the Big Bang, they postulated a continuous creation of matter from nothing. This matter was filling the empty space created by the expansion, continually forming new stars and new galaxies, maintaining the universe at constant density. This continuous creation entailed constant generation of an insignificant fraction of the mass of the universe and, for this reason, it had escaped observation. To those who objected that their model violated the Aristotelian principle of the conservation of mass (nothing can arise from nothing), they replied that the Big Bang violated it as well, creating all the matter at the beginning. They said that they were postulating an analogous process that was extended in time, a process that acted forever, in a universe that was infinite in time and space; this had the advantage of doing away with the beginning of time and the initial singularity. The Big Bang, they said, was much more absurd than the continuous creation.

One of the supporters of the idea of a stationary universe was the English astrophysicist Fred Hoyle, of whom we will read more in the coming chapters. To Hoyle, or better Sir Fred Hoyle (the knighthood was awarded for scientific merit), goes the involuntary credit of having invented the name "Big Bang." He chose it in a polemic with George Gamow, who developed this model in 1948, with the intention of ridiculing it. The name was so apt that it contributed to the success of the theory, and Big Bang is the name that everyone now uses.

Being a good scientist, Hoyle maintained that the question of the existence of the Big Bang should be resolved by observations. However, instead of invalidating the Big Bang model, he found elements in favor of its correctness. His studies of how the elements were created inside stars (which we will recount in Chapter 4) led him to calculate how much helium would have been created in the Big Bang; he showed that it would be much more abundant than that which would have resulted from nuclear reactions inside stars. The measurements then showed that the abundance of helium in the universe is exactly that computed by Hoyle, the one expected if the Big Bang had actually happened.

Helium, after hydrogen, is the second-most-abundant element in nature and is far more abundant than all the other elements taken together (Section 4.8

and Figure 4.11). To form helium by nuclear fusion, temperatures in excess of 20 million degrees Celsius (36 million degrees Fahrenheit) are required, enabling protons (which have the same electric charge and therefore repel one another) to come together close enough to permit the nuclear force to bind them into a nucleus of helium. High density is also required to make collisions between protons likely. According to the theory of the Big Bang, during the first three minutes of its life, the universe had the temperatures and the densities necessary for this reaction. Those three minutes were sufficient to transform one quarter of the mass of hydrogen into helium. These two elements today constitute, in a first approximation, three-quarters and one-quarter respectively of the ordinary matter present in the universe.

A stationary universe would not have produced so much helium, nor would any other known process, particularly the nuclear reactions that take place in the stars. Another blow to the theory of a stationary universe was given by the discovery of fossil radiation (Section 2.5) — a phenomenon that only the Big Bang was able to produce.

In spite of these results, Hoyle continued to explain the measurements using ever-more-sophisticated models of a stationary universe, assisted by other illustrious scientists, (like Geoffrey Burbidge), who were also skeptical about the Big Bang. Only toward the end of the last century, when faced with the numerous observational evidences, did he concede, although not completely. Until his death in 2001, Hoyle continued to maintain that "the Big Bang was only part of the truth but not the whole thing" and, after all, the new measurements showing that the expansion of the universe is accelerating and the existence of dark energy are proving him right.

When we discuss the origin and the development of life in Chapter 5, we will consider an argument that probably contributed to his skepticism. If we admit that evolution proceeded in a random (Darwinian) way, we find ourselves immediately confronted by the difficulty of explaining how a molecule of DNA could have formed from the random combinations of the bases. The molecule is too complex to have been formed "randomly" in the 4 billion years of the Earth's life. But if 4 billion years is too little to create so complex a molecule randomly, a universe that is only three times older (13.7 billion years) is just as incapable of building life elsewhere and bringing it to the Earth. We need a universe much older, practically one that has been here forever.

The observations are irrefutable, however. They show, as we shall see, that the galaxies around us are all moving apart, and that therefore they were closer together in the past. No one today doubts the expansion of the universe, but there are many scientists who refute the idea of an initial singularity, which seems to contradict the conservation laws of physics. String theory, which we shall mention at the end of the chapter, is an attempt to resolve the problem. Success in proving string theory (with measurements) would eliminate the initial singularity: the universe would begin its expansion from a finite size set by its previous existence. If we manage to demonstrate this, then we shall also have demonstrated that the universe has always existed and time has not have a beginning (Section 2.11). While we are waiting to find this "piece of the truth," as Hoyle said, we must stick with the models that give the best interpretations of the observations — the ones saying that our universe has a "year zero."

The idea that the universe is expanding and that it began with a great initial explosion is today accepted by everyone, although it took 50 years of battles and discussions (in the spirit of Section 1.3) to convince the most skeptical. The observations that have erased every doubt (without saying anything about a possible "before") have been made in the last 20 years; they show that the universe had a beginning and that its expansion started 13.7 billion years ago, as shown by the measurements of the American satellite *WMAP* in 2004. We devote the next sections to the evidences that the universe is expanding, starting with Olbers' paradox.

2.2   Olbers' Paradox

Among the arguments supporting the idea of an expanding universe, there is the solution it provides to Olbers' paradox, a problem that troubled scientists for almost two centuries. The German astronomer Wilhelm Matthias Olbers (1758–1840) took up an idea of Kepler[1] and asked himself why the night sky is dark. We are, of course, used to it being dark, but if we think about it carefully we conclude that if the universe is large and uniformly populated (as we will see, on a large scale the universe *is* uniform), the night sky should be luminous.

You can understand Olbers' reasoning by looking at Figure 2.1. In a sky uniformly filled with stars, the Earth (indicated with an E) is surrounded by a series of concentric spherical shells. If we look from the Earth in any direction containing a star, this point will no longer be dark, but bright, as is the stellar surface. If we consider only the stars in the inner sphere 1, the probability of finding a star in any given direction is quite small, but if we also consider the shell 2, the probability increases, and so on as we increase the number of shells. For those who recall a bit of algebra and physics, a more quantitative demonstration is given in the caption to the figure.

Olbers concluded by saying that if the universe were uniformly populated by stars, increasing its size would also increase the light that should reach the Earth. In the limit of an infinite universe, the whole sky should be as bright as the surface of an average star. From what we know about the universe, we

---

1 *Credit for the first formulation of the paradox goes to Kepler, who proposed it in 1610. The question was ignored by the scientific world, which at that time was completely engaged with the implications of the Copernican revolution.*

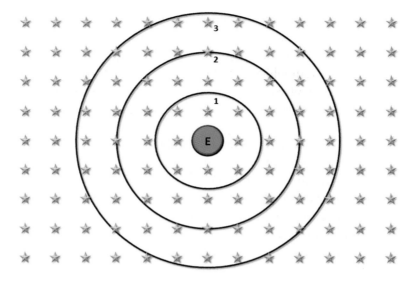

Figure 2.1: Olbers' paradox: Consider a series of concentric spherical shells around the Earth, E (not to scale), of which the figure represents a section. If the density of stars is uniform, their number will increase with the volume of the sphere, i.e. as D^3, where D is the distance from the Earth. The luminous flux from each star decreases as $1/D^2$, so that the total luminosity would grow as $D^3/D^2 = D$. For a sphere of infinite diameter, the flux of light arriving on the Earth would be similar to that of an average star surface.

should replace "stars" in the above discussion with "galaxies," but the same argument holds. However, everyone knows that even if the universe is enormous and uniformly populated with stars and galaxies (Section 2.4), our night sky is dark. So where is the error? There is no error. The reasoning is correct. The paradox was explained 150 years later by the simple discovery that the universe is expanding, and by the Doppler effect because of the reddening of the light of receding galaxies (Appendix A.3). The light of the farthest galaxies (which are the farthest because they are receding with the greatest velocity) is displaced toward infrared frequencies which the human eye does not perceive.

Today, however, there exist instruments with infrared and microwave capabilities, and their measurements show that the number of galaxies and the luminosity of the heavens increase at those wavelengths, as Olbers thought. The density of sources observed becomes so high (Figure 10.9) that in the field of view of the instruments there is more than one galaxy and the measurements of the farthest objects today are not limited by the sensitivity of the instruments but rather by their capability to separate one galaxy from another (which depends on the diameter of the telescope; Appendix A.5).

Fortunately, the luminosity of the sky which we observe at these wavelengths does not increase, as is expected according to Olbers' reasoning. In fact, even if we assume a universe is infinite in space, because it is finite in time (having come into being at the Big Bang), and because the velocity of light is finite (nothing can go faster than the speed of light), we can only see galaxies out to the distance from which light has reached us in 13.7 billion years: this is, the *horizon* of our observations (further explanation in Sections 2.6 and 2.7). Moreover, it turns out that the energy of traveling photons decreases with their redshift (increase in wavelength), and this energy reaches zero for the light departed from galaxies on the horizon. Therefore, the energy reaching us is much less than what we would have in a stationary universe.

With these reasons, Olbers' paradox can be considered resolved: the sky is dark since the farthest galaxies emit in the infrared which the human eye does not see and, because the energy carried by the photons decreases with distance, the total brightness of the night cannot be infinite.

2.3 Hubble's Constant

The fact that the universe is not static, and must be either expanding or contracting, is quite logical. Gravity causes all bodies to attract each other and therefore nothing can remain fixed. To conceive a static universe is like conceiving a stone that is suspended in the air: it is impossible — if we leave it alone, it will fall onto the ground; if we throw it up in the air, it will fly away. It will never remain suspended in the air. However, it was necessary to wait until the end of the 1920s to understand that a static model of the universe was not viable.

Even Einstein, when he wrote the equations of general relativity, found that they were incompatible with a static universe but, instead of imputing this to his image of the universe, he preferred to think that there was a mistake in his equations. He therefore introduced what he called *the "cosmological constant,"* which could maintain a static universe by countering the effect of gravity. When, in the early 1920s, a mathematician pointed out that he could eliminate the constant by supposing a universe that was either expanding or contracting, Einstein recognized the theoretical validity of the idea, but he rejected it because it was too strange. It was only in 1929, after a visit to Mount Wilson (the great observatory in the mountains above Los Angeles), and having discussed with Edwin Hubble (1889–1953) the measurements showing the redshift of the galaxies, that Einstein publicly recognized his error, and called the distrust he had shown of his own result the most colossal stupidity of his life.

Today, having accepted the idea of an expanding universe, it seems incredible that Einstein, who had come up with the theory of relativity, would not accept the idea of an expanding universe and therefore refused to admit that everything, including time, had an origin. Instead he preferred to modify his equations by introducing a term that had no clear physical meaning. This helps us understand how difficult it is to accept the idea of an origin of the universe and of time. We are all a bit Aristotelian.

The credit for the first proof of the expansion of the universe goes to the German astronomer Carl Wilhelm Wirtz (1876–1939), who published his result in 1922; this result was confirmed a few years later by the observations of the American astronomer Edwin Hubble, to whom, perhaps because he was American, now goes the principal credit for the discovery (the satellite in Figure 1.1 was named in his honor). Wirtz and Hubble noted that the spectra of distant galaxies were displaced toward the red, indicating that they were moving away from us (Appendix A.3). In spite of significant errors that were made in determining the distances, it was possible to show, beyond any doubt, that this displacement of the galaxies' spectra was greater in more distant objects. It was thus established that the galaxies were moving away from the Earth with a velocity that increases proportionally with distance, defining what is today called "*Hubble's constant,*" indicated by the symbol H_0. The first determination of H_0 was very rough, but it was enough to convince Einstein to eliminate the cosmological constant from his equations.

Today the value of H_0 is known with good accuracy: 20 ± 2[2] km/s for every million light years. This means that a galaxy 1000 million light years from us is receding at 20,000 km/s. The determination of H_0 with such precision has only been possible recently (the measurement is expected to be refined in the years to come, bringing the error to within 0.2 km/s, i.e. 1%). The measurement has been possible thanks to the discovery of some particular supernovae called type Ia. These all have the same intrinsic luminosity so that, by measuring the amount of their light that arrives at the Earth, we can derive their distances with good accuracy. (The flux of light diminishes with the square of the distance from a source.)

The mechanism by which all these supernovae explode with the same luminosity deserves to be described. Supernovae (Section 4.5) are massive stars which, at the end of their lives, explode with a luminosity of millions of Suns, comparable to that of the entire galaxy of which they are part and rendering them visible at great distances. The Indian Nobel laureate Subrahmanyan Chandrasekhar showed that stars which, like the Sun, have less than 1.44 solar masses at the end of their lives will become white dwarfs whilst stars that have more than 1.44 solar masses will explode as supernovae. The criterion for explosion therefore depends only on the mass of the star.

Supernovae of type Ia (Figure 2.2) belong to double systems, composed of two stars orbiting one around the other. The more massive of the two stars has

---

2 *The results of measurements are often reported in this way. It means that the value is more than 18 and less than 22.*

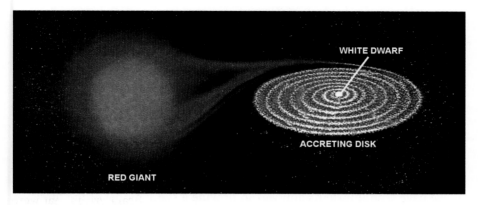

Figure 2.2: Supernovae of type Ia. They are binary systems consisting of a red giant orbiting around a white dwarf, two stars of similar masses but with very different diameters. The red giant has a diameter that can be greater than that of Earth's orbit, and it is very rarefied; the white dwarf has a diameter less than the Earth's, and it is very compact (Section 4.4 and Figure 4.8). The two stars are close enough that the gravitational field of the white dwarf attracts to itself the mass lost from the red giant, increasing its mass until it reaches the 1.44 solar masses necessary to implode under its weight and then explode as a supernova.

a mass slightly less than the Chandrasekhar limit of 1.44 solar masses. Because it is more massive, it evolves more rapidly (Section 4.3) and becomes a white dwarf; the other star, evolving more slowly, is still in the red giant phase, during which a star loses mass. The mass lost from the red giant ends up on the white dwarf, thereby gradually increasing its mass until it reaches the limit of 1.44 solar masses, whereupon it explodes as a supernova. Since type Ia supernovae all have the same mass of 1.44 solar masses, they all explode with the same luminosity and are the ideal instrument for measuring the distance of distant galaxies.

The measurements are not easy because of the difficulty of recognizing these supernovae, especially when they are far away. Patient work, lasting more than two decades, has made it possible to collect a sufficient number of measurements to show that all the galaxies around us are receding with a velocity proportional to their distance, thus removing any doubt about the expansion of the universe. The speeds measured are so high that the attractive forces between the masses which compose the universe are not enough to stop them. Therefore the expansion of the universe, initiated by the Big Bang, will never be stopped by the forces known today.

Measurements of the distances obtained from the supernovae indicate, however, that the universe is not only expanding, but that the expansion is undergoing an acceleration because of an unknown force, called the "dark force" or the "attraction of the vacuum"; it is generated by what cosmologists call "dark energy," an unknown energy that fills the whole universe. That is to say, the value of H_0 is not constant, but it increases as we move away from the Earth and, for the most distant galaxies, it may reach 30 km/s for every million light years.

At present, we do not know the nature of this energy, and it seems that, for the first 7 billion years, the expansion had been constant. Then the universe started accelerating, showing that the acceleration is variable in time (Section 2.9). When we understand the nature of this obscure energy, we may find that the expansion may one day slow down or stop or even reverse itself. In this case, the story of the universe would have to be rewritten and might be very different from what we believe now.

There is an amusing side to this: the acceleration of the universe can be described only by introducing a constant into the equations of general relativity: the cosmological constant which Einstein had proposed and then rejected. As we have seen, Einstein's aim was to provide a static model of the universe. In this case the constant not only keeps the universe from collapsing under its own weight, but it actually permits its expansion to accelerate.

Understanding the nature of this energy, and with it the attribution of a physical meaning to the cosmological constant, is one of the principal problems that cosmology and modern physics have to resolve today.

2.4   The Expanding Universe

How can the velocity of expansion of the universe be the same in all directions? If there was an explosion, there should be a center and the matter should move away from the point of the explosion equally in all directions; that is, it should spread out symmetrically. If there is a symmetric dispersal with respect to one point or any other point, which is not at the center of the explosion, we will not see the matter dispersing equally in all directions. However, observations show that all points in the universe are moving away from us symmetrically, with velocities that depend only on the distance from us (H_0 has the same value in all directions), as if we were at the point where the Big Bang took place. This is very difficult to accept because, among other things, the Sun and the Earth were created 9 billion years after the Big Bang; it therefore it seems impossible that, after so long a time, we should find ourselves at the center of the explosion. An explosion sends matter away; it does not hold it firmly at the center.

How is it, then, that we see all galaxies spreading out symmetrically around us? The answer lies in general relativity: the expansion is due to the dilation of space itself, so that every point in the universe sees all the other points spreading out equally in all directions; every point, therefore, seems to be at the center of the expansion.

An intuitive explanation can be given by imagining a two-dimensional universe spread out on the surface of a rubber balloon, with many points representing galaxies. If the balloon is inflated (Figure 2.3), the rubber is dilated uniformly. Each point sees the others as moving away in all directions and is

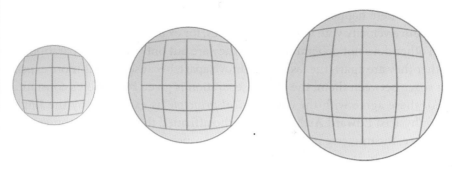

Figure 2.3: A model of the expanding universe. Imagine inflating a balloon; as the diameter increases, every point on the surface of the balloon sees the other points moving away as if it were in the center of the expansion.

justified in saying, "I am at the center of the universe." This is because one is living in a two-dimensional world (the surface of the balloon). As soon as the inhabitants of a point perceive what is happening in the three-dimensional space, in which the balloon is expanding, they understand (as general relativity has made us understand in the real universe) that all the points on the surface of the balloon appear to be at the center of the explosion.

The example of the expanding balloon helps us to understand the peculiar nature of this expansion. There is no true central point on the surface of the balloon, different from all the others, nor is there an external boundary or a direction toward which the surface is expanding. Moreover, the points on the balloon are firmly fixed, impressed upon the surface. They move apart because the surface is expanding. This is what happens in our universe. The objects that we see moving away from us are in fact fixed. The term Big Bang is a deceptive name that makes us think of an explosion, throwing pieces of the universe far from one another. That is not what is actually happening. The galaxies are fixed in space (apart from small random local movements), just as we are. It is the space separating the individual galaxies that expands with time, like the surface of the balloon when it is inflated.

In the example of the balloon, we used a third dimension: the volume into which the balloon expands. To explain the expansion of the universe, however, no new dimension is needed. According to the theory of general relativity, space–time expands, curves and bends. The description of the universe is contained in the properties of space and needs no extra dimension. Nor does it make any sense to ask what is the outside the expanding universe, because the curvature of space describes the entire universe; an "outside" does not exist.

The Big Bang is not, therefore, an explosion *in* space but rather an explosion *of* space, without a center and without an external boundary. It did not happen at any particular place, but at every point in the universe, each of which, in a sense, is at the center of the explosion. We are not in a privileged position; what happened here happened at every other point in the universe. This means that the relative velocity with which two galaxies, for example, are moving away from each other does not depend on where they are but only on their separation. The farther they are apart, the faster they move apart, because there is more expanding space between them; the rate of separation is given by Hubble's constant.

All these agree with the *cosmological principle* as expressed by the English astrophysicist Edward Arthur Milne, who affirmed that "at each moment of its existence, the universe must show the same aspect to all hypothetical observers at any point in the universe and what they observe does not depend on the direction in which they look."

Obviously the cosmological principle does not apply on a small scale between nearby objects. For example, our galaxy and Andromeda are moving toward each other at a velocity of 300 km per second. This is because these galaxies are relatively close, and their random proper motion is greater than the velocity given by Hubble's constant. Elsewhere, observations on large scales (great distances) confirm the cosmological principle without exception. In whatever direction one looks, the universe appears similar; that is, it is *isotropic*. And if the universe is isotropic around *every* point, it must also be homogeneous. It is easy to see that lack of homogeneity would mean anisotropy and vice versa.

2.5 Background Radiation

In 1964, two American radio astronomers, Arno A. Penzias and Robert W. Wilson, found themselves with a problem that many researchers would have ignored. They wanted to understand such a problem at a fundamental level, however, and their persistence won them the Nobel Prize. We are speaking of the problem of "noise," which is found in all telecommunication systems.

When we turn up the volume of a radio without having tuned into a station properly, we hear a uniform hiss that disturbs the radio signal with its "noise." From this hiss comes the name "noise," which is given to all random interference with telecommunications. We can also visualize it as that speckling, similar to a snowstorm, which appears on the screen of a television that is not properly tuned to the channel. Those involved in telecommunications fight against noise every day, because it is the noise that determines the weakest signal we can receive. When a signal is weaker than the noise, we can no longer distinguish it, just as it is impossible to hear a whisper in a noisy room; the words are uttered but we are unable to distinguish them. Improving the performance of a system of transmission/reception means reducing its noise to a minimum.

There are several sources of noise: some are produced by human activity (electric power lines, electric motors, electric switches); some are natural phenomena like thunderstorms; some are processes occurring inside the receiver itself, some finally are due to the background radiation, which we are about to discuss. They contribute between 1% and the 10% of the noise we see on a television screen (a kind of snow) when a station is not settled. But Wilson and Penzias were unaware of that.

In 1964, the two radio astronomers were using a radio antenna of the Bell Telephone Laboratories in New Jersey to measure the radio emissions of our galaxy. The antenna had been constructed to communicate with the satellite Echo and had the unusual form of a funnel (Figure 2.4), like the horns that were once used to amplify sounds, and it worked in a similar fashion. The entrance was a trapezium with sides of about 6 metres, sides, collecting the radiation that was reflected from the walls and concentrated on a detector at the end of the horn. To align it with satellites, the antenna was mounted on a mobile structure that permitted it to be pointed in all directions: it was an ideal instrument for radio astronomy.

Wilson and Penzias, trying to minimize the noise of their receiver, pointed the antenna toward a place in the sky where no signal was expected, but they found — at the wavelength of 7.35 cm (Appendix A.1), a band today used by television channels — much more noise than their calculations predicted. This noise did not depend on the direction toward which the antenna was pointed, suggesting that it was produced by the antenna itself and not by external sources. The two radio astronomers spent months checking every part of the system, even chasing away a family of pigeons who had spread, as they put it, a dielectric material on the surface of the antenna. They cleaned

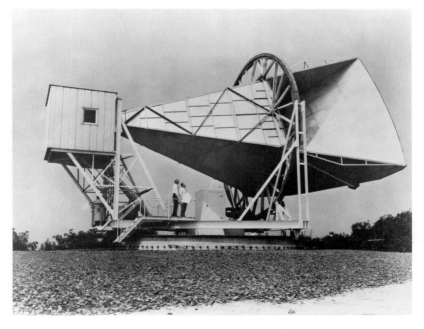

Figure 2.4: The Bell Telephone Laboratories antenna with which the fossil radiation, the residue of the Big Bang, was first discovered. In the picture, the two big gear wheels are visible, one horizontal and the other vertical, permitting the antenna to point toward any direction in the sky. (© GRIN.NASA database.)

the surface and checked the receiver but, in spite of their efforts, the noise was not reduced.

Many researchers would have given up trying to find the origin of that noise, considering such research to be a waste of time; they would have attributed it to an unidentified local source or to a small error in their calculations. Penzias and Wilson, on the contrary, did not give up but continued to seek the source of the noise until one day, when discussing the problem with some friends, they were introduced to James Peebles, a theoretician and cosmologist at Princeton University. He suggested that what they were measuring was not an excess noise from the antenna itself, but rather the "background radiation," the residue of radiation emitted by the fireball in which our universe was born. The following year, after much verification, they published the results of their work with the cautious title "Measurement of Excess Antenna Noise at 4080 Mc/s" while, in parallel, the Princeton theorists (Dicke, Peebles, Roll and Wilkinson) published a paper that interpreted the measurement as an observation of the residual radiation from the Big Bang. The results of Penzias and Wilson were confirmed in succeeding years by numerous observations and experiments. Today the existence of the background radiation is considered one of the most robust proofs of the existence of the Big Bang.

But what is the background radiation? In the first few moments of its life, the universe was very hot and filled with radiations; it was so hot, indeed, that for some time even nuclear fusion reactions were impossible, atomic nuclei being destroyed as soon as they were formed. With the expansion of the universe, the energy associated with this radiation was diluted into ever-greater volumes so that, as time went on, every cubic meter of the universe contained less energy and therefore became cooler. In fact, the temperature associated with this primordial radiation decreased in direct proportion to the dimensions of the universe. In Appendix A.2 you will see that to every temperature there corresponds a characteristic "color" of radiation, extending well beyond the range of the visible spectrum. As the temperature decreases, this color moves toward ever-longer wavelengths and lower frequencies. In 13.7 billion years, the temperature of the universe has come down to a few degrees above absolute zero, which corresponds to radiation in the microwave region of the spectrum (Appendix A.1). This is the background radiation that Wilson and Penzias measured and which led to their being awarded the Nobel Prize.

The discovery of the background radiation, together with the observed abundance of helium, was a mortal blow to the theory of a stationary universe (Section 2.1): only an initial fireball could have produced it. This meant that the universe had an origin and had always evolved with time.

In the following years, these measurements were repeated many times, always giving the same result. In the 1980s, the American satellite *COBE* was dedicated to such measurements, and achieved extreme precision. Figure 2.5

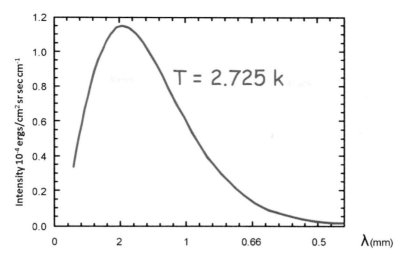

Figure 2.5: The fossil radiation measurement carried out in the 1980s by the American satellite COBE. The vertical axis shows the radiation intensity and the horizontal axis the wavelength which determines the color of the light. The curve is for a body at the temperature of 2.725° above absolute zero; the uncertainty, of ±0.002°, is within the thickness of the line on the graph.

shows the result of its measurements: a temperature of 2.725° (Kelvin) (with a precision of two-thousandths of a degree) above absolute zero and an emission peak (at $\lambda = 1$ mm) in the microwave region of the spectrum (Figure A.2).

The measurement showed, moreover, that the temperature of the sky is extremely uniform, so uniform that the difference in temperature between any two directions on the sky is less than a ten-thousandth of a degree. This is an extraordinary uniformity: no one on the Earth is capable of constructing an oven which has such a small difference of temperature in its interior. This presents a big problem for the cosmologists, because it is not easy to explain why the universe has a much more uniform temperature than a well-regulated oven. An explanation was also needed as to why the universe appeared to be "flatter" than it should have been, as we shall discuss in the next sections.

2.6   The Inflated Universe

As we have seen, the Big Bang theory explains very well how the universe expanded and cooled, the origin of the background radiation and the abundance of helium and hydrogen; it is the *only* model that succeeds in explaining all these phenomena. It does not explain other observations, such as the great uniformity of the background radiation that we mentioned at the end of the last paragraph. To attain such uniformity, the various parts of the universe must continually exchange energy (the only way to produce such a uniform temperature). And this energy would have to go from one part of the universe to another faster than the speed of light, which is physically impossible.

At lower velocities there would be greater differences in temperature than the one observed.

There is a second problem: measurements show that the universe is much "flatter" than it should be at its age. What do we mean by this? Consider again the balloon of Figure 2.3. If we were continuously measuring the curvature of its surface, we should find that it becomes less and less as the balloon is inflated, and the surface becomes more and more flat. In the limit of infinite inflation, it would be completely flat. Using a variety of observations, including those of the background radiation, astronomers have discovered that our universe is much flatter than it ought to be after expanding for 13.7 billion years at the rate deduced from *Hubble's constant* (Section 2.3). In analogy with the balloon, we see a universe much larger than it should be, given its age and the observed rate of expansion: a serious dilemma for the cosmologists.

The best explanation we have of this puzzle is furnished by the "theory of inflation," proposed in 1981 by Alan Guth of the Massachusetts Institute of Technology. According to this theory, the universe expanded very rapidly in its first moments of life, increasing its size enormously. To explain the observed curvature one must suppose that in its first 10^{-35} seconds of life the universe expanded by a factor of 10^{30} (1 followed by 30 zeros). To give an idea of how big this number is, we have to compare it with the actual dimension of the universe, estimated to be 10^{28} cm. If we divide this dimension by a factor of 10^{30}, we obtain 0.1 mm. We would come to the disconcerting conclusion that, without inflation, the universe would have a dimension of less than a millimeter in size. The model that explains the observation is therefore constrained to hypothesize that the highest rate of expansion happened in the first instants of life of the universe at an almost infinite velocity, passing from an infinitesimal size to a finite one (these are the peculiarities one encounters when dealing with a singularity). This is why the name "inflation" has been chosen for the theory.

The measurements seem to say that in its first few moments of life the universe underwent a very rapid expansion; an expansion so fast that it flew apart at speeds much greater than the speed of light — an idea that seems absurd, and an obvious contradiction of the principle of relativity, according to which the speed of light is the absolute limit at which anything can go. In reality this idea does not contradict the postulate, even if at first glance it leaves us perplexed. The limit on the speed with regard to the motion of an object in space does not apply to the *expansion of the space itself*, which is not a motion and therefore does not violate any hypothesis. It remains true that in the universe nothing can travel faster than the speed of light, even if we live in a universe in which space itself can expand faster than the speed of light. This means, moreover, that the regions of the universe moving away from us at a speed greater than the speed of light will never be reachable by us since we must move *through* space and therefore we cannot reach the speed of light.

The inflationary hypothesis is considered today the best explanation for a flat universe with the age of "only" 13.7 billion years. The theory explains why it is so uniform. The inhomogeneities of temperature that existed at the beginning have been "stretched out" by an enormous factor, and we could therefore live in a universe that is only a tiny part of something enormously larger (maybe infinite). With this expansion the inhomogeneities have been reduced to the tiny values we are observing (Section 2.10) and are those from which stars and galaxies were born.

This hypothesis is not simple conceptually, just as it is not for the idea of a beginning of everything connected with the Big Bang. However, it is today considered the best theory for explaining the measurements. The difficulties encountered in describing the first few instants in the life of the universe are evidenced by the lack of models that can represent it correctly. Close to the singularity the densities and the temperatures are so high that the equations we formulate no longer apply. A quantum theory of general relativity is needed, which today does not exist.

Observations of supernovae show that the universe is accelerating its expansion today (Section 2.3), giving a further confirmation of the theory of inflation. It shows that today we are living in a new era of "inflation" even if it is less rapid than the one that took place after the Big Bang. This is considered a confirmation of the theory and shows that there is still much to learn about the forces that regulate the universe and its evolution.

We should conclude that much must be done to discover the laws that govern the universe, and it is very probable that it is much larger than what we observe. In fact, because the velocity of light is finite and the universe is 13.7 billion years old, we can observe only those objects which have been able to send us their images in 13.7 billion years. This creates around us a "*horizon*" of which we have already spoken, and to which we devote the next section.

2.7 The Horizon of "Our" Universe

Light travels at 300,000 km/s, an apparently enormous speed, but it is still very small compared with the dimensions of the universe. The star closest to us, Alpha Centauri, is 4.28 light years from the Earth. This means that the light of the star takes 4.28 years to reach our planet. The stars of the Orion constellation are 1500 light years from us; when we look at it, the light we observe left those stars 1500 years ago, when the Western Roman Empire had just fallen. The center of our galaxy, the Milky Way, is about 30,000 light years from the Sun, so the light we observe started out when the Earth was inhabited by Neanderthal man.

Therefore, if the light from a distant galaxy has taken a million or a billion years to come, what we see is not how that galaxy is today, but how it was a million or a billion years ago when the light was emitted, even though in the meantime that galaxy has aged as we did. Equally (in accordance with the cosmological principle), an inhabitant of one of those distant worlds who looks toward us would not see the Earth as it is today, but as it was a million or a billion years ago, according to the distance. The farther one looks out into space, the younger are the objects seen.

If the universe had a beginning in time, there is a limiting distance beyond which we cannot "see." It is the distance light traverses in 13.7 billion years, traveling at 300,000 km/s. This means that we are surrounded by a "horizon," beyond which we cannot see; it is called the "particle horizon." We can suppose that there is something beyond that horizon, but we cannot observe it directly. The theory of inflation says that the universe is gigantic and a large part of it should be beyond that horizon. The manner in which the universe expands changes that horizon; it changes what we can see today and what we will see in the future.

To clarify this concept, imagine living in a *universe that is very large, but not expanding.* If its age is 13.7 billion years (it was born already big) the most distant objects we can see are those at a distance of 13.7 billion light years from us, the maximum distance that light can have traveled from the birth of the universe until today. That is the horizon; if in this static universe there were other galaxies beyond the horizon, we could not see them today, but our descendants will see them after the time necessary for light to arrive. With the passage of time the sphere will enlarge, taking in new galaxies so that the number of objects we could observe increases with time.

Now imagine a galaxy 5 billion light years away. The inhabitants of its planets would also be able to see a sphere with a radius of 13.7 billion light years, centered on them. Because their sphere is displaced from ours, they would see objects that we have not yet seen — those at which the light has arrived, but has not yet arrived at the Earth. After 5 billion years we too will see the objects they see now. Every galaxy would therefore observe a spherical region around itself, whose size is determined by the age of the universe.

In *a universe that is expanding* with a velocity that increases with distance (Section 2.3), things get more complicated, because while the photons travel toward us, the space (keep in mind the balloon of Figure 2.3) continues to expand. This means that when a photon arrives, the galaxy that emitted it is farther away than it was when the photon left it. Models show that when we receive a photon that has traveled 13.7 billion light years, the source that emitted it is about 45 billion light years from us. Therefore, we see today all the objects that are in a sphere 45 billion light years in radius. That is our horizon.

As in the previous case, the inhabitants of a galaxy 5 billion light years from us would see all the objects in a sphere displaced from ours, but now the radius of that sphere would be 45 billion light years. Each galaxy would observe a spherical region around itself with a radius larger than the distance light has traveled from the beginning of the universe until today. As before, the sphere (the particle horizon) always grows larger with time, incorporating new galaxies, so that the number of objects that can be seen increases with time. It does not increase indefinitely, however. It will increase only until the distance at which the speed of expansion of the universe (which increases with the distance from us) reaches the speed of light. The galaxies that are farther away than that point are moving away faster than the speed of light and will never be visible from the Earth. We have seen, in the preceding section, that the existence of galaxies moving away from us faster than the velocity of light does not violate the postulate of the special theory of relativity, because we are speaking of an expansion of space, which is not a movement "on space" and does not violate the postulate. Therefore, galaxies we will never see could exist (according to the theory of inflation, there are many more than what we do see). The photons they emit toward us will never reach our part of the universe. They move away from us together with the galaxies which had emitted them. With the passing of time and the expansion of the universe, more and more objects will exceed that limit and disappear from our view.

The third case, which measurements have by now excluded, is that of *a universe in which the expansion is decelerating*, as was thought possible until recently. In this hypothesis, with the passing of time one would see an increasing number of objects: those that were beyond the particle horizon and were going away from us faster than the speed of light but, because of the deceleration, entered our horizon.

The last case, *a universe in which the expansion is accelerated*, as shown by the latest measures (end of Section 2.3), is not very different from the case of a universe that expands with a velocity proportional to the distance from us, except that galaxies exceed the particle horizon more rapidly, leaving our universe forever, never to become visible again.

An expanding universe consists therefore of many parts that are not visible to one another: a *multiverse*. In fact we observe only the part of the universe inside our particle horizon. A hypothetical observer who found himself at a distance twice that of our horizon would see a universe that would never be visible to us, while an observer at an intermediate distance would see a part of the universe visible to us and a part that we will never see. Proceeding in the same way, one can imagine a sequence of universes that are never in contact with one another (at least as long as space–time continues to expand). There is no reason to believe that these universes are different from ours; they could therefore contain stars like the Sun and planets like the Earth, with scientists who are asking what could exist beyond the particle horizon.

Even if we cannot observe the entire universe, we can see very distant objects, those which are moving away from us with the largest velocity (Section 2.3). Because of the Doppler effect the radiation they emit is also that with the greatest displacement toward the red (Appendix A.3). The radiation emitted during the Big Bang (observable inside our particle horizon) is displaced so much in wavelength that it peaks in the microwave (Figure 2.5). But is it possible to photograph the Big Bang? Is it possible that the background radiation contains a trace of its image? We will speak of this in the next section.

2.8   The Image of the Most Distant Source

Can the Big Bang be photographed? In reality, no, but we can come very close. In the first years of its life the universe was a very hot cloud of atomic nuclei (of hydrogen and helium) and electrons which, because of the very high temperature, could not bind with the nuclei to form atoms. When electrons are free (i.e. not in orbit about an atom), they are very efficient at scattering light in different directions, like fog does, and it does not allow us to see inside it.

Between us and the Big Bang, then, there is a dense cloud of free electrons. We can see the surface of the cloud as it was 380,000 years after the Big Bang, when the universe had expanded and cooled down to a temperature of 3000° (Kelvin). At that temperature the electrons were captured by the hydrogen and helium nuclei that had been formed in the first minutes of life of the universe. The ability of the electrons to scatter light was thus reduced; the "cloud thinned," becoming transparent, so that light began to travel freely through the universe. From that point on, the universe is transparent. The oldest and most distant thing we can observe, therefore, is the surface of that cloud of fire — 380,000 years after the Big Bang.

Photographing that surface has been the ambitious goal of a group of researchers, led by Paolo De Bernardis of the University of Rome and Andrew Lange of the California Institute of Technology, since 1998. To achieve this goal, they launched a stratospheric balloon to an altitude of 36 km, where the atmosphere is particularly transparent to microwaves. On the gondola under the balloon, they attached a telescope and sensitive instruments capable of taking images in the submillimeter range (Appendix A.1). For 10 days, the instruments on board "photographed" the universe as it appeared 380,000 years after the Big Bang, making use of a characteristic of the south-polar launch site: stable, high-altitude winds that blow for months along a circular path around the pole. The balloon, subject to these winds, returned after 10 days to the point where the flight started, allowing its precious cargo to be recovered. The experiment was named *Boomerang*, because of its circular flight path.

Figure 2.6 shows the balloon before the launch. The sky has been replaced by the image of the primordial universe obtained by this difficult measurement; the various colored areas indicate regions of different temperatures and

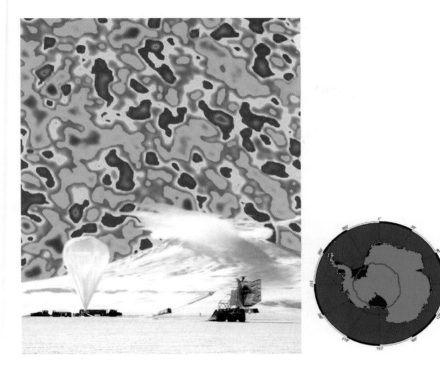

Figure 2.6: A picture of the Boomerang experiment before its launch. The sky has been replaced by the result of the measurement — the image of the universe as it was 380,000 years after the Big Bang. The trajectory of the balloon's path is shown on the right. [Images courtesy of Professor Paolo De Bernardis (University of Rome) and the Boomerang team.]

densities with their real dimensions, which are of the order of one degree of arc, about twice that of the Moon. As we said earlier (Section 2.5), the greatest difference in temperature between any of the points is less than a ten-thousandth of a degree. This image represents the initial inhomogeneity of our universe which is accounted for by the theory of inflation. This inhomogeneity subsequently increased with time and, after hundreds of millions of years, gave rise to the first great condensations of matter that formed the first stars and then the first galaxies.

The image in the figure is the most distant and, in a certain sense, the oldest picture we can have of our universe, the beginning of everything. The matter would have had to cool for many years before organizing itself into structures dense enough to form the first stars — the processes which we shall discuss in the following section.

2.9 Dark Matter and Energy

In 1933, the Swiss-American astronomer Fritz Zwicky, while studying two nearby clusters of galaxies, made a discovery that even today presents a challenge to physicists, astronomers and cosmologists. Galaxies (the one to which the Sun belongs is the Milky Way; see Figure 8.2) are generally found in clusters of hundreds or even thousands of members, each one consisting of hundreds of billions of stars, bound together by the gravitational attraction of the total mass of the group. Zwicky, studying the galaxies of the Virgo and the Coma Berenices clusters, discovered that the velocities with which the galaxies moved (determined by the Doppler effect; Appendix A.3) were too high for them to have been confined to the clusters by their self-gravity. In other words, all the galaxies had more than the escape velocity of the cluster and should, therefore, have rapidly dispersed into space. As we shall see in Section 4.3, the mass of a star is easily derived from its luminosity, so that it is possible to determine the total mass of a galaxy from the luminosities of all the stars it contains. Zwicky found that the mass needed to hold the galaxies together was 100 times larger than that deduced from their luminosities. In other words, the mass of the stars constituted only 1.1% of the total mass of the group. Thus was born the concept — and the problem — of "missing mass," the major component of a galaxy's mass that had "escaped" in the measurement of luminosity.

In the following years, it was discovered that the problem of missing mass existed even inside individual galaxies. They rotated about their centers as if their mass were much greater than that calculated from the radiation they emitted. Figure 2.7 shows the results of measurements made on the stars in our

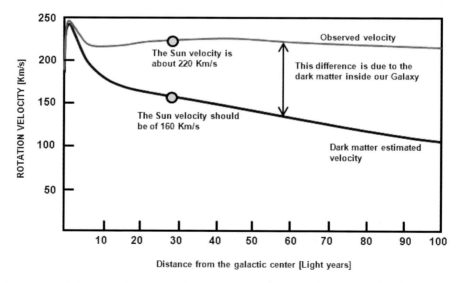

Figure 2.7: *Rotation curve of our galaxy. The upper, red curve represents what is measured and is, practically, constant when moving away from the galaxy center as if the galaxy were a solid body. The lower curve is the rotational velocity as deduced from the luminosity of the stars. The Sun rotates around the center of the Milky Way at about 220 km/s, while its speed, with respect to the mass measured by the visible matter, should be 160 km/s. The mass of the Milky Way, estimated by the radiation emitted by its luminous matter, is about 5×10^{10} times the mass of the Sun, whilst the one derived from the rotation is around 3×10^{12}. The difference is caused by dark matter.*

galaxy: the rotational velocity is much higher than what would be derived from the luminosity of the stars and the difference increases toward the outer parts of the galaxy. In this case the discrepancy is about a factor of 10.

For years astronomers have searched for this missing mass. They hypothesized that it was due to clouds of gas and dust (Section 3.4), to white dwarfs (Section 4.4), to exhausted pulsars or to black holes (Section 4.5), of which they did not know the existence until recently. All objects are difficult to observe, but have a notable mass. With the passage of time and the progress of observations and models, it has been possible to estimate the contributions of all these objects and find that they make a negligible contribution to the "missing mass." It has to be due to some new entity which has eluded the physicists up till today, and is called "*dark matter,*" in the sense that it is *dark* because it does not emit any detectable radiation (from x-rays to the optical to the infrared and to radio waves). Its presence is revealed only by the gravity that its large mass imposes on "*ordinary matter*" (the kind of matter that emits radiation we observe which, in the past, was thought to constitute the entire mass of the universe).

Today it is estimated that the mass of "ordinary matter" constitutes about 5% of the mass of the universe, 1% of which is in optically visible stars, the remaining 4% consisting of dust, gas, white dwarfs, neutron stars and black holes — objects that either emit no visible light or are faint enough to escape observation. The next component, making up about 25%, is "dark matter", whose nature we know little; the only certainty is that it is everywhere within the galaxy and therefore also around us. Dark matter has thus entered prominently the physicist's world — above all, the world of particle physicists who wish to identify the fundamental components of the matter that constitutes the universe. Experiments done in the particle accelerators and in the subterranean laboratories like the Gran Sasso in Italy may give us in the future an understanding of the nature of this matter.

The remaining, approximately 70% of the mass of the universe, is today believed to be in the form of a mysterious energy called *dark energy,* of which we have spoken at the end of Section 2.3. This energy is responsible for the *dark force* that accelerated the expansion of the universe. (If the universe is accelerated it needs energy, just as a car needs energy to accelerate.) With the laws of physics we know today, we cannot explain this energy, which was positive for a brief period at the beginning (inflation), then disappeared, to reappear after 7 billion years and give a new acceleration to the universe. This energy can be converted into mass *via* Einstein's formula for equivalence of mass and energy, the famous $E = mc^2$, and is shown in Figure 2.8 along with the dark and ordinary matter.

We can therefore say that our universe is dominated by dark matter and dark energy, and that all the structures we see are due to their effects on

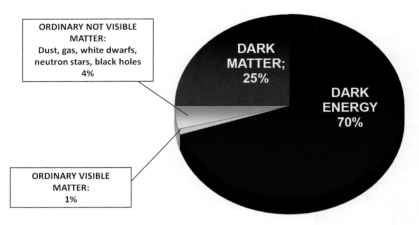

ORDINARY NOT VISIBLE
MATTER:
Dust, gas, white dwarfs,
neutron stars, black holes
4%

DARK
MATTER;
25%

DARK
ENERGY
70%

ORDINARY VISIBLE
MATTER:
1%

Figure 2.8: *Contributions to the mass of the universe: 95% is dark matter and dark energy, neither of which is understood today. Visible matter, consisting of the stars illuminating the sky, is about 1% of the total mass.*

ordinary matter. Every new theory that treats the origin of the universe and its evolution has to take into account these obscure components and their effects on ordinary matter.

2.10   After the Big Bang

In Figure 2.9, we represent schematically what has happened in the universe from its origins up until the present time. Along the horizontal axis of the figure there is a timeline with the most important events that occurred since the Big Bang. None of the dimensions in the figure are drawn-to-scale with any physical dimension. The conical form of the design serves to remind us that the universe is expanding, so its dimensions are increasing as time passes.

To the left of Figure 2.9 is the Big Bang — a singularity and a point at which the temperature and the density rise to infinity, which is physically nonsensical. The history of the universe since the Big Bang is described by models that try to reconstruct, according to the known laws of physics, the state of matter at the beginning and how it evolved to what we observe today.

These models begin a hundredth of a second after the Big Bang because, as we have said, near the singularity, the temperature and the density were so high that the physics we know does not apply. To describe matter in these conditions would require, as we already said, a quantum theory of general relativity, which does not exist today. In that first hundredth of a second, as we have seen, many things happened; there was the inflation and matter was probably already divided into "ordinary" and "dark."

We will begin our story, then, from a hundredth of a second after the Big Bang and follow the evolution of ordinary matter according to the laws we know. The universe had a temperature of around 100 billion degrees (Kelvin) and a density 4 billion times that of water. In these conditions atomic nuclei

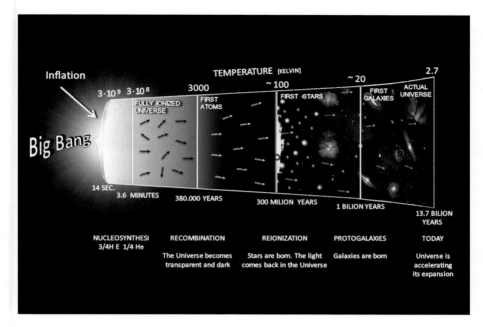

Figure 2.9: A space–time diagram of the universe. On the left, the rapid space–time expansion after the Big Bang is shown (Section 2.6), whilst the slight divergence of the conical structure on the right indicates the present acceleration phase of the universe (Section 2.3). Above the cone, the temperature of the universe is shown; below, the age of the universe and the main events since the Big Bang are indicated.

could not survive; as soon as they formed they were destroyed by the radiation. The only particles that existed, apart from photons, were electrons, neutrinos and their antiparticles.

A tenth of a second after the Big Bang, the temperature had already descended to 30 billion degrees and the density to 30 million times that of water. At this temperature protons and neutrons were formed.

At 14 s, the temperature was reduced to 3 billion degrees and the first nuclei of deuterium were formed (an atom of hydrogen with one proton and one neutron). They broke up immediately, however, without forming heavier atoms.

At about 3 min, nucleosynthesis began: the temperature was down to a billion degrees (70 times the temperature inside the Sun), low enough to permit the existence of stable nuclei of deuterium. Thus began the first nuclear fusion reactions that transformed protons (hydrogen nuclei) into nuclei of deuterium, and then into nuclei of helium (Section 4.1).

At 3.6 min, nucleosynthesis ended: the temperature had descended to 300 million degrees and the density to about 0.1 g/cm³, which is too low for further nuclear fusion to occur. In those 0.6 min, ¼ of the hydrogen nuclei were transformed into helium nuclei. No other atomic nuclei were formed because, when there was enough helium around to form carbon nuclei (needing four helium nuclei), the density was too low to overcome what the physicists call the beryllium (^8Be) bottleneck, which we will discuss in Section 4.1. The universe remained populated by ¾ hydrogen and ¼ helium, proportions that changed only a little in the 13.7 billion years that separate us from the Big Bang. Nuclear fusion reactions that have occurred within stars (Section 4.6) since then have transformed only 2.3% of that matter into nuclei heavier than helium.[3]

After the end of nucleosynthesis, the universe continued to expand but the temperature was still too high for the electrons to bind together with the nuclei of the hydrogen and helium present to form the first atoms. The free electrons, as we said in the previous section, scattered the light, making the universe opaque.

380,000 years after the Big Bang, the temperature of the universe fell to below 3000° and recombination occurred: electrons were captured by atoms and *the universe became transparent*. The most distant image that we have (Figure 2.6) is that of the so-called "last scattering surface"; when going back in time the universe changed from being opaque to being transparent.

The universe continued to expand and cool. When the temperature dropped to below 1000°, the universe became dark, because there were no longer photons with wavelengths in the visible band (Appendix A.1). With the drop in temperature, the atoms of hydrogen (^1H) began to bind in groups of two, forming the first molecules: those of H_2 made-up of two nuclei of ^1H and two electrons.

300–400 million years after the Big Bang, atoms of helium and hydrogen began to accumulate in clouds ever larger denser and cooler. When these clouds reached a temperature of between 100 K and 200 K (−170°C and −70°C), the clouds began to collapse under their own weight and the first stars were formed, and light come back into the universe (the reionization in Figure 2.9). The first stars, as we shall see, were very massive, with masses hundreds of times that of the Sun. They lived less than a million years and exploded as supernovae, leaving their residue as a black hole. *The stars and black holes were the first inhabitants of the universe.*

In the interior of the first stars, nuclear fusion reactions transformed the helium and hydrogen produced by the Big Bang into nuclei heavier than helium, such as carbon, oxygen, nitrogen, silicon and iron. That is to say, they formed all the atoms that constitute the Earth and biological organisms, without which life would not exist.

About a billion years after the Big Bang came the first galaxies, the great structures consisting of hundreds of billions of stars we see today in the sky. Many current theories suggest that the first galaxies were formed by the condensation

---

[3] *For those who wish to learn more, we suggest a book that is rather old but still largely up to date: The First three Minutes, by Steven Weinberg (1977).*

of matter around the black holes left by the first stars. This would explain why black holes of enormous mass are found at the centers of all galaxies (the one at the center of our galaxy has a mass of 3 million Suns!). Their masses are large because they have been capturing matter for more than 12 billion years!

Today, we have only a theoretical description of the way in which the first stars and galaxies were born. Although we can observe the surface of the primordial fire cloud (that of Figure 2.6), we have not yet constructed instruments capable of seeing the first stars and the first galaxies.

The space telescope Hubble was built for that purpose but did not find them in the visible spectrum even though its sensitivity was more than sufficient to observe them.[4] It was this lack of success that led to the hypothesis that the large galaxies we observe today were born later than stars, after nuclear reactions within the stars had produced enough heavy elements to form grains of dust. Clouds of these grains then hid the galaxies from the HST's observations in the visible.

In the near future, space telescopes with large diameters capable of working in the infrared and submillimeter bands — the only ones that can penetrate the clouds of dust and gas surrounding the primordial galaxies — will be built (Appendix A.4). They will be assigned to find the first stars and galaxies that populated the universe. The first of these instruments is the *Herschel* satellite (Figure 2.10) with a 3.5 m telescope (the largest telescope ever sent into space), dedicated to studying the universe in the far-infrared and submillimeter wavelengths. Launched on 14 May 2009, *Herschel* will probably not have the necessary spatial resolution to observe all the first galaxies, but it will probably see the most luminous ones, giving the first evidences of the hypothesis discussed above.

In 2018, the James Webb Space Telescope (JWST), the successor to Hubble, will be launched. The satellite (Figure 2.11) will harbor a 6.4 m deployable telescope (its diameter being larger than that of the launch rocket) with near- and mid-infrared instrumentation. Its assignment is to discover the first stars that illuminated the universe and the first supernovae (Section 4.5). The intense ultraviolet emission of these objects will be displaced by the Doppler effect into the near-infrared, where the JWST will operate.

In years to come there will be a new generation of satellites (the *SAFIR* satellite, a NASA project, is shown in Figure 2.12) and interferometers, operating in the far-infrared, with diameters sufficiently large to resolve spatially all the first galaxies and confirm the story we have presented.

After the formation of the first galaxies, the universe continued to expand and evolve, producing in 13.7 billion years what we observe today, and we shall discuss this in the coming chapters.

---

4 *The primordial galaxies must be very bright in the ultraviolet and x-rays which, because of the Doppler effect, move to the optical and near-infrared bands which Hubble is observing.*

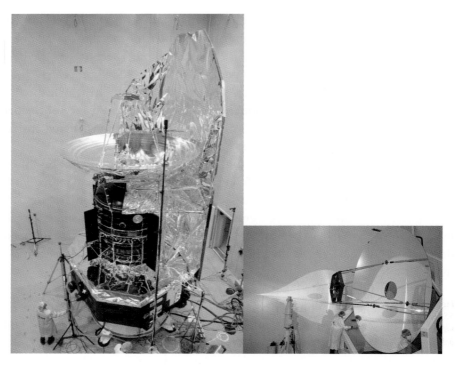

Figure 2.10: On the left is the Herschel satellite (with height 7.5 m, width 4 m and launch mass 3400 kg) during the acoustic tests. On the right is the mirror, 3.5 m in diameter, the largest ever sent to space. This mirror is a great technological achievement; made of ceramic (silicon carbide), it has a weight of only 300 kg, one-third of the HST's mirror (Figure 1.1).

2.11   Before the Big Bang

Until a few years ago many scientists believed that the story of the origins of the universe had been essentially settled and that future study would serve mainly to understand the details better. Some even said that, within 20–40 years, everything about the beginning of the universe would be understood, and that cosmology would have outlived its purpose.

Now, with the discovery that the expansion of the universe is accelerated, the necessity of again inserting a cosmological constant into Einstein's equations (Section 2.3), the problem of dark matter and dark energy (Section 2.9), and the need for new theories like inflation, are showing that our picture of the origin and evolution of the universe is in need of new measurements and perhaps new theories.

Among the theories on the origin of the universe, a model that is acquiring importance is the "string" theory, which proposes a scenario different from that of inflation. According to this theory, all the particles are formed from a single entity — a string. These strings are microscopic bits of energy, hidden

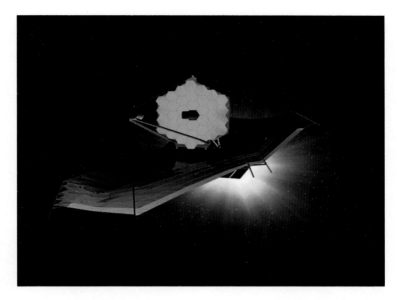

Figure 2.11: The James Webb Space Telescope. Hubble's successor will be launched in 2012. Built in a NASA–ESA collaboration, it will carry three instruments in the focal plane, all working in the infrared. The big shield, visible under the mirror, will protect the telescope from solar radiation, allowing the instrumentation to be below −240°C. Among the mission objectives is the measurement of the first stars born in the universe. (© ESA.)

in the heart of matter, that could unify the forces of nature; this is why the theory is sometimes called the "theory of everything" (TOE).

This theory predicts that the universe must have laws that avoid the initial singularity and keep it from collapsing below a certain size (just as there are laws in quantum mechanics that keep an orbiting electron from falling into a nucleus). If the theory succeeds in explaining the observations — it has not yet — we will no longer need inflation or a beginning of time. The universe would be a consequence of what existed before the Big Bang — a symmetric universe characterized by a "big crunch" which arrived at a finite, even if tiny, size and then was followed by a "big bounce" from which the current expansion began.

The strength of the theory is the elimination of the initial singularity. The weakness is that the theory, whilst avoiding inflation, does not explain the acceleration we see today, nor does it predict the uniformity of temperature (end of Section 2.5) or the flatness of the universe, all the observations that led to the theory of inflation (Section 2.6).

The scientists who sustain this theory are not frightened by these differencs; they are convinced that when the model has been refined by measurements, it will succeed in explaining everything. They maintain that it is important to eliminate the singularity of the Big Bang, which is intrinsically irrational.

According to string theory, if the universe exists today, it is derived from one symmetric in time that should have left traces in the fossil radiation; such

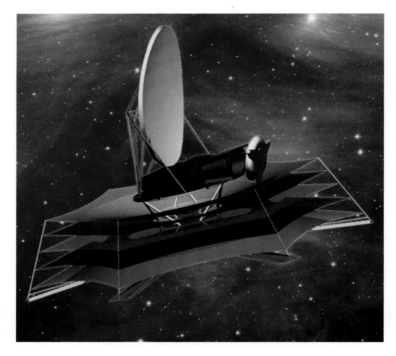

Figure 2.12: NASA's SAFIR Single Aperture Far-Infrared Observatory, a 10 m space telescope that represents the new generation of such instruments, will operate in the infrared and microwave regions of the spectrum. As in the JWST, the big shields in the satellite's upper part will protect the telescope from the solar light, allowing the mirror to go down to low temperatures. The big telecommunications radio antenna is visible above the shield. Such instruments are needed to study the first galaxies ever created. (© JPL NASA.)

traces could be seen by the European satellite *Planck* (launched 14 May 2009). If they are seen, it means that the hyper-expansion envisaged by inflation did not occur and what we observe is a consequence of the properties of the universe from which we come. Also, the gravitational waves could demonstrate the correctness of this theory. These waves have the advantage that they are not scattered by ionized free electrons (Section 2.8); therefore, they contain information about the universe at an earlier stage than the image of Figure 2.6 and allow us to see the first few instants of the universe's life. The problem is that of sensitivity; we may need decades to develop sufficiently sensitive antennae. (Such measurements will probably be made by the European Space Agency's *LISA* mission, using spatial interferometry.)

If these measurements confirm the string model, within a few decades cosmologists will be studying what existed before the Big Bang — a question that in the 1990s would have been meaningless in the spirit of what we said in Section 1.2. If this happens, it will mean that the pendulum of the origin of everything will swing back to the Aristotelian conception: nothing can be created out of nothing; if time exists, then it has always existed.

For now, we keep to the simpler interpretation of the measurements — in which the Big Bang is the origin of space and time — and take the next steps along the path of understanding our origins. We shall move in the diagram of Figure 2.9 to the birth of the stars, the first objects that, with nuclear reactions occurring in their interior, have brought light back to the universe. As we will see, the birth of stars is also that of planets; both are born in the same process.

Origins of Stars
and Planets

<u>Chapter 3</u>

3.1 The Stars and the Planets

One of the major topics of research in the last few decades has been to understand how stars and planets are born, and the connection between these processes and the origin of life in the universe. In fact, the stars furnish the energy that enables life to exist and the planets are the places where life can be born and evolve.

As we shall see in the following chapters, not all stars can host forms of evolved life in their surroundings, even though they all contributed to providing the elements from which life arose (Chapter 4). Among the planets, few can host evolved forms of animal life; in the solar system, only the Earth can do so. To estimate the probability of finding life in the universe as a whole, it is necessary to understand how probable the lucky combination of a planet like the Earth and a star like the Sun is.

We know that stars like the Sun are among the most common in the universe, but we can say very little about the frequency with which planets like the Earth arise. This is because our current models and observations do not provide a clear-enough picture of the circumstances giving rise to planetary systems.

Our present picture of the solar system, with the Sun at the center and the planets orbiting around it, is comparatively recent. Until the 16th century, the dominant picture was the Ptolemaic one, with the Earth at the center of the universe and all the heavenly bodies moving around it. Greek astronomers had considered the idea that the Earth might be moving, but their observations showed that stars, over the course of the years, remained firmly fixed in their relative positions in the sky. This led them to deduce that the Earth was fixed and the heavenly spheres were rotating around it, with the planets following the very special trajectories. They thought that if the Earth were in motion, the stars at various distances from the Earth would have to change their relative positions (imagine looking at a forest: the relative positions of the trees change as we move). They had no idea of the size of the universe and could not conceive that movement of the stars might be imperceptible to them. We can see how accurate their observations were if we realize that Hipparchus, in 150 BC, established that an Earth year lasted 365 days and 6 hours[1] and discovered the precession of the equinoxes (a motion that occurs with a period of 26,000 years; Section 10.1). The precision of the measurements, however, was not sufficient to reveal the motion of the Earth around the Sun; such an observation was possible only in 1800, after the great improvements made to the astronomical instrumentation. Supporting the idea of an Earth immobile in space was the observation that objects fall vertically, a fact that was erroneously considered to be impossible if the Earth were in motion.

Placing the Earth at the center of the universe was therefore based on measurements, showing that even rigorous methods can lead to erroneous results if the limits of the observations are not understood (Section 1.2). Not all the Greek philosophers accepted this conclusion: both Archimedes and Plutarch reported the Aristarchus as believing that the planets rotated about the Sun. Unfortunately, we do not know how he arrived at that conclusion, since the original text is lost.

The ideas and the observations of the Greeks were collected together three centuries later by Ptolemy, a great admirer of Hipparchus, who lived in Alexandria in the second century BC. He described this model of the universe, and the measurements supporting it, in the *Almagest*, a text that formed the basis of all Arab astronomy in the following centuries. It is this text which the Arabs carried to Europe, where it became the point of reference for all medieval astronomy.

The Ptolemaic vision of the universe faced a crisis in 1540, when the Polish astronomer Nicholas Copernicus wrote *De Revolutionibus Orbium Coelestium*, where he showed that the description of the planetary motions could be simplified if one allowed the Earth and planets to rotate about the Sun. He also maintained that it was unthinkable that the Earth should be the only immobile heavenly body, and that it was much more sensible to think of the Earth as rotating about its axis rather than that all heavenly bodies rotating around the

---

1 *The exact period is 365 days, 6 hours, 9 minutes and 9 seconds.*

Earth. In presenting this idea he was in some sense a precursor of the cosmological principle (Section 2.4): there are no privileged points in the universe!

Copernicus' data were no different from those used by the Greeks 1700 years earlier. Nor did he have the physical explanation of the motions that was discovered by Newton 140 years later. His intuition was only the result of a different interpretation of those data. According to legend, during his long stay in Italy, he was put on the right track by reading the text of Aristarchus, but there is no proof that this text survived until the 16th century. He had one great advantage over the Greeks — the greatly increased facility of calculation provided by the algebra that came to Europe via the Arabs.

The Copernican revolution was the beginning of modern astronomy and very soon there began to emerge the first recognizably modern theories of the universe and of the origin of the stars and solar system. Changing the perspective from which one looks at the world led, very quickly, to revolutionary ideas.[2]

A few years later, Johannes Kepler (1571–1630) adopted the Copernican idea that the Sun was at the center and the planets were in motion. Being a good mathematician, he searched for a geometric scheme to describe the motion of the planets. In this way, he established his famous laws that the planets move in elliptical orbits about the Sun and that their periods and velocities are related to the major axis of the ellipse (roughly the distance from the Sun). He did not know about gravity, which explains his law on physical grounds, and it was therefore described as celestial harmony.[3]

Around that time, Galileo Galilei (1564–1642) invented the telescope and used it to study the planets. He discovered that Venus had phases similar to those of the Moon: an observation compatible only with the assumption that it is orbiting around the Sun. This was the first direct evidence of the rotation of the planets around the Sun and it was confirming Copernicus' intuition.

Galileo also discovered the four brightest satellites of Jupiter, which he called the Medicians,[4] in honor of his patron. His observations showed that they rotated about Jupiter, reproducing on a slighter scale the motion of the planets around the Sun. The four satellites were also a direct proof that there are objects in the heavens that do not have the Earth as the center of their motion. The only regret of Galileo was that of not having found the "key evidence" of

---

2 *The word "revolution," which originally indicated the movement of a planet around a star, has now taken on an additional meaning. It has been used to describe phenomena or ideas that completely change the way things are conceived; that is why one talks of the French or the Russian revolution.*

3 *He maintained that God was the architect and musician of the universe — that music and the position of planets were expressions of a common divine harmony. His "laws" were so named because they were manifestations of a divine will.*

4 *Today, the four major Jovian moons are known as the Galilean satellites, in honor of their discoverer, who gave them the mythological names Io, Europa, Ganymede and Callisto.*

the Earth's motion about the Sun: its movement with respect to the "fixed stars"; a measurement done only in 1837 by the German mathematician–astronomer Fredrick Wilhem Bessel.

In 1636, Galileo published the *Dialogo sui Sistemi Tolemaico e Copernicano* ("Dialog on the Ptolemaic and Copernican Systems")[5] and came into conflict with the Inquisition and with Pope Urban VIII. His trial was convened in Rome and he was condemned in 1633. To avoid the punishment he made his famous retraction. Seventy years old and nearly blind, he was forced to go on his knees in front of his judges and retract everything he had said in contradiction to the Ptolemaic theory. In a place that for him, a believer, was the place of the truth, he was obliged to read a document that began with the words "With a sincere heart and pure faith I swear, curse and detest….," followed by an affirmation he knew to be untrue.

Anyone looking at the sky with a telescope would have found proof of Galileo's thesis. It is said that many of his judges refused to look, that others looked and pretended not to see, and that yet others found absurd explanations of what they saw for not being in conflict with the Pope. Only three cardinals (among them Francesco Barbarini, nephew of Urban VIII) refused to sign the condemnation.

Galileo's abjuration has often been criticized but it was an act of great wisdom; the laws of the universe would not be proven by his sacrifice, and humanity would have stupidly lost a genius. He avoided the stake but his life was destroyed. Confined to his house at Arcetri, he died few years later, alone and embittered. Those last years were not inactive, however: abandoning astronomy, he laid the foundations of modern mechanics.

The Inquisition persecuted him even after his death, prohibiting the public funeral that the Medici wanted to give him in recognition of his genius and to redress in part the wrongs he had suffered. From his first works Galileo had affirmed the principle that science should be based solely on observation and on what can be deduced from them (Section 1.2). In that, he indirectly had his revenge on the Inquisition and he pioneered the basis of modern scientific thought. Many of our reflections in Chapter 1 are based on the thoughts of Galileo.

---

5 *With the accession of Urban VIII (1623), there seemed to be more freedom for new ideas and so Galileo decided to publish his results. The book, initially written with the authorization of the Church and the Pope, who was a great admirer of Galileo, was in the form of a dialog between friends (Salvati and Simplicio) who argued about the merits of the two systems by showing them to another friend (Sagredo). Galileo chose this approach to present the two pictures without personal commitment to either. It did not succeed in hiding his ideas, however. Readers immediately identified Galileo with Salvati and the Pope with Simplicio, the one who came out badly in the dialog. Indeed, the Pope had discussed many of Simplicio's theses with Galileo before he became the pontiff. This brought Galileo into conflict with him and the Inquisition. The book was a great success as it was written in vulgar Italian, simple and readable by anyone, and it was translated into many languages, including Chinese.*

The story of Giordano Bruno (1548–1600) must have influenced Galileo's decision not to defend his convictions *ad extremum*. Bruno, a monk, philosopher and scientist, had preceded Galileo in the chair of mathematics in the University of Padua. In 1584, he supposed that the solar system was not unique and that there could be other stars, which, like the Sun, were surrounded by planets. He also postulated the existence of other earths, populated by other forms of life that might be intelligent. Unlike Galileo, he chose to defend the principle of freedom of thought. He resisted Cardinal Bellarmine (1542–1621), refused abjuration and suffered a horrible death: he was burned alive in Rome at Campo dei Fiori in February 1600, 60 years after Copernicus' publication.

In the same year that Galileo died, Isaac Newton (1642–1727), one of the greatest geniuses of science, was born in England. In 1687, he published the *Principia*,[6] in which he identified gravity as the force that regulates the motions of the heavenly bodies. He evaluated the effect of gravity on planetary motion, estimated the mass of the Earth and some of the other planets, and explained the tides. He also built the first reflector telescope, overcoming the chromatic problems of lenses. He gave a solid theoretical basis to the intuition of Copernicus and to the laws of Kepler. He was the first in science to find a simple principle (in this case gravity) from which many complex phenomena can be derived. He published the *Principia* only on the insistence of his great friend and admirer, Edmond Halley (1656–1742), the discoverer of the eponymous comet, overcoming the scruples of his profound faith in doing something disagreeable to the Church. With Newton, the Copernican revolution can be considered complete: after 147 years, the world had accepted the Copernican picture even though the Roman Church persisted with its opposition for many years.

Because of these extraordinary discoveries, the 1600s are celebrated as the century of rationalism, in which the capacity of the intellect to confront any problem was affirmed. The new ideas were opposed at length by the dogmatic vision of the church, even though not one of those scientists, in particular Newton, Galileo and Bruno, ever doubted the complete compatibility of their ideas with their profound religious faith: our intellect came from God, and to use it honestly to understand the world what He had created cannot be in contrast with His will. In spite of this, it was only in 1737, a century after his death, that Galileo's remains could be buried in the Church of the Sacred Cross in

---

6 The *Principia* *is a true mathematicophysical treatise in three volumes. The first deals with the foundation of mechanics and gravity. The second deals with fluid dynamics and demonstrates the dependence of the speed of sound on the density of the medium. The third deals with the effects of gravity in the universe: on the movements of planets around the Sun; on the Moon's orbit; and on the orbits of comets, including their possible return in the future. On the basis of the universal law of gravitation, the masses of the Earth and Jupiter were calculated, the tides were explained as an effect of the Sun's and Moon's attraction, and the precession of the equinoxes was computed.*

Florence, where they repose next to those of Michelangelo. His *Dialogue* remained on the Index[7] until 1822.

If it had not been for the Inquisition, the life of Galileo might have been longer. In spite of efforts of the intellectual world, even of many Catholics, only in 1992 did the Church criticize the way Galileo was treated. On the contrary, Cardinal Bellarmine, one of the principal architects of the Inquisition, was proclaimed a saint in 1930.

In 1992, Pope John Paul II proclaimed Galileo's thesis to be right. The Pope also recognized that "…Galileo, sincere believer, was more perspicacious than his theologian opponents" in the interpretation of the Holy Sciptures.

Opposition to the Copernican revolution was a long affair that caused useless suffering to people of great genius and good faith; it interfered with the progress of human thought and is a good example of the absurd positions to which preconceived ideas (that have nothing to do with religious sentiments) can lead when they replace scientific reasoning.

3.2   The Placental Cloud

The following century, the 17th, was that of the Enlightenment: people believed less in the absolute power of reason, which could sometimes lead to wrong conclusions, but they affirmed the need to use it, free from any imposition, to illuminate the path of humans.[8] Kant (1724–1804), in his *Critique of Pure Reason*, speaks of a tribunal in which reason was both the judge and the accused; with the Enlightenment the approach to the problems became more thoughtful and critical.

It was in fact Kant, in his *Natural History and Theory of the Heavens*, published in 1755, who suggested for the first time that the universe and the stars could have been born from a rotating nebula. This theory was conceived in parallel by the great French mathematician Pierre Simon Laplace (1749–1827), who in 1796 published it in his *Exposition of the World System*. According to this theory, the solar system was born from the collapse of a rotating nebula which, with the decreasing of its diameter, was rotating ever more rapidly, forming a disk with the Sun at its center. Then the planets were born from the inhomogeneity of the disk. This hypothesis has been confirmed by the recent observations.

In only 250 years, we have passed from Ptolemy's geocentric concept to the formulation of the processes that gave rise to stars and planets. This extraordinary development arose from the different ways in which the universe was

---

7 *The* Index Librorum Prohibitorum *("List of Prohibited Books") was a list of publications* prohibited *by the* Catholic Church. *It was formally abolished on 14 June 1966 by Pope Paul VI.*
8 *It was a return to the medieval thinking of Saint Thomas, who had great faith in reason — which he considered as God's gift — but he was wondering where it might take us….*

viewed and the increased confidence of humans in the power of reason. The Copernican revolution was a true "revolution," and with it the modern era was born.

After the original intuition of Kant and Laplace, it was necessary to wait until our times before a model, explaining the birth of stars and planets, could be formulated. What were lacking, in all this time, were observations capable of testing the correctness of the theory. Until the 1980s, no one knew which of the objects in the sky were protostars: stellar nuclei accreting matter. For this reason protostars were called the Holy Grail of astrophysics, something that everyone was searching for but no one could find.

Nowadays, there are many protostars in the catalogs of celestial objects, even though many problems remain unsolved: we do not know, for example, the precise mechanism by which the mass of a star is determined. The mass is a very important property, governing the entire evolution of a star and its possibility of hosting inhabitable planets (Section 4.3). There remain questions about planetary formation too: we do not know how planets come to have certain masses or why they are born at particular distances from their stars. In fact, we have only a qualitative picture of the processes by which stars and planets are born. We lack a model that would permit us to generalize the results and help us to understand how rare a planet like the Earth, which orbits around a star like the Sun, is.

Today, there are still many problems that cannot be solved with the existing instruments. Stars and planets are born inside nebulae, similar to those proposed by Kant and Laplace (Figure 3.1), which are, unfortunately, impenetrable by visible light. They have to be studied using radio and infrared radiation (Appendix A.4), with techniques that have existed for less than 40 years. Moreover, to make things more difficult, these phenomena take place in regions that are tiny as compared with their distance from us (Section 12.2), for which instruments with great angular resolution in the infrared are required; these instruments are not available today, but they will be in the future (Appendix A.5).

In Figure 3.1, there is an example of a cloud of dust and gas within which stars are born. It is an image obtained in the visible band by one of the great telescopes of the European Southern Observatory (ESO), installed in a Chilean desert at a height of 2600 m (Figure 12.11) — a site where the air is particularly dry and the best atmospheric conditions for astronomical observations are found. The image shows a dark cloud, a zone of the sky with no stars, that astronomers identify with the abbreviation B68.[9] We are aware of its existence only because the cloud is relatively close to the Earth and obscures the light

---

9 B68 (abbreviated from Bernard 68) has a lower density than the clouds from which stars are normally born (Figure A.5), but they all appear like B68 in optical wavelengths.

Figure 3.1: The dark cloud B68 in the Ophiuchus constellation. Four billion years ago, there was a similar cloud where our Sun is now. These clouds are opaque in the visible but they can be penetrated by light of infrared and radio frequencies. The same cloud is shown in Figure A.5 in the infrared bands where it becomes transparent. (Picture taken January 2001. © VLT-ESO.)

from the stars behind, creating the effect of a black patch in the sky. Clouds further away are less easy to identify, because they appear smaller and are indistinguishable from the dark zones that separate stars from one another.

Five billion years ago (about one-third of the age of the universe), there was a similar, but denser, cloud where the Sun is now, containing all the elements from which the solar system was built. Clouds like that in Figure 3.1 are called *placental clouds*. They are also called *dense clouds*, because they are much denser than the interstellar matter, even though their density is modest by terrestrial standards — millions of times lower than that of the best vacuum obtainable on the Earth.[10]

The cloud is about 7 million times larger than the Sun and, at the speed of light, it would take several months to traverse it. If we are to understand how

10 *A cloud of this type, in its densest regions, can reach densities of 100,000 atoms per cm^3, compared with about 10 atoms per cm^3 for interstellar matter. In comparison, the density of the Earth's atmosphere is greater than 10 quadrillion atoms per cm^3 (10^{19}) and the lowest achievable vacuum in the laboratories is of the order of hundreds of billions of atoms per cm^3 (10^{11}).*

stars are born from such clouds, we need to identify the process by which they can be compressed by a factor of 7 million to reach the size of a star. It is not a simple process and there are several opposing forces. The enormous compression has an important effect on planetary formation: as the cloud collapses, it rotates faster and faster and, as it nears stellar size, develops a flattened circumstellar structure, like a disk, that will evolve into planets (the process proposed by Kant and Laplace).

Stars and planets have, therefore, a common origin, being born by the same process. We should therefore expect planets to exist around almost every star, even though no one knows today how many of them are similar to the Earth.

3.3 From the Cloud to the Star

In the heavens there are large stars and small stars; they come in all sizes, from less than a hundredth of the mass of the Sun to a hundred times its mass. The mass of a star is a most important parameter. As we shall see, it influences the evolution of the star and defines its capacity to host planets on which life might develop (Sections 4.3 and 8.6). Today, we cannot say with certainty what mechanism determines a star's mass; the process is complex because, in order to allow some of the cloud matter to collapse into the star, a larger mass has to be spewed out into space, a process that is still an object of research. However, it is possible to give a simple, intuitive and qualitative answer: the mass of a star is, to a first approximation, proportional to that of the placental cloud. Larger clouds build larger stars, and smaller clouds build smaller stars. The reason is that when, in a dense cloud like that of Figure 3.1, a protostellar (star-forming) nucleus begins to accrete matter, the process continues until the cloud's matter is entirely accreted by the star or is spewed out into space. Otherwise stars would still be immersed in their placental clouds, and would not be shining in the heavens. The problem of the mass of a star is therefore turned into that of the mass of its placental cloud. What could it depend on? Why is one cloud smaller and another larger? The answer, also qualitative, is that the mass of the placental cloud depends on its temperature: little stars are born from small cold clouds, while larger ones are born from clouds that are more massive and warmer.

This answer, which is confirmed by observation, might seem surprising. How could a simple parameter, like the temperature of a gas, define the mass of a star? In reality, it is not a strange result. In a gas, the motion of the particles increases with the temperature (the temperature of an ideal gas is proportional to mv^2, where m is the mass and v is the velocity of the particles); therefore in a hot gas the particles move more rapidly than in a cold one. If the mass is too small, the particles will have sufficient velocity to escape from the gravitational attraction of the cloud and will disperse into space. For each temperature of the gas, therefore, there exists a minimum mass (called the *Jeans mass*,

Table 3.1 Power radiated by a surface of 1 m² in interstellar space (sheltered from the heat of the Sun), as a function of its temperature. The first column gives the temperature in degrees Celsius, and the second in absolute temperature (in kelvins), whilst the third gives the power radiated at that temperature. This radiated power is proportional to the absolute temperature of the body raised to the fourth power. Therefore, the lower the temperature's, the slower the body cools. The table shows that, while at −73°C the power radiated by 1 m² is *a little less than 100 W, at a temperature of −253°C (necessary for forming stars like the Sun) the power radiated* is 10,000 times lower. If clouds could only cool by radiation, the Sun would never have been born.

Temperature (°C)	Temperature (K)	Power radiated: $W = \sigma T^4 (\text{W/m}^2)$
−73	200	91
−123	150	29
−173	100	5.7
−223	50	0.35
−253	20	0.009

after the English physicist Sir James Jeans, who proposed it in 1926) for a cloud to remain gravitationally bound. The Jeans mass increases with the gas temperature. Moreover, because massive stars are built in warm clouds where particles move faster and therefore hit the protostellar nucleus more frequently, we have the paradox that massive stars grow faster than stars of lower mass (Section 3.10).

To obtain stars like the Sun, we need relatively small clouds with temperatures not much above 10 K (−260°C).[11] If the temperature is higher, the cloud needs to be more massive and larger stars are made. Therefore, to build a star like the Sun, a cloud has to be cooled to a very low temperature. This is not easy, because it is not easy to remove heat from a body already cold. We are used to seeing bodies cooling themselves by radiation, a mechanism that does not work with cold objects because, as shown in Table 3.1, the capacity of a body to radiate heat diminishes as the fourth power of the temperature.

Fortunately, there are molecules inside clouds (Section 4.10 and 5.1). They cool the cloud, lowering its temperature to that necessary for forming small stars like the Sun. When the molecules collide with each other, they become excited. That is to say, they convert small amounts of their energy of motion into vibrations of the atoms inside the molecule or rotations of the whole

---

11 *Measured in solar masses, the Jeans mass is given by* Mj = 40 (T³/n)^{1/2} , *where T and n are the temperature (in kelvins) and the density (in particles/cm³) of the gas in the cloud. Mj decreases with decreasing temperature and increasing density. If we substitute the typical density of 100,000 particles/cm³ for a placental cloud, we find that a temperature of about 10° is needed to obtain Mj = 1 (one solar mass).*

molecule. This reduces their velocities. To understand how this mechanism works, we can refer to a similar and more familiar example, that of a hammer striking a tuning fork. The hammer, having struck the tuning fork, bounces off with a velocity smaller than it had before the bump, the difference in energy being trasferred to the vibration of the fork. The energy of this vibration is then radiated away in the form of sound waves — a form of energy different from the mechanical motion of the hammer. Finally, this sound energy is absorbed by the surroundings, heating them.

Molecules react similarly: they are struck into excited states, absorbing a bit of the energy of their motion, they slow down (and thus decrease the temperature of the cloud), and then they re-emit the absorbed energy in the form of electromagnetic radiation (Appendix A.1). If this radiation manages to escape from the cloud without being absorbed by other particles, the cloud cools and shrinks. The electromagnetic waves that are emitted vary from molecule to molecule and range from infrared to radio frequencies (Figure A.1), just like different tuning forks emitting different notes at different frequencies.

Among the molecules of the placental cloud, one of the most important is that of CO (the dangerous carbon monoxide emitted by stoves when they are malfunctioning), the most efficient in cooling the cloud.[12] With every collision, molecules (especially CO) extract energy from the atoms or molecules of the clouds which have their velocity reduced.

The temperature of a cloud depends therefore on the abundance of the molecules capable of cooling it. The abundance of these molecules in today's universe allows the cooling a large fraction of the placental clouds to temperatures of about 10 K, which, as we have seen, is the temperature needed to form stars like the Sun. This is why stars of solar mass are so common in the universe.

Collision after collision, molecules slow down all the gas of the cloud, thus cooling it. Losing their speed, the particles under the influence of gravity fall inward and the cloud collapses, and becomes denser and denser. At the center of the cloud the density increases until it forms a nucleus, on which matter accretes to form a star. The collapse continues until the initial matter of the cloud has either collapsed onto the star or dispersed into space. The process is quite fast: for a star like the Sun it is estimated to last for a few hundred thousand

---

12 *CO is the most efficient molecule in cooling the cloud for four reasons. The first is that CO is the most abundant molecule in the universe after hydrogen. The second is that the CO molecule is very robust; it survives to temperatures and radiation that would destroy other molecules (it exists even on the surface of the Sun). The third reason is that the molecule can be excited by impacts which are too small to excite most of the other molecules, especially hydrogen, the most abundant. The final reason is that the CO molecule becomes de-excited by emitting energy in the form of (millimetric) radio waves which, unlike visible light, can exit the cloud (Figures 3.1 and A.5) and disperse into space. Other molecules are also excited by small impacts and emit in the radio region but they have a so small abundance that their contribution to the cloud's cooling is negligible.*

years (Section 3.10) — short compared with other astronomical phenomena or with the biological processes from which life has developed on the Earth.

Approaching the nucleus, the matter is constrained to rotate ever more rapidly, forming a disk (Section 3.6), from which planets will be born. Before considering this process we shall look at the giant molecular clouds where placental clouds are formed, and at different types of stars born in different eras of the universe — different because the chemical composition (and the molecular content) of the clouds has changed with time.

3.4 The Giant Molecular Clouds

The energy emitted by molecules can be measured by radio telescopes on the Earth (Figure 3.2) because, when they de-excite, they emit radiation at radio frequencies. Radio telescopes have thus been able, in the last few decades, to map all the gas and dust of our galaxy, even the most tenuous and thin, which, unlike the gas of the placental clouds of Figure 3.1, absorb little light and are therefore transparent.

These great tenuous clouds, transparent to visible light, are called *giant molecular clouds*. One of them is near us and is superimposed on the Orion constellation. On the left of Figure 3.3 is a picture of the constellation, familiar to any keen observer. In the southern part of the constellation lies the famous Orion nebula (shown on the right of Figure 3.3), where massive luminous stars are being born, coloring the cloud red (Appendix A.4). The red of the newborn

Figure 3.2: *The IRAM radio telescopes are examples of the instruments used to study molecular emissions. They were built in the French Alps, north of Grenoble, at an altitude of 2560 m above sea level to reduce the atmospheric absorption of millimetre wavelength radiation. (© IRAM, Granata.)*

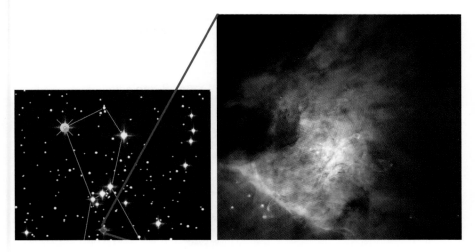

Figure 3.3: On the left is an image of the Orion constellation as seen from the Earth. On the top left, and in red, is the red giant Betelgeuse (Figure 4.3). On the right is an enlargement of the small red spot that marks the position of the Orion nebula. Superimposed on the constellation is the giant Orion molecular cloud, traced by the CO map in Figure 3.4. The Orion nebula is at the center of region 1c. The three central stars of the belt correspond to the stars of region 1b in Figure 3.4. (Picture taken by the HST. © NASA & C. R. O'Dell and S. K. Wong, Rice University.)

stars inspired the Greeks to name the constellation after the legendary hunter Orion; the three stars at the center of the constellation represent the belt from which his bloody sword hangs, and the point of the sword is the nebula with his reddish color.

If our eyes were sensitive to the light emitted by the CO molecule, the molecular cloud of Orion would appear even larger than the constellation itself. What the radio telescopes see is shown in Figure 3.4, which can be read like a topographic map. The contours show bands of equal abundance of CO; the darker points are where the abundance is greatest and correspond to the regions in which there are dense clouds and in which stars are being born. The clouds extend for over 16° on the sky, which is 32 times the diameter of the full moon. If the CO emission were visible, the Orion nebula would appear as large as 32 full moons side by side!

Molecular clouds do not live very long, being tenuous and lightly bound (compared with placental clouds); they are destroyed when they cross the plane of the galaxy, which happens within 10–100 million years of their formation. Therefore, the birth of stars has to occur much more rapidly (Section 3.10) or there would be no stars in the sky.

Having discovered these thin clouds, by their molecular radiation, it was realized that interstellar matter, before forming dense clouds, arranges itself in larger structures, which have dimensions of hundreds of light years (in contrast with the light months of the placental clouds), containing as many as a

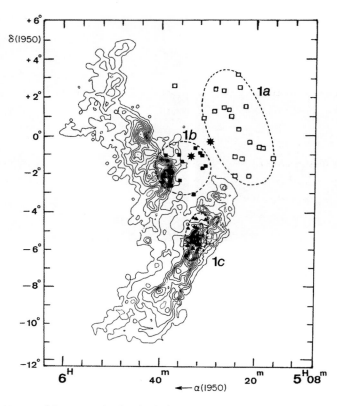

Figure 3.4: A CO map of the giant molecular cloud of Orion. In the past, the cloud was much bigger than it appears today, because part of it has been destroyed by earlier generations of stars. Today, at least three generations of stars are identified: the oldest stars, contained in region 1a, were born 12 million years ago, comparatively recent in evolutionary-biological times; dinosaurs had already disappeared 50 million years earlier. Once born these stars destroyed the cloud around them (Section 3.8). The second generation of stars, in region 1b, were born 7 million years ago; among them we recognize the stars of Orion's belt (Figure 3.3, left). The last generation are in region 1c, where stars are still being formed. The crowding of the lines of constant light indicates that 1c is the region of greatest density in the cloud. At the center of 1c, there is the Orion nebula (Figure 3.3, right) illuminated by the newborn stars. (© H. Zinnecker et al., Protostars and Planets III; E. Levy and J. I. Lunine, eds., 1993.)

million solar masses (in contrast with the few solar masses of the placental clouds). These tenuous clouds are made up mainly of the molecules of hydrogen (H_2) and carbon monoxide (CO) — the second-most-abundant molecule in the universe. As we have seen, although CO is 10,000 times less abundant than H_2, it is nevertheless the principal coolant of the clouds. There are many other molecules (H_2O, NH_3, CS_2, etc.; Section 4.10) that contribute to the cooling of the cloud, although they are less abundant.

Cloud fragmentation. Observations indicate that stars are not born in isolation, separated one from another. In the darkest zones of Figure 3.4, stars are born in groups of tens or hundreds — in some cases thousands — of stars. The most numerous of these groups are called "clusters."

We do not yet have a satisfactory theory of how the effects of turbulence, magnetic fields and rotation within the clouds contribute to the origin of these groups. Nevertheless, astronomers believe that the majority of stars are born in groups because they observe that all the protostars are multiple and that about half of the stars in the sky belong to groups of two or more objects. Because with the passing of time stars tend to separate each other not to bind, it is reasonable to believe that even stars that are now isolated, like the Sun, were in a group when they were born.

Clusters of stars are divided into two groups, *open clusters* and *globular clusters* (Figure 3.5); born in different epochs, from clouds of different chemical composition, they differ in their properties.

An example of an *open cluster* is the Pleiades (Figure 3.5, left), easily visible to the naked eye because it is close to the Earth. Such clusters are called open because their stars are not gravitationally bound. They appear as groups of tens, hundreds, or sometimes thousands of stars asymmetrically distributed. They are all in the galactic plane (Figure 8.2) and their stars are among the youngest of our galaxy. They have a chemical composition similar to that of the Sun (which was probably formed in a similar cluster, then moved away from it). Their ages range from hundreds of thousands of years to billions of years.

Globular clusters are much older than open clusters; they were formed, along with our galaxy, about 12 billion years ago, when stars still did not have all the heavy elements that exist today. These clusters are much larger than the young ones and contain hundreds of thousands of stars; the largest (Ω Centauri) has a mass of five million Suns. The stars of these clusters are gravitationally bound

Figure 3.5: Examples of star clusters. On the left are the Pleiades: an open cluster easily visibile to the naked eye. It is a young cluster (less than 100 million years old) of about 3000 stars, 400 light years away from us. On the right is M13, a globular cluster, about 12 billion years old — the age of our galaxy. Open clusters like the Pleiades are in the plane of our galaxy (Figure 8.2). Globular clusters, on the contrary, are spherically distributed about the galactic center, because, when they were formed, the matter making up the galaxy had not yet flattened into a disk. (Left: © Bert Katzung, 2004, San Rafael California. Right: © Marco Fazzoli Grosseto, 2003, http://www.marcofazzoli.com.)

and for that reason they have spherical symmetry (Figure 3.5, right). Globular clusters are distributed within a sphere about the galactic center because they were formed when the matter from which the galaxy was born had not yet flattened into a disk but had a spherical structure. (Our galaxy originated from an enormous placental cloud over 100,000 light years in diameter!) The chemical composition of the stars in these clusters is very different from that of our Sun, having far fewer heavy elements. For this reason, if they have planets, the rocky ones are probably absent or rare because of the lack of heavy atoms necessary for building them.

The common origin of all the members of a cluster allows astronomers to understand the physical mechanism that controls the functioning of stars. Since the stars of a cluster all have the same age and the same chemical composition — that of the cloud from which they were born — the differences between them can depend only on one parameter — their mass. It has therefore been easy to interpret the observed differences as a function of mass. Having worked out how their evolution is influenced by their mass, it becomes possible to study the influence of other parameters. Our understanding of many of the properties to be discussed in the next chapter comes from studying stars in clusters.

The way in which the stars are born can be summarized thus: the interstellar matter, enriched by matter expelled from earlier generations of stars (Sections 4.4–4.6), first condenses into *giant molecular clouds* (Figures 3.3 and 3.4), transparent to the visible light. Within these molecular clouds several dense clouds are formed: the *placental clouds* (Figure 3.1). Inside these dense clouds, stars are born in more or less numerous groups. With the passage of time, some stars escape from the gravitational attraction of the cluster and become isolated objects like the Sun. This isolation is believed to be essential for the birth of habitable planets, because the mutual interaction of stars within a cluster is thought to render improbable life on their planets.

3.5   Populations of Stars

The great differences we observe between the old clusters, like globular clusters (born in a universe with few heavy atoms), and the young open clusters that were born recently, show how much star formation has changed with time and how it has been influenced by the heavy atoms produced in the previous generations of stars (Section 4.2). Stars born in different epochs differ so much from one another that astronomers classify them into three *populations*. The oldest is called *population III* and the youngest, to which the Sun belongs, is called *population I*. The numbering is inverted with respect to chronology because it follows the order in which they are identified.

The stars of population III were the earliest; they were born in the primordial universe, before the formation of galaxies. At that time, only hydrogen and helium existed (Section 2.1) and the only molecule capable of cooling the cloud was hydrogen (H_2), which is not excited below 100 K. Without CO, the clouds could not cool themselves and remained at temperatures of hundreds of kelvins (as opposed to the 10–30 K of clouds cooled by CO). They are therefore much larger (Section 3.3) than those of Figure 3.1, hundreds of times more massive than the Sun. These were the stars of Figure 2.9 that brought light back into the universe. They were very luminous and consequently had short lives (Section 4.3), a million years at the most. Thereafter they exploded as supernovae, expelling the heavy atoms that had been produced (Section 4.5) and leaving a black hole around which the first galaxies might form (Section 2.10). Stars of population III are no longer seen, but we believe that they existed, because we observe the matter they produced, out of which our galaxy was born.

The globular clusters were born from the matter expelled by the first stars; they constitute *population II*. They have a lower proportion of heavy elements than the Sun and are distributed spherically around the center of the galaxy because they were born before the galactic matter had flattened into a disk. From the beginning, these clusters were dominated by massive and luminous stars that did not last very long. Later, thanks to the matter produced by those stars, some smaller ones — even smaller than the Sun — were born, living long enough to remain brilliant now, after 13 billion years (Figure 3.5).The birth of globular clusters can still be observed today in the Magellanic clouds[13]: two small young galaxies near ours, in which heavy elements are much less abundant than in the Sun and which are similar to our galaxy in its early years.

With the passage of time, the matter of the protogalactic sphere flattened to form the plane of our galaxy (the Milky Way; Figure 8.2), and successive generations of stars were born — all in this enormous disk (Figures 8.1 and 8.2). These stars have continually enriched the interstellar matter with heavy elements, leading to the abundance of molecules of the population I stars of the open clusters.

Population III is therefore characterized by massive stars that did not live long and for which we observe only the chemical elements they produced; population II is made up of great clusters of stars that have existed for 13 billion years; population I consists of young stars with ages ranging from millions to billions of years. To the last population belong those stars that, although born in clusters, have since liberated themselves and become single stars like our Sun, and they are the best candidates for harboring habitable planets.

The formation of stars changes with time because the chemical composition and the abundance of molecules in the placental cloud change with time; both

---

13 *The Magellanic clouds are two young galaxies discovered by the great navigator Ferdinand Magellan in the Southern Hemisphere; they are easily visible to the naked eye and look like two luminous flocks in the sky.*

play a fundamental role in the process of star formation. It is moreover interesting to note that the existence of stars like the Sun, necessary for life, depends on a small molecule like CO. If nature had not provided such a molecule in sufficient quantities, the universe would be dominated by massive stars that, as we will see, explode as supernovae and do not live long enough to allow life to develop.

3.6 Disks

When Kant and Laplace put forward the hypothesis of a protosolar nebula, they were seeking an explanation of the heliocentric structure suggested by Copernicus and explained by Newton — a star surrounded by orbiting planets. They asked themselves what could have generated this nebula and guessed that the initial configuration might have been something similar to the cloud of Figure 3.1. A structure similar to the solar system could arise from the contraction of such a cloud because, in contracting, the cloud would rotate faster and faster according to the law of physics known as "conservation of angular momentum" (Figure 3.6). This law states that *if you reduce the diameter of a rotating object by a certain factor, its rate of rotation has to increase by the same factor*. In order to become a star, the diameter of the cloud decreases by a factor

Figure 3.6: *The law of conservation of angular momentum states that in an isolated system, such as a cloud, the rate of rotation multiplied by the diameter of the system remains constant. This physical law is well known because it is easily observed in everyday life. One example is the very rapid rotation that ice skaters can achieve. To execute this maneuver they first spread their arms and legs as far as possible (increasing their effective diameter) and rotate as fast as they can. Then, they quickly reduce their diameter, by bringing in their arms and legs, increasing proportionately their rate of rotation.*

of 10^7. Rotating ever faster, the cloud would flatten, forming a disk-shaped structure, out of which planets could be born.

This increase in rotation has two effects: first, the cloud breaks into pieces; second, as predicted by Kant and Laplace, the matter flattens out and forms a disk. Both these effects can be observed in the sky.

The first effect contributes to the fragmentation of the cloud, which we discussed in a previous section. The cloud divides itself into smaller clouds from which stars are born. This is the origin of the clusters of stars seen in the sky.

The second effect leads to the formation of disks: the rotating matter tends to settle into a trajectory where the gravitational attraction toward the center of the cloud is balanced by the centrifugal force tending to throw the matter outward. Think of a merry-go-round. As soon as it begins to rotate, its seats arrange themselves in an "imaginary" disk about the axis of rotation — held in place by the chains with which they are attached to the rotational axis. Thus the same is for the matter of the disk, which, depending on the velocity of rotation, disposes itself at a distance from the center so that the centrifugal force is balanced by gravity.

Astronomers have searched for disks around protostars since they were first proposed by Kant and Laplace.[14] It was only in the 1970s that we began to find indirect proofs of their existence, by measuring their emissions at infrared and millimeter wavelengths. Finally, in the 1990s, a direct proof of their existence was discovered. Credit goes to the HST (Figure 1.1), which managed to photograph them, thanks to its extraordinary spatial resolution. Imaging the disks is not easy, because their dimensions are small compared to their distances, demanding very high spatial resolution.

In Figure 3.7 we see an HST image of a region of Orion where stars are being born. In part (a) of the figure we see a region of the nebula illuminated by distant stars which are out of the field of view. We can see two protostars, which are enlarged in (b) and (c). In (b) we can clearly distinguish a disk with a star being born at its center. The star appears red because its light passes through dense layers of gas and dust which scatter the blue light but let the red through, just as the atmosphere does to solar light at sunset (Appendix A.4). In (c) we see an even younger object, an oblong cocoon, the residue of the parental cloud (Figure 3.1) which, having lost most of its matter, has become transparent and permits us to see the black circumstellar disk. The disk is so dense that it obscures the light of its protostar, even though the latter is more luminous than the Sun. In (a) we also see five protostellar objects that are probably the result of an earlier fragmentation.

---

14 When the first planetary nebulae were discovered, it was thought that they were protostellar disks, hence their mistaken name. They are not newborn stars but, as we shall see, dead stars (Section 4.4).

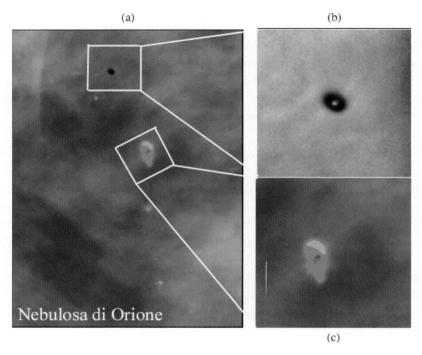

(a)　　　　　　　　　　　　(b)

Nebulosa di Orione

(c)

Figure 3.7: Star formation in the Orion nebula. (a) A region in which five condensations of matter are identi-
fied. (b) A disk with a red protostar at the center; (it is red for the same reason that the Sun is red at sunset
(Appendix A.3). (c) A still-younger object with a residue of the parental cloud; the central black spot is the
protostellar disk. (Picture by the HST. © NASA & C. R. O'Dell, Rice University and M. J. McCaughrean, MPIA.)

In Figure 3.8 we see two disks, also photographed by the HST. The one on
the left is seen in profile but we can just see, above the disk, the red light from
the protostar. The other disk is seen face-on and clearly shows the protostar at
the center. (The circular white zone is where the detector has become saturated
by the intensity of the light.)

3.7 Outflows

The hypothesis of Kant and Laplace, whilst explaining how circumstellar disks
and therefore the planets were formed, did not explain how the cloud's material
could reach the surface of the star. As explained in the previous section, matter
would have settled where centrifugal force was balanced by gravitational
attraction. In this state of equilibrium, the disk would orbit about the protostar
forever, as indeed the planets around the Sun have done for billions of years.

Since the stars do exist, there must be other mechanisms for slowing the
orbiting matter enough to let it fall into the central nucleus. These mechanisms
must act on the gas and dust alone, allowing the planets to remain in their
orbits. They are still not completely understood. They are complex, involving

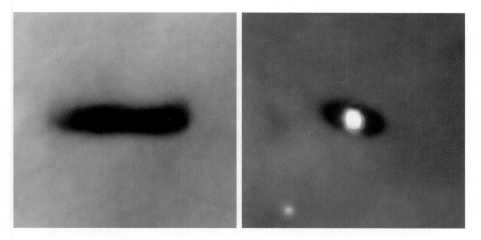

Figure 3.8: Two examples of disks in Orion. (Photographed by HST. © NASA & C. R. O'Dell, Rice University and M. J. McCaughrean, MPIA.)

magnetic fields and viscosity — friction between the inner parts of the disk, which orbit rapidly, and the outer parts, which, according to Kepler's law, orbit more slowly. Because of this friction, matter near the star slows down until it falls into the star, whilst that further away is accelerated until it reaches escape velocity and disperses into space. To get an idea of how much the orbital velocity can differ in the various parts of the disk, we may consider the planets of the solar system. They orbit with the same velocity as matter in a disk at the same distances from the Sun. Mercury, which is 60 million km from the Sun, makes a complete orbit about the Sun in 87 days. Pluto, which is 6 billion km out, does it in 247 years. Imagine the effect of some friction between matter in those two orbits!

The least-understood of these mechanisms produce violent outflows in a direction nearly perpendicular to the disk, without disturbing the motion of the matter moving toward the star in the plane of the disk. The outflows are very powerful jets about which little is known. When they were discovered in the 1960s, they surprised astronomers who were searching for indications of matter flowing into the protostar, with motion for which there is still little evidence even today. Instead, the astronomers found jets carrying matter away! Outflows are found in all the protostars, and observations indicate that the power of the outflow increases with the size of the disk.

The energy in these jets is so large that if it were transferred to the parental cloud, it would quickly destroy it. But, as one can see in Figure 3.9, the jets are very narrow, and therefore the energy transferred to the cloud is small. The jets transit the cloud and continue for tens of light years. Sometimes they strike other clouds nearby, compressing them enough to trigger the birth of new stars.

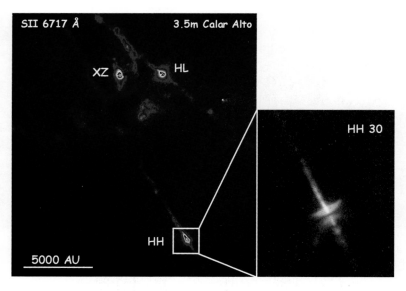

Figure 3.9: Image of a jet. On the left is a photograph taken by a ground-based telescope at Calar Alto, Spain; on the right is one taken by the HST. (© Left: R. Mundt MPIA Heidelberg. Right: NASA & T. Ray.)

On the left of Figure 3.9 is an image of a region where stars are just born, obtained by a ground-based telescope; on the right is an image taken by the HST. In the latter image we see a large region of diffuse luminosity (blue because it is scattered light; Appendix A.4) crossed by a dark band. The light is from the protostar, and the dark band is the disk.

Perpendicular to the disk is a jet; it is so energetic that it strikes and excites atoms along its way (Appendix A.2 and Section 3.3); during their subsequent spontaneous de-excitation, these atoms emit light. The image gives an idea of how narrow a jet can be, similar to a laser beam. The weak blue strip crossing the entire image on the left tells about the extension of the jet, the scale being given in AU (astronomical units) (Appendix B) at the bottom. Above, one sees a similar object (indicated by the initials HL), producing a jet of similar length. The jets seem to be well over 5000 AU, but only part of them is made visible by collision with interstellar matter; in reality they extend even further. Jets have been measured extending more than 5 light years from the exciting stars, a distance equivalent to 30–50 times the diameter of a parental cloud.

In summary: the rotation of the cloud is slowed down at the expense of its own material, which is expelled from the cloud. In the end only a small fraction of the mass of the cloud will reach the star; the rest returns to the interstellar medium to condense into new giant molecular clouds, where it will form new dense nuclei from which new stars will be born.

Figure 3.10(a) illustrates a protostar schematically: a stellar nucleus is surrounded by a disk onto which the matter of the parental cloud falls following

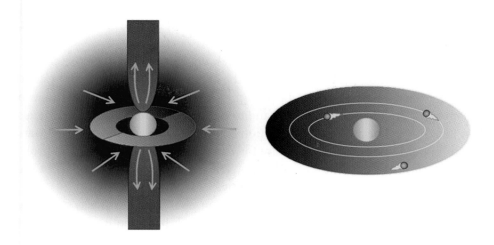

Figure 3.10: (a) Schematic picture of a protostar: the blue sphere represents the parental cloud; the arrows represent the matter falling from the cloud onto the disk. By losing angular momentum, the material in the disk spirals into the protostar. The jets of material, i.e. the outflows, are indicated by the two red cones extending perpendicular to the disk. (b) A close-up of the disk, with some protoplanets growing out of the dust and gas in the disk.

the directions indicated by the arrows. Jets of material leave the disk in the perpendicular direction.

3.8 The Planets

The mechanisms we have just described are very efficient at slowing down the microscopic particles of gas and dust in the disk. They do not act efficiently on larger objects, and this explains how planets can form. If a solid nucleus is formed in the parental cloud or in the disk, and if it exceeds a certain mass, there is a good probability that it will survive and continue to accrete mass without being braked by the mechanisms that affect the gas and dust. After all the dust and gas of the parental cloud has either collapsed onto the star and planets, or been expelled by the outflow or blown away by the violent stellar winds that accompany the beginning of stellar nuclear activity, the planets will remain in their orbits, just as they do in the solar system today.

According to models of the birth of the solar system, the process began when the density of the disk was high enough to make collisions between particles very likely. These colliding particles aggregated the grains of dust that grew from one collision to another until they reached sizes of a few centimeters. On these grains, the gas and dust condensed ever more easily until a collection of solid bodies called *planetesimals* [from the Latin words *planeta* and (infinit)*esimal*] was formed, with dimensions ranging from millimeters to tens of kilometers. Some, those nearest to the Sun, were rocky; others, in the cold outer regions of the solar system, were mainly composed of ices — above all, those of H_2O, CO_2 and NH_3.

Planetesimals continued to collide, break up, and reaggregate in a chaotic process that produced the embryos of the actual planets of the solar system. Once a relatively large object (a protoplanet) is built up, it will continue to accrete gas and dust from the disk and to capture the remaining planetesimals in its vicinity. The asteroid belt (Section 8.3) found between the orbits of Mars and Jupiter is evidence of the failure for a planet to form in that region, probably because of the proximity of Jupiter. The planetesimals in that part of the disk have therefore survived until our time.

Just at the end of the planetesimal accretion phase, the Earth was struck by a large asteroid, comparable in size to Mars, which has a tenth of the mass of the Earth. The resulting mass of ejected debris then aggregated into the Moon. The impact occurred when the Earth had already formed and when its ferrous matter had sunk into the center, and therefore our satellite is made up of matter from the exterior parts of the Earth. This explains why the Earth has an iron nucleus, whilst the Moon has not, and why the average density of the Moon is similar to that of the Earth's crust, well below the average density of the whole Earth.[15] A similar mechanism probably accounts for the birth of Charon, the largest satellite of Pluto.

The residue of this hierarchical growth is still evident in the solar system, most noticeably in the asteroid belt, where we find objects of all sizes (Section 8.3), from grains of dust up to objects hundreds of kilometers in diameter. They are similar to the rocky planetesimals that built the Earth and that formed the planets in the inner, hottest parts of the disk, where the lightest elements (mainly H and He) have evaporated. Other planetesimals that have survived are the comets, which have kept most of their light elements because they were formed in the outer, coldest part of the disk where we find the giant planets. They are composed mainly of water ice, carbon dioxide, methane and ammonia. Examination of the water content of comets shows that they were formed in the outer regions of the solar system,[16] at temperatures between $-230°C$ and $-250°C$, close to absolute zero ($-273°C$). Only far from the Sun can such temperatures be found.

The efficiency of the process by which matter aggregates to form the planets of the solar system depends on the frequency with which the particles collide

---

15 *The Moon's average density is 3.34 times that of water, a value similar to the 2.8 of the Earth's crust and quite different from 5.52, the mean for the Earth as a whole.*
16 *According to the relative spin directions of its hydrogen atoms, the water molecule has two forms called "ortho" and "para," which are recognizable spectroscopically (Appendix A.1). The ratio between the two forms depends, in a sensitive way, on the temperature of the gas from which the ice is formed. From its emitted light, the proportions of the "ortho" and "para" forms of water in a comet can easily be calculated. From the ratio of the two, the temperature of the gas from which the comet was formed billions of years ago can be obtained.*

with one another; this increases with the density of the gas and with its temperature, which is proportional to the square of the velocity of the particles. The planets are thus born quicker in the inner part of the disk (where the gas is denser and hotter) than in the outer parts (where it is less dense and colder). Even the chemical composition of a planet depends on its distance from the Sun; light atoms like hydrogen and helium are retained only by planets which are a long way from their star. Thus, we have two different classes of planets in the solar system: the inner ones, which are close to the Sun and are small and rocky; and those more distant, which are massive and are composed of light elements.

The inner planets (Mercury, Venus, the Earth and Mars) took about a million years to form near the Sun, which was then 20–100 times more luminous than it is today. They are called rocky or terrestrial-type planets, because they consist predominantly of heavy atoms, especially iron and silicon. They were formed of rocky planetesimals from which the light elements, at least those which were not chemically bound to the heavier atoms, had mostly evaporated.

The outer planets (Jupiter, Saturn, Uranus and Neptune) took longer to form, probably tens of millions of years, because the density and temperature in that part of the disk were lower. The first part of the process by which the giant planets were formed was similar to that of the inner planets — the aggregation of planetesimals. When a planet, similar to the Earth in size, was built up from planetesimals, it began, unlike the inner planets, to accrete light gaseous atoms — mostly hydrogen and helium — from the disk, because its large mass and low temperature were sufficient to trap them. This accretion of gases continued until inside the Sun sustained nuclear reactions started producing an intense flow of particles (solar wind[17]) that swept away the residual gas of the disk. The external planets are therefore more massive than the internal ones because they have been able to retain the hydrogen and helium of the disk, which provide most of their mass. For this reason they have many more light atoms than the rocky planets.

Finally, we note that these massive planets, because of their mass and their distance from the Sun, took up most of the angular momentum of the solar system (Section 3.6), thus allowing other disk material to spiral into the star. Indeed, at the end of the process, 98% of the angular momentum of the disk will be held by these planets, the remaining 2% being absorbed into the Sun.

Outside the region occupied by the giant planets, the density and temperature of the gas were so low that the aggregation process was very slow and unable to form bodies larger than planetisimals, with very few of them exceeding 100 km in diameter. These fill a large band encircling the solar system — the Kuiper belt (Section 8.2), which extends from 39 AU to about 500 AU from

---

17 *The solar wind consists of particles flowing away from the Sun at high velocity.*

the Sun. From time to time, the orbits of these objects is perturbed by the giant planets and they enter the inner solar system. They are the short-period comets whose periodicity was discovered by Edmond Halley, Newton's great friend.

Pluto (Section 8.3), which is no longer officially a planet, is one of these objects. It is much smaller than the outer planets, having roughly the same mass as our moon, and is rocky like the inner planets. The process leading to its formation was probably so slow that it did not manage to form a nucleus large enough to accrete the light elements of the disk before the solar wind blew them away.

The planetesimals, in contrast to the gas, survived the solar wind because they were sufficiently massive to resist its pressure. Most of them were aggregated into planets in the first phase of solar system's evolution. A few, those whose disordered orbits carried them close to the giant planets (especially Jupiter and Saturn; Section 10.2), were deflected into the Sun or out to the edges of the solar system, forming the Oort cloud (Section 8.2). In a relatively short time, the solar system assumed the orderly aspect that we see today, even though an occasional asteroid or comet in a disordered orbit passes near the Earth.

Some models say that very massive stars,[18] of more than 5–10 solar masses, cannot have disks. All these stars are distant from the Earth and are formed so rapidly that we have no observational data to allow us to determine with certainty whether or not they possess disks.

Massive stars, however, constitute less than 5% of all the stars in the sky, so we may conclude that almost all (if not all) stars have disks because without disks they cannot form. We must therefore expect to find planets around some 95% of the stars in the sky. How many among these planetary systems are similar to ours? The limitations of current observations prevent us answering that question.

The picture of the formation of the stellar systems described in this section, with its hierarchical growth of planetesimals, explains how the solar system was born. It does not, however, explain the formation of those recently discovered systems that have a giant planet near the star. It is now believed that there are at least two processes leading to the formation of planetary systems: the first is the one we have just described that creates systems like our solar system; the second, which will be discussed in the next section, is very fast and allows for giant planets to exist near their stars but rules out the possibility of habitable planets in this area with the exception of the satellites of the giant planets.

---

18 In Chapter 4, we will see that these stars are very important for the chemical evolution of the universe but they are completely unsuitable as hosts to biological life because they do not live long enough.

3.9 The Discovery of the First Planets

From what we have said up to now, we should not be astonished by the discovery of many planets in recent years. On the contrary, because planets are believed to exist around most, if not all, stars, we might be astonished that they were not seen earlier; the reason is that, it is difficult to detect a planet because it is visible only by reflected light, that is a small fraction of the total emission of its parent star. Given its great distance, the planet appears very close to the star, which dazzles astronomical instruments and renders the planet not observable. Separating the planet's light from that of the star is practically impossible today but, as we will see in Chapter 12, it should be feasible in the coming years. Up to now, scientists have discovered planets using techniques based not on the direct observation of planets but on secondary effects.

The first extrasolar planet was discovered in 1994 by the Swiss astronomer M. Mayor and D. Queloz, using the Doppler effect. The laws of physics demand that both the planet and the star do orbit around their common center of mass.[19] If the planet is relatively massive, and is close to the star, this common center of mass can be appreciably displaced from the center of the star, which therefore orbits around a point between the planet and the star (Figure 3.11).

To an observer, represented in the figure by a telescope, the star is seen alternately approaching and receding, depending on which part of the orbit it is in. According to the Doppler effect, described in Appendix A.3, the observed spectrum of an object is displaced toward the blue end of the spectrum when it is approaching, and toward the red end when it is moving away. These displacements can be measured on the Earth, and the lower part of Figure 3.11 shows the spectrum of the star 51 Pegasi, about 45 light years from us. The horizontal axis of the figure represents time, whilst the vertical axis gives the velocity of the star as it alternately approaches and moves away from us. From these observations, Mayor was able to deduce the existence of a planet orbiting 51 Pegasi.

From these measurements the mass of the planet can be obtained. The technique is obviously not very sensitive and it does not work well for giant planets that are far from their star. There are two reasons for this. First, in the case of a distant planet, because the displacement of the center of mass is small, the displacement of its spectrum is small and difficult to detect. Secondly, even supposing that the sensitivity of the instrument is high enough, it will be necessary to follow the star for at least half an orbit in order to produce data like those in Figure 3.11. But planets far from their stars have

---

19 A fact well known to car drivers is that if a wheel of their car is not well balanced, it will vibrate because the center of mass, about which it tries to rotate, is not its axis.

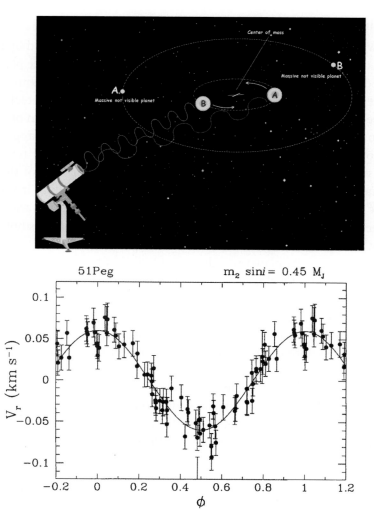

*Figure 3.11: The measurement of the first extrasolar planet. Top: The star and the planet are orbiting about their common center of mass and they are shown in two positions: A and B. A ground-based observer (who does not see the planet) sees the star approaching for half the orbit, and moving away during the other half. If the planet is massive, and close to the star, the speed at which the star approaches or moves away from the observer can be measured via the Doppler effect (Appendix A.3): the starlight shifts toward the blue end of the spectrum when the star is approaching the Earth (position B of star and planet), and toward the red end when it is moving away (position A). Bottom: The approaching/receding speed of the star 51 Peg, as measured by M. Mayor in 1994. This was the first clear observational evidence for the existence of an extrasolar planet. © Top: Geoff Marcy, University of California. Bottom: M. Mayor and D. Queloz, Nature **378**, 355 (1995).*

long orbital periods and it is necessary to wait 10–30 years to accumulate significant statistics. Moreover, a planet like the Earth can never be discovered with this technique because its mass, compared with that of the star, is too small to cause measurable displacements, especially if more massive planets are also present.

From 1994 to 2011, 771 planets belonging to 559 planetary systems were discovered using this method. The number of discovered planets is higher than the one of planetary systems because using the Fourier an of the signals, it was possible to detect more planets in a single system.

Starting in 2007, another search technique was added to the radial velocity: while passing in front of a star, a planet causes a reduction in the observed stellar brightness, which can be detected by very precise photometric measurements. Such measurements can be done both from the ground using very large telescopes (8 m diameter or larger) and from space with a mission dedicated to this research. The first of these satellites was the European *COROT* (COnvection, ROtation and planetary Transits), launched in December 2006; the second was *Kepler* of NASA, launched in March 2009.

The advantage of this technique is that it can detect planets as small as the Earth; the disadvantage is that it can only be applied to planetary systems whose orbits are edge-on to our line of sight, and these are estimated to be a few per thousand of all planetary systems. Up to 2011, 230 new planets belonging to 196 planetary systems have been detected with this technique.

The number of known planets has increased with dizzying speed after the discovery, of the first in 1994: by 1996, two years later, 20 planets were discovered; by 2000, 60 planets; in 2003, 120; in 2004, 150; in 2005, 200; and 2007; and finally about 800 by 2011 (some planets have been detected with both methods). Given the low sensitivity of these methods, the relative large number of detections confirms our belief that planetary systems are common.

Most of the planet's discoveries have been made with the method of radial velocity (Figure 3.11) and, as we have said, these observations can only discover massive planets that are closer to their stars than is the Earth to the Sun. Such planets are not consistent with the picture given in Section 3.8, in which giant planets are born far from the star where they can directly accrete gas from the disk, whilst small and rocky planets are born near the star, aggregating planetesimals from which the lighter elements have evaporated.

To explain the existence of massive planets close to a star, a mechanism different from that which gave rise to the solar system is needed. This mechanism has to be fast, because it must form a giant planet before the light elements (mainly H and He) evaporate as they did in the rocky planets of the solar system. Once a planet has acquired a large mass, its gravitational field can retain the lighter elements, even in the inner regions of a planetary system near the star. The suggested mechanism is based on "disk instability," which produces concentrations of matter at various distances from the center of the star (like sand dunes in a desert), which are then rapidly transformed into giant planets.

However, if disk instability explains how giant planets are born near stars, it cannot at the same time account for planetary systems like ours with small rocky planets near the star. As result of these observations, we have to accept that there are at least two mechanisms involved in the formation of planetary systems: a fast process, like disk instability, which rapidly produces planets made of light elements near the stars (unsuitable for hosting life), and a slow process, such as that which formed the solar system, with a slow accretion of planetesimals to form rocky planets in the right place to sustain life.

By 2020 we may have acquired good statistics on stars that have giant planets far from their star, with long periods like that of Jupiter (11 years) and Saturn (30 years). These planetary systems are the only ones that can host planets like the Earth and are therefore the most interesting in the search for habitable worlds.

To conclude, because the techniques used so far are sensitive to large planets orbiting close to a star, therefore, the resulting detections have statistical significance only for large planets. The systems detected up until now might only be exceptions and most planetary systems could still be like the solar system (Section 3.8). We cannot say today how many planetary systems have rocky planets. That question will only be answered by making the observations described in the last chapter of the book.

3.10   Timescales

The timescales on which stars and planets form are short, of the order of millions of years for stars like the Sun and even less for more massive stars. They are short compared to the 10–billion-year lifetime of a star like the Sun (Section 4.3) and the 3.5 billion years that have elapsed since life on the Earth began (Section 6.3).

How do we estimate these timescales? The method is very simple: it is based on Newton's famous law of universal gravitation, telling us that the particles of a cloud attract each other with a force proportional to their masses and inversely proportional to the square of the distance between them. Contraction under that force releases an amount of energy given by $0.6(GM^2)/R$,[20] where M and R are the mass and radius of a star respectively and G is a universal constant determined experimentally. Putting solar values (Appendix B) into this simple formula tells us that if the Sun had been formed in a single year, its luminosity would have been equal to that of 20 million Suns. As the time needed to build it increases, a star's luminosity decreases in proportion. For example, if the Sun had been constructed in 10 years, its luminosity would have been that of 2 million Suns, because the product between the luminosity and the number of years must always be 20 million. This allows us to calculate the duration of the process

---

20 *Because there is a relation between the mass and the radius of a protostar, this formula can be written as $2 \times 10^7 [M/M^0]^2$, where M_0 is the mass of the Sun.*

approximately[21] by observing the luminosity of the protostars. These observations have to be made in the infrared (Appendix A.4), because the protostars are immersed in the placental cloud (Figure 3.1) from which only infrared radiation emerges.

Stars that, to a first approximation, will evolve into something like the Sun have luminosities of 20–100 Suns, so they take 200,000–1,000,000 years to form. More massive stars form quicker and less massive stars take longer. In any case, these times are very short compared to those of the physical and biological processes discussed in this book. We will summarize in the following table:

	Million Years
Time needed to form a star	0.01–1
Time to form a rocky planet in the solar system	1–5
Age of *Lucy* (Section 6.9)	3.5
Life of a 10-solar-mass star	10
Time to form a giant planet in the solar system	10–50
Time elapsed since the disappearance of dinosaurs	65
Life of giant molecular clouds (Section 3.4)	100
Time elapsed since the origin of the species (Section 6.6)	543
Time elapsed since the beginning of life on the Earth	3600
Time elapsed since the birth of the Sun	4500
Life of a star like the Sun	10,000
Age of our galaxy, the Milky Way	13,000
Age of the universe	13,700

3.11 The End of the Cloud

Sometimes, near the clouds where stars are born, we find massive stars whose enormous luminosity can be more than a million times that of the Sun. In such cases the flux of energy is so high that it evaporates the cloud. The result is seen in some of the most beautiful images captured by the HST, the most famous of which is shown in Figure 3.12 (an enlargement of it is in Figure A.7), where we see the nebula Aquila in the constellation Serpens. The picture looks like a series of columns produced by erosion and, in fact, it is due to the erosion of the cloud by photons which produced the very luminous stars. The luminous

---

21 *The birth of a star does not take place at a constant rate, so the star's luminosity varies during the various stages of its formation. It is the average luminosity that the above reasoning provides, whereas we observe the instantaneous luminosity. However, the order of magnitude of this estimate is correct.*

Figure 3.12: *The nebula Aquila in the constellation Serpens. Radiation from massive stars (outside the field of view) is evaporating the cloud, except in very-high-density regions where stars are being born (see text). (Picture-by the HST. © NASA & J. Hester and P. Scowen, Arizona State University.)*

stars do not appear in the picture; they are out of sight over the upper right but their presence is revealed by the huge luminosity of the upper parts of the columns.

What is happening in Aquila is analogous to what happens on the Earth in regions where rainfall causes severe erosion. If a rock falls onto the ground, rainfall gradually washes away the soil around the rock, but not that underneath it. This results first in a mushroom-like structure and then in a column of earth holding the rock upright. The same happens in the Aquila region; the intense radiation of the stars (the rain) is washing away the cloud except that part of it (the ground under the rock) that is protected by material which is denser and opaque to the light (the rock).

What protects the cloud from the erosion of light? At the top of the column, when the luminous stars began their destructive work, there are few protostars with their disks that are impervious to the flux of photons and that did not evaporate because they are too dense. These forming stars at the top of the columns have protected the gas beneath them.

What will happen in the future? Gradually, the clouds will be dissipated. Anything that remains will be swept away by the intense radiation and the energetic particles produced by the luminous and massive stars when they will explode as supernovae (Section 4.5). If one of these stars were to have an Earth-like planet, could any form of life survive? The answer is no. Every biological form would be destroyed by the flux of high-energy particles, x-rays and gamma rays emitted by the explosion.

The majority of stars, in particular those originating in our region of the galaxy (Section 8.1), are born with more tranquilly, far from luminous stars like those of Figure 3.12. They will follow the scenario described in the preceding section. Little by little, as the stars and their planetary systems are formed, the cloud will dissolve, dispersing part of its matter into space. We have seen that this process can give rise to planetary systems but we cannot answer the question posed at the beginning of this chapter: Does this process build planets like ours? At the moment, there is not much we can say. We have seen that massive planets do exist close to stars and that they are probably not rocky like the Earth. There could, however, be exceptions selected by bias in the method of measurement (Section 3.9). We must conclude by saying that, today, we cannot answer the question of how common Earth-like planets are. We will have a better answer in 10–20 years' time, when we have better statistics on these giant planets, and we will have the result of the new observational techniques to be discussed in Chapter 12.

The Origin
of the Elements

<u>Chapter 4</u>

4.1 The Primordial Abundances

The placental cloud from which our solar system was born had everything it needed to build the Sun and the planets, including all the elements found on the Earth — oxygen, carbon, nitrogen, silicon, iron, etc., as well as heavy elements like lead and uranium. Although these atoms are very common on the Earth, they make up a very small fraction of the elements in the universe as a whole. In fact, as we have seen, they were nonexistent in the primordial universe and were formed, during the 13.7 billion years that separate us from the Big Bang, by nuclear reactions occurring in the heart of stars. Even so, they are rare: in a dense cloud like that of Figure 3.1, 97 out of every 100 grams consist of hydrogen and helium. Only the remaining 3 grams are made up of the heavier atoms created within stars.

On the Earth, however, the abundances are reversed — for every 100 grams of matter, 96 are made up of heavy atoms and only 4 of hydrogen and helium. On the Earth, the presence of the heavy elements is the constituents of mountains, seas, plants and, finally, of our own bodies. Heavy atoms are very abundant on the Earth because the light atoms evaporated early in the Earth's history, the temperature of our planet being too high and its mass too small to bind them gravitationally.[1] Even today, when the small quantities of underground

---

1 *The kinetic energy of the gas, which increases with temperature, is proportional to* mv^2, *where* m *is the atomic mass and* v *its speed. Thus, at a given temperature, the lighter atoms move faster than the heavier ones and reach the velocity of escape from the Earth's attraction.*

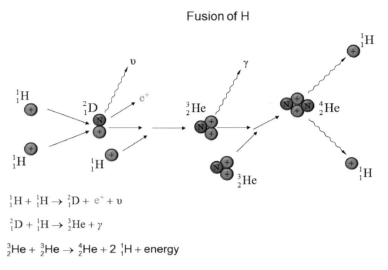

Fusion of H

$$^1_1H + ^1_1H \rightarrow ^2_1D + e^+ + \upsilon$$

$$^2_1D + ^1_1H \rightarrow ^3_2He + \gamma$$

$$^3_2He + ^3_2He \rightarrow ^4_2He + 2\ ^1_1H + energy$$

Figure 4.1: The proton–proton cycle. Four hydrogen nuclei ($_1^1$H) fuse into an atom of $_2^4$He, via two intermediate elements, $_1^2$D and $_2^3$He. The blue circles labeled "N" are neutrons, which have no electric charge. The red circles, labeled with a plus sign, are protons (hydrogen nuclei) with a mass nearly equal to that of neutrons, but with a positive charge. The e⁺ are positrons, particles with a mass equal to that of electrons, but with a positive charge. The number at the top left of an atomic symbol is the atomic weight, i.e. the total number of neutrons and protons together; the number at the bottom left is the atomic number, i.e. the number of protons contained in the nucleus, corresponding to the number of electrons (e⁻) the atom needs in order to be neutral.

hydrogen and helium are liberated into the atmosphere, they disperse into space because the temperature is high enough for them to reach escape velocity against the Earth's gravity. Other planets, like Jupiter, Saturn and Uranus, which are much more massive, being farther from the Sun, are cold, and can hold on to the light elements on their surface. As a result, they consist mainly of hydrogen and helium.

Let us try to reconstruct the story of nuclear reactions, starting from the Big Bang. After its birth, the universe had an exceedingly high density and temperature and consisted of subnuclear elementary particles. As it rapidly expanded and cooled, the first subatomic particles began to form, and then the protons that are the nuclei of hydrogen (H) atoms. After that, the first nuclear fusion reactions took place, in which four protons fused into a helium (He) nucleus with a consequent release of energy.

The process (Figure 4.1) consists of a sequence of three nuclear fusion reactions. In the first, two protons ($_1^1$H) fuse into an atom of deuterium ($_1^2$D); in the second, the deuterium, colliding with a proton, forms an isotope² of

---

2 *An isotope is an atom having a different number of neutrons in its nucleus than does the normal atom. This changes the atomic weight of the atom without modifying its chemical characteristics, which are determined by the electric charge of the nucleus and, therefore, the number of electrons in the atom. Figure 4.1 shows two examples: $_2^3$He is an isotope of $_2^4$He that has the same number of protons but one neutron less; deuterium is a hydrogen isotope with one more neutron.*

helium, $_2^3$He. When the density of $_2^3$He becomes high enough to make collisions between them probable, the third fusion reaction occurs, producing an atom of $_2^4$He and liberating two protons, $_1^1$H. This process, called the proton–proton cycle, is still active in the Sun and together with, another process called the CNO cycle[3], are the principal sources of energy in stars. In the primordial universe this reaction lasted a little more than 3 min before the density became too low to sustain it.

The next element to be formed was carbon, $_6^{12}$C, consisting of three atoms of $_2^4$He, via a reaction that needs temperatures of the order of 100 million degrees to proceed.[4] In this reaction, two atoms of helium collide first to form an atom of beryllium ($_4^8$Be), and then a third atom of helium arrives, producing an atom of $_6^{12}$C. There is a problem, though: $_4^8$Be, unlike the $_1^2$D and $_2^3$He in the reaction of Figure 4.1, is unstable — it breaks into the two original atoms of $_2^4$He in the extraordinarily short time of 1/10,000,000,000,000,000 s (16 zeros, or 10^{-16})! To form a carbon nucleus, the third $_2^4$He nucleus must hit the other two (bound in $_4^8$Be) in a shorter time, and this requires a very high density. When the fusion of hydrogen had produced enough helium to make the fusion of two atoms of $_2^4$He into one of $_4^8$Be possible, the density was already too low; the third atom of $_2^4$He always arrived too late, after the beryllium had already decayed. It is estimated that, in the first 3 min after the Big Bang, less than one atom of carbon was formed for every 100 billion atoms of hydrogen — a totally negligible number.

As we shall see in the next section, temperatures and densities high enough to form carbon and other heavy atoms are found only within stars. If we could construct an instrument capable of detecting biological forms out to the edge of the universe, and therefore looking backward in time, we would find a "dead" zone, in which there were no planets and where no life existed because the stars had not had time to make atoms heavier than helium.

4.2 The Origin of the Elements

Stars are much more efficient than the Big Bang in creating heavy elements because they maintain the conditions of high temperature and high density necessary for nuclear reactions over long times. In fact, while expanding after the Big Bang, the matter became colder and less dense, and the opposite happens in the core of stars: with the passage of time, the material in the star becomes hotter and denser as it is compressed under its own weight. This contraction stops only when nuclear reactions begin and the resultant energy,

---

3 *The cycle is known as CNO, because it involves carbon, nitrogen and oxygen. These elements did not exist when the first stars (population III stars; Section 3.5) were formed.*
4 *The temperature is higher than is needed for hydrogen fusion because the electrostatic repulsion between $_2^4$He nuclei, which has to be overcome by the colliding nuclei, is quadruple that of $_1^1$H.*

making its way to the surface, pushes the matter outward and balances its weight.

In a certain sense, the history of a star is that of processes opposing its contraction under its own weight. The story begins with the contraction of the placental cloud and the effect of angular momentum (Section 3.6). Contraction then continues in the protostar until its center reaches the temperature of 10 or 20 million degrees necessary for fusing hydrogen (Figure 4.1). At that point, the beginning of nuclear reactions marks the birth of a new star, and the contraction gets halted by the huge outward flux of energy. Then follows the longest phase in the life of a star, which is also the one that produces the most energy. For example, the fusion of hydrogen began 4 billion years ago in the Sun and will finish in about 5 billion years.

When a star has burned up all the hydrogen in its core, where only the helium produced remains, the fusion stops. The stellar matter, no longer supported by a flux of energy, begins contracting again, the temperature thereby increases until it reaches 200 million degrees, and the density necessary for three atoms of helium to form one of carbon (Figure 4.2), achieving what the Big Bang could not.

The size of the star remains constant while it burns helium. Then, when most of the helium is spent, it contracts onto its carbon core until it reaches the temperature needed for subsequent nuclear reactions ($C + He \rightarrow O$, $C + C \rightarrow Mg$, etc.).

The process continues with a series of contractions alternating with periods of nuclear fusion. In each new fusion reaction, the energy produced is less than

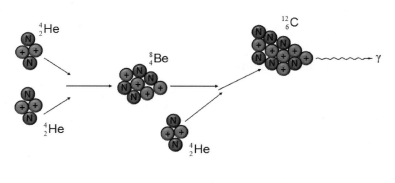

Fusion of Helium – Triple Alpha Process

$${}^{4}_{2}He + {}^{4}_{2}He \rightarrow {}^{8}_{4}Be$$

$${}^{8}_{4}Be + {}^{4}_{2}He \rightarrow {}^{12}_{6}C + \gamma$$

Figure 4.2: *The triple-alpha process. Three atoms of ${}_{2}^{4}He$ fuse into an atom of carbon, ${}^{12}_{6}C$. The cycle requires very high densities, because the element ${}_{4}^{8}Be$ is unstable and has a very short life. (The nomenclature is the same as in Figure 4.1.)*

the preceding energy. For example, the combustion of hydrogen (Figure 4.1) produces about four times as much energy as the combustion of helium (Figure 4.2). Consequently a star like the Sun will take about 10 billion years to burn its hydrogen but only about 1 billion years to burn the produced helium. This reduction in the produced energy with the increase in the mass of the nuclei continues until it is produced iron (Fe), which has 56 time the mass of hydrogen. From this point on, nuclear reactions not only fail to produce energy, but absorb it, thus cooling the star and halting any further nuclear reactions. Elements heavier than iron cannot, therefore, be produced with this mechanism.[5]

With the energy produced ever-decreasing, the star has more and more difficulty in supporting its own weight. When it can no longer do so, it begins its final contraction. As we shall see, the result of this phase, and the manner in which a star ends its life, depend on its mass. The less massive stars, those that began their life with less than about 7 solar masses, will end their life as white dwarfs, weak little stars that are very dense and hot. They cool gradually and quietly for the rest of the life of the universe. The more massive stars will first collapse under their own weight and then explode as supernovae. The enormous gravitational energy liberated by this collapse is used in part to produce elements heavier than iron.

4.3 The Luminosity of the Stars

The stars use gravity to resolve a problem that humans have yet to overcome — how to confine a hot dense gas for nuclear fusion to take place. If, one day, we succeed in solving this problem, we shall have an inexhaustible source of energy by using hydrogen, extracted from water, as fuel.

In the laboratory, matter has been heated to millions of degrees, but it is not yet possible to reach the density needed for nuclear fusion.[6] It is not possible to build a container, since no material can survive at such high temperatures: refractory materials survive temperatures thousands of degrees, but not the millions needed for fusion. The alternative is to try confining the gas with magnetic fields but instability in the plasma has defeated all attempts thus far.

In stars the "container" is gravity, i.e. the weight of the overlying layers of the star, which succeeds in confining the very hot and dense gases. The greater the mass of a star, the higher the temperature and density that can be reached at its center. *The heaviest atoms a star can produce therefore depend on its mass.*

---

5 *To build an atom heavier than iron, energy has to be supplied. The reverse of this is seen in a nuclear power plant where heavy nuclei, uranium for example, break up by fission and release energy. Nuclei lighter than iron, on the contrary, release energy when they fuse together. This is what we seek to achieve in nuclear fusion power plants — to produce energy using the process shown in Figure 4.1, as does the Sun.*
6 *So far, hydrogen fusion has only been achieved in H-bombs, where the explosion lasts a fraction of a second, and confinement of the gas is not necessary.*

Stars like the Sun have produced helium (^{4}He) and carbon (^{12}C) when finishing their lives as white dwarfs. To produce all elements from hydrogen to iron, a star needs to have 5–6 solar masses at least.

The luminosity and lifespan of a star depend therefore on its mass. Nuclear fusion occurs in fact in the central region of a star and the size of this region increases with the star's mass. Above a certain mass, one can even have more than one fusion process taking place simultaneously. For example, a star of a certain mass can burn helium to produce carbon in its center, where the density and temperature are higher, while it is burning hydrogen to produce helium in an external layer where the temperature and density are lower.

It is not surprising, therefore, that the luminosity of stars increases rapidly with their mass; in fact, it can be shown that it increases as the 3.5-th power of the mass ($M^{3.5}$). This means that a star that has twice the mass of the Sun emits 11.3 times its light, one with 10 times the solar mass will be 3162 times more luminous and a star of 100 solar masses will be 10 million times more luminous than the Sun.[7]

This increase in luminosity has a dramatic effect on the life of a star. If a star has 10 solar masses, it is 3162 times more luminous than the Sun. It therefore fuses 3162 times more hydrogen every second. But since it has only 10 times more hydrogen than the Sun in the first place, its life will be 10/3162 times shorter. The life of a star therefore decreases with its mass as $1/M^{2.5}$. Going back to the previous examples, a star with twice the mass of the Sun will live one sixth (=2/11.3) of the Sun's life, whilst one with 10 solar masses will live one three-hundredths (=10/3162) of the Sun's life. These massive stars started burning hydrogen while still inside the placental cloud, because they have not had time to disperse the energy.

Massive stars are like cars which have been given greatly improved performance and higher fuel consumption but without a corresponding increase in the size of the fuel tanks: they do not "last" long. This phenomenon is relevant to the search for life. As we shall see in the next chapters, life needs billions of years — times much longer than the lives of these stars — to evolve from bacteria to more complex forms, and extraterrestrial life, if it exists at all, will only be found around stars of four solar masses or less, the only ones that live long enough for life to develop on their planets.

The opposite effect is also true: a star whose mass is half that of the Sun will live 6 times longer than the Sun, while one with a tenth of the Sun's mass will live 300 times longer. In these stars everything is slower, including their birth.

---

7 *An example is the Pistol star, tens of million times more luminous than the Sun and probably the most luminous object in our galaxy. In spite of its luminosity, the star is not easily visible. Born a million years ago, it is still shrouded by its placental cloud, not having had the time to disperse it. It therefore emits most of its energy in the far-infrared. Only the HST has succeeded in measuring the weak optical radiation from the cloud.*

This fact is equally relevant to the search for extraterrestrial life, because protostars that have less than a quarter of the mass of the Sun have not had enough time to become stars and then allow life to develop around them.

4.4 White Dwarfs and Red Giants

In 1850, a German astronomer, Wilhelm Bessel (the same person who measured the motion of the Earth relative to the fixed stars in 1837; see. Section 3.1), observed that the motion of Sirius, one of the brightest stars in the sky, was irregular. It seemed to be affected by the presence of another, nearby star but, in spite of his efforts, he could not find such an object. This was surprising, because the object would need to be fairly massive to influence the motion of Sirius. This mysterious companion was only discovered in 1862, by a telescope maker who constructed an instrument that, for its time, was of exceptional quality. He found a small star, 10,000 times weaker than Sirius, which he called Sirius B. This was the first discovery of a white dwarf, a star similar to what the Sun will become in 6 billion years.

To understand what these small stars are like, we will take up the story of stellar evolution from the moment they begin to burn hydrogen. For about

Figure 4.3: Sirius A and B. The larger star is Sirius A, which has 2.4 solar masses and a radius of 1.2 million km. The smaller one, on the lower left, is Sirius B. It has one solar mass and a radius of 5700 km, smaller than that of the Earth. The progenitor of Sirius B was originally more massive than Sirius itself, with five solar masses. Being more massive, it evolved faster (Section 4.3), losing 4/5 of its mass in various red giant phases. (Photo by the HST. © NASA.)

10 billion years, a star like the Sun continues in this phase, increasing its luminosity by only a factor of about 2. At the end of this phase, the star has consumed all its central hydrogen, and finds itself remaining with a core of helium. No longer supported by a flux of radiation, this core contracts, the temperature of the star increases and hydrogen begins to be consumed in a shell around the helium core. As time goes on, the helium core gets bigger and the hydrogen-burning shell is displaced toward the surface of the star.

During this phase, the luminosity of the star increases dramatically, because much more hydrogen is burnt in the shell than was originally burnt in the core. The large amount of heat released through these reactions pushes the stellar material outward, increasing the diameter of the star enormously until it becomes hundreds of times that of the Sun. At this stage of their evolution, stars are called red giants and may be recognized by their extraordinary luminosity and their reddish color. Betelgeuse, a red supergiant in the Orion constellation (Figure 4.4), is an example of this type of star and has a diameter large enough to contain the orbit of Jupiter.

As the hydrogen-burning shell moves ever closer to the surface of the star, it eventually reaches a zone where the temperature and density are too low to support the fusion of hydrogen. The star's luminosity decreases and, itself no longer supported by the energy produced in its interior, the star shrinks. As a result of this collapse in size, the central temperature increases until it is high

Figure 4.4: Betelgeuse, a red giant in the Orion constellation, 420 light years from the Earth. It is one of the few stars whose diameter can be measured directly. This diameter is a little higher than that of Jupiter's orbit, shown in the picture together with the Earth's orbit. Betelgeuse is more massive than Sun and will end its life by exploding as a supernova. (Picture taken by HST. © NASA & A. Dupree, CfA.)

enough to fuse helium into carbon. A new phase begins, similar to the previous one, but this time the star is less luminous because the energy produced by the fusion of helium is less than that produced by the fusion of hydrogen (Section 4.3).

After some time, a core of carbon is formed at the center of the star and helium begins to fuse in the surrounding shell. The star's luminosity, and consequently its diameter, increase and it enters another red giant phase until nuclear fusion ceases again. Depending on the mass of the star, there may be several of these red giant phases. For our sun there will only be two, because it is not massive enough for its core to reach the temperature needed for fusing atoms heavier than helium. The first red giant phase will occur 5 billion years from now, when hydrogen is exhausted in the core. When that happens, the Sun will be big enough to envelope the Earth in a fiery inferno that will destroy our planet.

Massive stars can go through many red giant phases during their lives. Each time, they swell so much that they lose a large part of their mass to interstellar space, enriching it with the heavy elements produced by nuclear reactions.

Stars initially of less than about 7 solar masses are reduced, by mass loss during their red giant phases, to below 1.44 solar masses: the Chandrasekhar limit[8] (Section 2.3). They become white dwarfs, dense stars that squash a solar mass within a diameter of a few thousand kilometers (Figure 4.3). Because they are small, they are not very luminous even though their surface temperature can be 10 times that of the Sun. They are destined to cool slowly for all the life of the universe. There are many such stars near the Earth, which have ended their lives quite recently. We can see the material they ejected during their final phases; some of the most beautiful HST pictures are of planetary nebulae — Figures 4.5 and 4.6 show two of them. The white dwarf is the luminous point at the center, surrounded by the concentric circles of ejected matter that are responsible for planetary nebulae, being so named. Although they appear as circles in the images, they are, in fact, concentric shells of matter. The material is moving away from the star at velocities of 20–30 km/s.

Within a few thousand years, the gas and dust of the planetary nebula will have dispersed into space, forming new giant molecular clouds (Section 3.4) and new stars. Only the white dwarfs will remain and, like the one orbiting Sirius (Figure 4.3), may be difficult to see. All the planetary nebulae we see are

---

8 In 1930, the Indian astrophysicist Subrahmanyan Chandrasekhar noted that a white dwarf's radius decreases with increasing mass. A white dwarf with the mass of the Sun has a radius of 6000 km, whilst one of half a solar mass has a radius of 20,000 km. This inverse dependence arises from the properties of the dense stellar material and its ability to support the weight of the star. Chandrasekhar showed that 1.44 solar masses was the maximum mass that could be supported and that stars that are more massive would collapse first and later explode like supernovae (Section 4.5), leaving behind either a neutron star or a black hole.

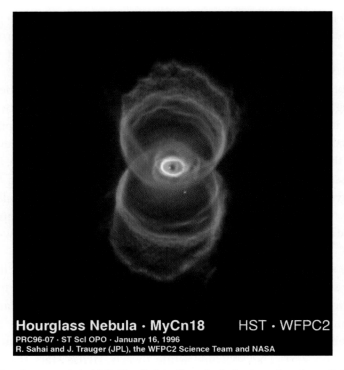

Hourglass Nebula · MyCn18 HST · WFPC2
PRC96-07 · ST ScI OPO · January 16, 1996
R. Sahai and J. Trauger (JPL), the WFPC2 Science Team and NASA

Figure 4.5: The planetary nebula MyCn18, called the "hourglass" after its curious shape. Each circle corresponds to matter expelled by the star during its red giant phases. At the center of the smallest circle we see the white dwarf remnant of the star. (Picture by the HST, January 1996. © NASA & R Sahai and J. Trauger, JPL.)

therefore very young. Within 5 billion years, the Sun will also finish up in the same way, having most of its mass, and that of its planets, dispersed to form the next generation of stars.

The Earth will die during the Sun's first red giant phase, when it will be swallowed up by the fiery solar atmosphere. Our planet will not disappear immediately; it will continue to orbit the Sun for thousands of years while its outer layers melt and evaporate. With the passage of time, the friction caused by the solar atmosphere will slow the Earth, causing it to spiral toward the center of the Sun. When it approaches within about a million kilometers of the center, it will encounter a temperature of half a million degrees, and the last vestige of our planet — its metallic core — will evaporate.

The name "planetary nebulae" was given to these objects immediately after their discovery, because they looked like little stars surrounded by the nebulae — the gas and dust — predicted by Kant and Laplace to be the birthplace of planetary systems. Only later it was discovered that they were, on the contrary, dying systems. Stars and planets are born of dust, and to dust they will return, feeding new generations in their turn, like biological organisms. "Dust unto dust, ashes unto ashes."

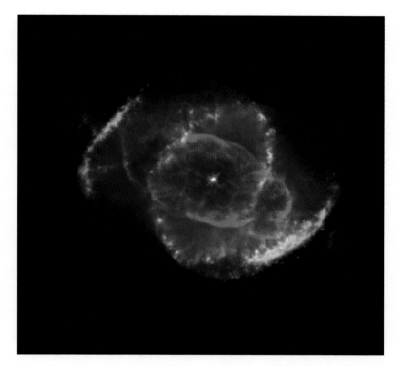

Figure 4.6: The planetary nebula NGC 6543, called "cat eyes" because of the curious shape of the material expelled by the star. The luminous dot at the center is the white dwarf. (Picture by the HST, January 1995. © NASA & J. P. Harrington and K. J. Borkowsky, University of Maryland.)

4.5 Supernovae

In 1054, a Chinese astronomer wrote in his annals that a new star had suddenly been born in the constellation Taurus. It was more brilliant than any star that had ever been seen, being four times brighter than Venus, the most luminous body in the heavens. It was so bright that he was able to observe it in daylight for 23 consecutive days. Then its luminosity began to fade and, after a few months, it was no longer visible, even at night.[9] This is considered to be the first observation of a supernova explosion. Other cases were reported in historical records, especially in China, but they were not as precise and their identification with supernovae is less certain. Now we know that the brilliant star seen by the Chinese in 1054 is associated with the Crab Nebula (Figure 4.7), 6300 light years distant, which has a neutron star at its core. This neutron star is also a pulsar and is one of the most intense x-ray sources in the sky.

Only two other supernovae have been observed in our galaxy since 1054 — one in Cassiopeia, recorded by Tyco Brahe in his annals of 1572, and the other

---

9 *Strangely, nobody in Europe noted the event — perhaps a proof of how "dark" those centuries were in the western world. Such an outstanding phenomenon must have been observed but nobody recorded it.*

The Crab Nebula in Taurus   (VLT KUEYEN + FORS2)   ₊ES₊

Figure 4.7: The Crab Nebula. The remains of a supernova whose explosion was registered in the annals of a Chinese astronomer in 1054. The star is 6300 light years from the Sun, contains a pulsar at its core, and is a bright x-ray source. (Picture taken in November 1999 by the VLT telescope, Chile. ©ESO.)

in Ophiucus, observed 32 years later by Kepler in 1604. Since then no supernovae have exploded in our galaxy, notwithstanding the keen expectation of astronomers. In spite of this, supernovae can be studied in other galaxies because they are so luminous (one exploded in a nearby galaxy in 1987). Indeed, we now have a reasonable sample of these objects. As we saw in Section 2.3, it is by means of supernovae that we have been able to establish the cosmological distance scale and measure precisely the rate of expansion of the universe.

For a relatively short time — a matter of hours — a supernova generates an enormous amount of energy that can outshine an entire galaxy. The luminosity then decreases and, after 200 days, it is 100 times less bright.

The progenitor stars, which are born with more than 7 solar masses, evolve in a manner similar to that of smaller stars but their life is much shorter (Section 4.3). They pass through the sequence of nuclear fusions and all the red giant phases, during which they lose most of their mass. In the final fusion stage, these massive stars create an iron core that grows with time. When the core passes the Chandrasekhar limit of 1.44 solar masses, its structure can no longer support its weight, and it rapidly collapses, generating an enormous amount of energy. Part of this energy is used to produce elements heavier than iron, part to feed its great luminosity, comparable to billions of Suns, and part to expel material enriched with the heavy elements the star has produced. The expulsion velocities are very high, reaching 10,000 km per second, a 30th of the speed of light! (The highest velocity recorded by a man-made vehicle is 17 km/s, set by the Voyager satellites.)

In those first few hours, nuclear reactions within supernovae produce the elements heavier than iron (cobalt, nickel, copper, zinc, arsenic, gold, lead, platinum, uranium,[10] etc.) Fusion of lighter elements that were not burnt in a previous phase also takes place and contributes to the energy of the explosion. Supernovae are therefore the major source of both light and heavy atoms. In Figure 4.7, the luminous filaments that are flying away with vertiginous speed from the center of the explosion contain all the elements produced by the supernova; they will enrich the chemical composition of a future giant molecular cloud. The energy stored in the transferrous atoms can be recovered by splitting them again. This phenomenon is exploited in nuclear "fission" plants where very fast particles, crashing into a heavy nucleus like uranium, break it up and release the energy stored in it by an ancient supernova billions of years ago!

While the exterior of the star explodes, the center implodes, cooled by the endothermic nuclear reactions that are building the heavy elements. What remains is an object with the density of an atomic nucleus, the only structure that can bear the immense weight. This object, called a *neutron star*, has so high a density that a solar mass is contained within a diameter of less than 20 km.

Some of these stars are also called pulsars, because their electromagnetic emission pulsates in intensity. The first pulsar was discovered in 1967 by Jocelyn Bell, a PhD student at Cambridge, when she was making radio observations of an object today known as CP1919. She noticed that the object emitted pulses with a period of 1.33728 s. Because this period was so precise and constant (to a hundred-thousandth of a second), it was at first thought that it was a signal from some extraterrestrial civilization. It was then discovered that objects of this type are associated with the remains of supernovae — neutron stars — and that the energies involved were so large that no civilization would be able to control the emissions. It would be like having to switch the Sun on and off every second.

Further observations showed that the pulses contain radiation at all wavelengths, like the radiation emitted by the particle accelerators called synchrotrons.

---

10 Some of the trans-ferrous nuclei are radioactive (after radium, the first element in which the phenomenon was studied), with decay times that can exceed 10 billion years. Measuring the proportion of decay products in the solar system's rocks allows us to determine their age very accurately (Section 6.1). The energy produced in these decays is the source of heat in the Earth's interior, feeding earthquakes and volcanoes and sustaining continental drift (Section 9.9).

In these accelerators, particles are accelerated in a magnetic field to speeds close to that of light, and they emit *synchrotron radiation*, which extends from radio waves to x-rays (Appendix A.1).

These observations led to a model which could explain a signal that was so precisely periodic. It was suggested that the neutron star rotates with exactly the same period of the pulses. (As we will see in Section 9.8, every periodic phenomenon is associated with a mechanical motion.) The star's magnetic field naturally rotates with the same period and accelerates particles like a synchrotron, emitting radiation within a narrow cone. Every time the cone sweeps the Earth, we see a flash, just as with a lighthouse (Figure 4.8). As we have seen, synchrotron radiation is emitted at all frequencies; this holds for neutron star pulses. Whilst pulsars were originally discovered in the radio region of the spectrum, they have since been observed in the infrared, visible and x-ray bands. The rotational velocity of pulsars is very high, the longest observed period being about 4s.

With the passage of time, pulsars lose energy and slow down their rotation. The particles are less accelerated and the rotating beam of the star (Figure 4.8) becomes less luminous. After 10 million years of emission, a pulsar will be so attenuated that it should no longer be detectable, so that there should be many

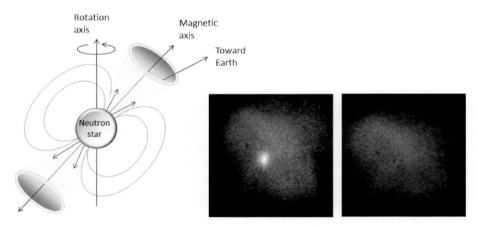

Figure 4.8: *On the left is a schematic drawing of the pulsar model. The pulsar's emission is explained by supposing that a neutron star rotates rapidly around an axis that is not aligned with its magnetic axis. The particles evaporating from the star's surface are constrained to move along the magnetic lines of force and are accelerated to speeds close to light by the rotating field. These accelerated particles emit intense synchrotron radiation within the cones shown in the figure. The two cones rotate with the star so that the resulting effect is that of a lighthouse. We observe the light each time the beam sweeps the Earth in its line of sight. The period of the "lighthouse" is that of the star's rotation and can have a value going from a fraction of a second to a few seconds. On the right are two images of the Crab Nebula (also shown in Figure 4.7), viewed in x-rays by the Einstein satellite, whose "shutter" was opened and closed with the same period of the radio signal. The image on the left, in which the star — the white spot — is visible, was obtained by adding together images taken in phase with the radio signal. The image on the right, in which the star is not visible, was obtained by adding antiphase images. (© NASA & F. R. Harden, Harvard Smithsonian Center for Astrophysics.)*

"extinct" pulsars around us, whose existence we do not even realize.[11] Up till to today thousands of pulsars have been discovered, allowing us to estimate the number of "extinct" pulsars, which were born during the 10-billion-year lifetime of the galaxy. Their number should be a little smaller than a billion. To these extinct pulsars is attributed the diffuse x-ray radiation coming from all directions of the sky. It is believed that it is produced by the interstellar matter that falls onto these stars, accelerated by its enormous gravity (an x-ray is obtained by striking a surface with very energetic particles).

Some of the intense gamma ray emissions observed in the sky are attributed to extinct pulsars belonging to binary systems. It is believed that these neutron stars, orbiting about each other, are losing angular momentum (Section 3.6), and will eventually fall into one another. These gamma beams could be their final emission before they coalesce into a black hole. These are phenomena that, fortunately, do not occur today in our galaxy; the energy emitted is so great that it could destroy every form of life on the Earth. (It is thought that some of the great extinctions of the past might have been caused by these beams, but there is no fossil evidence to support this hypothesis.)

If the mass of a star is more than 4 solar masses when it explodes as a supernova, nothing can stop the collapse and the core will become a black hole. Physicists had predicted the existence of black holes long ago but observational evidence was missing as they are not easy to find. The gravitational field of a black hole is so strong that practically nothing can escape — not even light. Only in the past 30 years has the observational proof of their existence been obtained from the gravitational perturbation produced on the motion of nearby stars.

Numerous black holes have now been discovered, mainly at the center of galaxies; for example, at the center of our own galaxy there is one of 3 million solar masses. When they devour other stars, they emit beams of energy similar to those emitted during the collapse of two pulsars. These beams can destroy any form of life around them (Section 8.1). Fortunately the black hole at the center of our galaxy is a "quiet" one; its emission is weak and poses no danger to us.

4.6 The Cycle of Matter

As we have seen, the stars are similar to biological organisms; just like them, they are born, they live and they die. At the end of their lives, most of their mass is recycled and used to make new stars, just as with biological organisms.

---

11 *Some supernovae have certainly exploded near our placental cloud, a short while before the birth of the Sun; the radioactive materials inside the Earth that produce its internal heat are due to these supernovae (Chapter 8). The decay times of radioactive elements allow us to estimate that these explosions came a few hundreds of millions of years before the birth of the Sun (Section 9.9).*

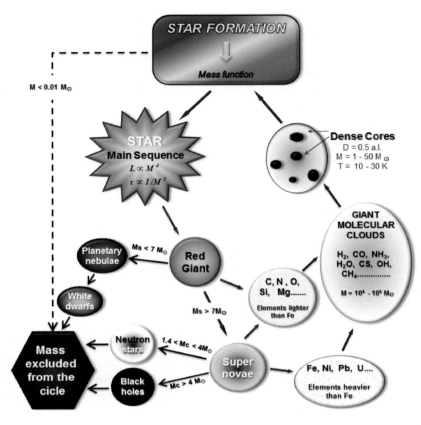

Figure 4.9: *The cycle of matter since the origin of the universe. In every cycle, matter is enriched with heavy elements and part of the mass is lost to the evolutionary cycle. M_s indicates the mass of the star at birth and M_c the mass remaining after it has lost mass.*

It is a process of continuous evolution because, in each successive generation, the chemical composition of the matter from which stars are born is different (Section 3.5).

In all of this process, the mass of the star plays the most important role. As we have seen, the mass influences the luminosity of the star, its lifetime, how many heavy elements it produces, how it dies and, finally, whether it can host planets that support life.

Figure 4.9 shows schematically the cycle of matter in the universe. We begin at the top with stellar birth, the process discussed in the last chapter. In this phase, the fragmentation of the placental cloud defines the mass of the stars which form — the mass function. If the fragments are less than eight hundredths of a solar mass (80 times the mass of Jupiter), they will not reach the temperatures and densities needed to fuse hydrogen in their centers. They will evolve into the small objects known as "brown dwarfs" that will wander around the universe forever. The mass locked up in these objects is lost to the cycle and finishes in the reservoir of "blocked mass" (on the bottom left of the figure). Fragments weighing more than eight hundredths of the Sun will have luminosities proportional to $M^{3.5}$ and lifetimes that decrease as $M^{-2.5}$ (Section 4.3). Only if they have masses between ¼ and 4 solar masses will they have planets capable of supporting life. Fragments of less than ¼ of a solar mass have not become stars in our galaxy, whilst stars with more than 4 solar masses will not live long enough for life to develop.

Having burnt up their hydrogen, stars of less than 0.7 solar masses become white dwarfs after their first red giant phase because they are not big enough to fuse helium. Stars of more than 0.7 solar masses can enter the helium-fusing phase to produce carbon and, depending on their actual mass, may go on to produce all the elements up to iron. Each phase ends with an increase in the energy production that transforms the star into a red giant (Figure 4.3). In the red giant phase, the star begins to lose mass and, in so doing, enriches interstellar space with elements that are ever heavier, from helium to iron.

Those stars that began their life with less than 7 solar masses will lose sufficient mass during their red giant phases to finish below the Chandrasekhar limit of 1.44 solar masses (Section 4.4). Then they will end their lives as white dwarfs within the planetary nebulae of Figures 4.5 and 4.6, with their mass being inside the "blocked mass" reservoir.

A star with more than 7 solar masses will end its fusion sequence with more than 1.44 solar masses, even after successive periods of mass loss. It will not be able to sustain its own weight but will first implode and then explode as a supernova (Section 4.5), producing many new elements, including the transferrous nuclei which can only be created in this way. A small fraction of less than 1/10 of the original mass will end up in the blocked mass reservoir as a black hole or a neutron star.

All matter expelled from stars will build the giant molecular clouds (Section 3.4) which can be up to tens of thousands of solar masses (Figure 4.9, left). Inside these clouds there will be denser and more compact clouds, the placental clouds of Figure 3.1, which will form new stars and initiate another cycle. Each cycle enriches the interstellar matter with heavy elements and locks up a little more "lost mass."

By these processes, stars — large and small — contribute to the life of the universe. Stars like the Sun moderately enrich the universe with heavy elements and live long enough for life to form in their environs. Heavier stars do not live long enough to allow life to form, but they enrich the universe with elements that are indispensable for building the planets and the biological forms they can host.

4.7 Fuel for the Stars

The first stars formed in the primordial universe were of hundreds of solar masses (Section 3.5) because of the lack of molecules, particularly of CO (Section 3.3). Their lives were short and they exploded as supernovae, producing not only the first heavy elements but also the first black holes, which were probably the seeds for galaxy formation.

Successive generations of stars produced all the elements — other than hydrogen and primordial helium — that make up the universe today. This process continues for a long time, until the nuclei that can be burnt in nuclear reactions finish.

Figure 4.10 shows the proportion by mass of the principal elements, as they were after the Big Bang and as they are today. The 92 elements of the periodic table (Figure 5.10) are created inside stars, but in very modest quantities compared with the total mass of the universe. None of these heavier elements account for as much as 1% of the mass of the universe. The most abundant element after helium is oxygen, which does not make up more than 8 parts per thousand.

In Figure 4.11 the proportions are shown in terms of the number of atoms. The heavy elements are represented by smaller spheres than in Figure 4.10 only because they *are* heavy. An atom of iron, for example, weighs as much as 56 atoms of hydrogen; it is counted as 56 in Figure 4.10, representing its mass, while it counts as only 1 in Figure 4.11 because it is only one atom. The production of this small proportion of atoms within stars has furnished the energy to heat the universe since the first stars were born, a little more than 13 billion years ago.

Looking at Figures 4.10 and 4.11, one notices how little hydrogen has been consumed since the beginning of the universe — merely 2.3% of its initial mass.

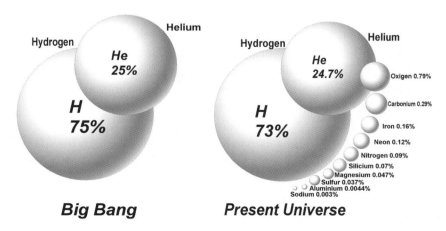

Figure 4.10: *The relative proportion of elements by mass, as they were immediately after the Big Bang (3/4 of hydrogen and 1/4 of helium) and as they are today.*

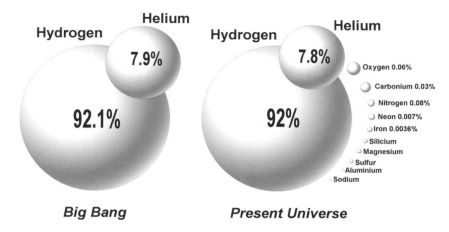

Big Bang   Present Universe

Figure 4.11: This figure is similar to Figure 4.10, but with the relative abundances shown in terms of the number of atoms. Figure 4.10 tells us that, in the present universe, there is about 1g of oxygen for every 100 grams of hydrogen. This figure tells us that there is about one atom of oxygen for every 1600 atoms of hydrogen, one oxygen atom weighing approximately as much as 16 hydrogen atoms.

It follows that the universe is still young and that the reserves of hydrogen and helium are abundant: stars will continue to be born and to die for hundreds, if not thousands, of billions of years. If the universe continues to expand forever, a decline in star activity will occur, not because nuclear fuel is exhausted but because matter is too diluted. In the very distant future, matter expelled by individual stars at the end of their lives may not be able to agglomerate to form new giant molecular clouds and thence new stars.

4.8   The Abundances of the Elements on the Earth

The abundances of the elements on the Earth are very different from those seen in the universe. This is a result of the process by which the Earth was formed (Section 3.8) and which, in this region of the solar system, favored the heavier elements over the lighter ones. As we have seen, the lighter elements evaporated from the planetesimals that formed the Earth. We noted that, even today, the temperature of the Earth is high enough to evaporate into space any hydrogen and helium that escapes into the atmosphere.

Hydrogen, the most abundant element in the universe, constitutes only about 1 part per thousand of the Earth's mass and is mainly in the form of water, where it makes up only 2/18 (being in H_2O, the mass of $H = 1$ of $0 = 16$) of the weight of the molecule. Helium is even rarer; being a noble gas, it does not react with other atoms to form heavy molecules as hydrogen does. It therefore exists only in atomic form and, being light, ascends in the atmosphere as soon as it reaches the Earth's surface and eventually escapes into space. The few quantities we have are trapped in rocks that are slowly outgassing. There are large quantities of hydrogen and helium in the giant planets — Jupiter, Saturn and Uranus — because they are cold and massive.

Table 4.1 The abundances of the elements in the Earth's crust, in descending order of magnitude. For comparison, the abundances in the universe are shown in the last column.

Element	Symbol	Atomic weight	Abundance by mass (%) (in the Earth's crust)	Abundance by mass (%) (in today's universe)
Oxygen	O	16	45.5	0.7664
Silicon	Si	28	27.2	0.0681
Aluminum	Al	27	8.3	0.0044
Iron	Fe	56	6.2	0.0030
Calcium	Ca	40	4.66	0.0058
Magnesium	Mg	25	2.76	0.0471
Sodium	Na	23	2.27	0.0030
Potassium	K	39	1.84	0.0003
Hydrogen	H	1	0.152	73.000
Phosphorus	P	31	0.112	0.0007
Sulfur	S	32	0.034	0.0374
Carbon	C	12	0.018	0.2960
Chlorine	Cl	34	0.012	0.0012
Nitrogen	N	14	0.0019	0.0940

It is estimated that the Earth is mainly composed of a few elements by mass: 35% iron (mostly in the core), 30% oxygen, 15% silicon and 13% magnesium. All the other elements contribute only to the 7% of its mass. Of the 92 elements that exist in nature, four contribute to 93% of the mass of the Earth, the other 88 making up the remaining 7%. This 7% is precious, though, because the existence of life depends on it.

The composition of the crust of the Earth (Section 9.2), the thin layer on which we live and from which we extract the elements necessary for life, differs from the *average* composition of the Earth. This resulted from the descent of the heavier elements into the Earth's core, which occurred at the beginning, when our planet was a mass of molten rock. Table 4.1 shows the crustal composition by weight. For a comparison, in the last column, the composition of today's universe is given, the values being those of Figure 4.10.

Again, a few elements make up more than half the mass of the crust: of the 92 naturally occurring elements, only 8 are present at concentrations of over 1%. The most abundant is oxygen, followed by silicon and aluminum; hydrogen contributes to only 0.15%; carbon, the basic element of life, is rare, contributing 0.018%. In Chapter 9 we will see that carbon could easily disappear from the Earth's crust, trapped under the depths of the sea; should this occur, life on our planet will be impossible. As we shall see, its presence on the planet's surface is the result of volcanic activity (Section 9.5).

The elemental abundance changes once again in biological organisms (Table 4.2) because of their particular chemistry. In biological organisms, of primary importance are carbon and oxygen, and most of their mass again consists of only six elements: oxygen, carbon, hydrogen, nitrogen, calcium and phosphorus. Of these, only oxygen and calcium are among the most abundant elements in the Earth's crust; other elements, which are present only in traces in biological organisms, are nevertheless essential for the complex chemistry of life. The last column of Table 4.2 shows a few of these rare elements and summarizes their role in an organism. Without iron, for example, blood could not transport oxygen to the cells; without phosphorus and nitrogen, the DNA chain could not be built; for this reason, the latter are the main components of fertilizers in agriculture.

The food chains of living species have the effect of concentrating the elements of Table 4.2. At the base of the chain, there are the vegetables which find rare chemical nutrients from the soil (with their roots) and from the air (with their leaves). They provide food for the herbivores, but this food is still low in the concentration of the elements for life. Consequently, herbivores spend most of their time eating. They provide food for carnivores, which, being at the

Table 4.2 The abundance, by weight, of the most common elements in living things. The last column lists their function in biological organisms. The last three elements are representative of many others that, despite their presence in minute traces, play a fundamental role in biological activity.

Element	Symbol	Atomic weight	Abundances by mass (%) in living things	Essential function
Oxygen	O	16	65.0	Water and cellular respiration
Carbon	C	12	18.5	Basic for organisms
Hydrogen	H	1	9.5	Water and energy
Nitrogen	N	14	3.3	Proteins and DNA
Calcium	Ca	40	1.5	Bones
Phosphorus	Ph	31	1.0	DNA and energy transport
Potassium	K	39	0.4	Body fluid and nervous system
Sodium	Na	23	0.2	Body fluid and nervous system
Sulfur	S	32	0.2	Protein
Chlorine	Cl	34	0.2	Negative ions
Magnesium	Mg	25	0.1	Enzymes and energy transport
Iron	Fe	56	Trace	Hemoglobin
Copper	Cu	29	Trace	Component of many enzymes
Zinc	Zn	30	Trace	Component of many enzymes
Iodine	I	53	Trace	Thyroid hormone

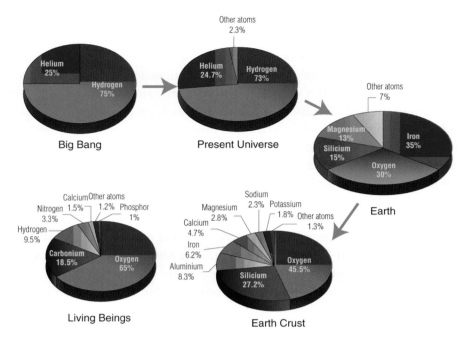

Figure 4.12: Shown clockwise are the average abundances of the elements after the Big Bang, in the present universe, in the Earth, in the Earth's crust and in biological organisms. The two figures at the top use the same data as Figure 4.10 but in a pie chart, where the slices are proportional to the amount of the element present (instead of the volumes of Figure 4.10).

end of the food chain, have a diet rich in the necessary elements and can therefore devote a relatively small amount of time to eating.[12]

Figure 4.12 summarizes the proportions discussed in the last few paragraphs. Except for a tiny amount of hydrogen, the elements that constitute the Earth and biological organisms did not exist immediately after the Big Bang. They are the 2.3% of the present universal abundances that have been made within stars.

The tiny quantities of heavy elements that stars have produced are extremely precious: without them, planets would not exist, nor would the complex chemistry that regulates life.

4.9 The Spectra of the Stars

In this and the previous chapter, we have discussed the chemical composition of atoms and molecules. Before continuing, we shall describe how we *measure* the chemical composition of a planet, a star or a distant galaxy. We saw in Section 3.3 that the molecules of a gas, in colliding with one another, become excited. That is to say, they transfer part of their energy of motion into the rotation of

---

12 *Flours, rich in animal proteins, are often used to feed herbivores. As a consequence these grow faster and produce more milk (up to 50 litres a day). This is how the "mad cow" disease has spread.*

the molecule as a whole and into the oscillation of the atoms of which they are composed. Then they de-excite, emitting characteristic photons which differ from molecule to molecule. Similarly with atoms, although, in this case, the energy of collision is stored in the motion of the electrons around the nucleus.

This emitted light has "colors" that depend on the structure of the atom and of the molecules. They are characteristic of each element, allowing astronomers to analyze the chemical composition of stars and galaxies. As long as we have instruments capable of measuring their light, we can study chemical reactions everywhere in the universe.

As discussed in Appendix A.1, everyday experience tells us that the light of a sodium lamp is orange and that of a mercury lamp is blue, whilst a neon lamp is red. These colors are emitted when those particular atoms are excited by an electric discharge inside the lamp. In Section 1.2 we saw that the color of light is determined by the frequency of the corresponding electromagnetic wave or, alternatively, by its wavelength, which the eye recognizes, so that it is making it possible for us to distinguish one color from another.

There are instruments called *spectrometers* that split light into its constituent colors, just as drops of rain do to produce a rainbow in the sky. Using these instruments, we can measure the intensity of light corresponding to each wavelength with great precision. The results can be presented on a graph, in which the horizontal axis represents the wavelength or frequency which identifies the color and the vertical axis represents the intensity of that color. When an atom or a molecule is particularly abundant, the graph will have peaks corresponding to the wavelength emitted. Figure 4.13 shows a

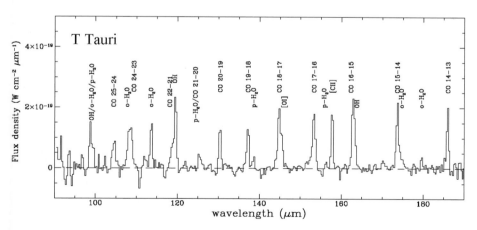

Figure 4.13: *The spectrum of the very young star, T Tauri, whose name is used to categorize all stars that are in this evolutionary phase. The heights of the emission peaks are proportional to the abundance of the atoms or molecules producing them. Note the presence of water (H_2O), carbon monoxide (CO), neutral oxygen ([OI]) and ionized carbon ([CII]). This spectrum was obtained with an LWS spectrometer mounted on the ISO satellite. (© A&A Luigi Spinoglio IFSI-INAF & the ISO LWS team.)*

measurement, by the European satellite *ISO*, of a very young star called T Tauri, which is similar to our Sun 4 billion years younger, when the solar system began to form.

Each peak in the graph is labeled by the chemical symbol of the species that produced it, showing that these species are abundant around this star. One can recognize water (H_2O), carbon monoxide (CO) and oxygen (O). A graph of this kind is called a *spectrum* and researchers studying spectra are called *spectroscopists*. From the graph, they can determine not only the abundances of the various species but also the temperature and density of the gas containing them. Each atom contains many electrons, each of which can be excited to a particular mode; the same is true of molecules. For this reason a given atom or molecule can emit various colors or wavelengths (Appendix A.1). In Figure 4.13, for example, there are eight peaks emitted by H_2O and nine by CO. Each set of peaks constitutes an identity card for the species that emits it, a sort of bar code made up of peaks. Another example of a spectrum is shown in Figure A.3, in this case for a single chemical component — neon — in its gaseous state. The lines are emitted in the optical and infrared regions of its spectrum.

The peaks we observe are therefore the sums of all the characteristic photons emitted by the atoms and molecules to be found in the direction at which the instrument is pointing. If there are sufficient atoms and molecules, the peak is high enough to be detected and measured on the Earth, just as in Figure 4.13. The more the molecules of a given type, the stronger the peak they produce. From the distribution of the peaks one can determine the abundance of each species of atom and molecule.

Great sensitivity is needed to see the rarer atoms and molecules, which can be very important for understanding the chemical processes that are taking place, or for studying the most distant objects in the furthest parts of the universe. To obtain greater sensitivity, larger and more powerful telescopes are constructed. Furthermore, a technique has been invented for the increase in this sensitivity. It allows us to employ several telescopes simultaneously, by making them function as a single telescope.

The most ambitious of such projects is being pursued by Europe, the United States, Canada and Japan, which are constructing an observatory at Llano de Chajnantor in the Chilean Andes to study the millimeter emissions of atoms and molecules. This project is so extraordinary that it is worth devoting a few lines to it. Llano de Chajnantor is in the Atacama desert, at an altitude of 5000 m, a place that, apart from the normal discomfort of high altitude, is still one of the most arid deserts in the world: they say that it rains once every 100 years! Though location is so hostile, every means is sought to reduce human intervention in its localities: most of the operations are executed by remote control from a base at a lower altitude, as if the telescopes were orbiting a

Figure 4.14: The ALMA millimeter and submillimeter interferometer, built at Llano de Chajnantor, Chile, in the Atacama desert. The system has 64 telescopes — each 12 m in diameter — that function as a single telescope with a total collecting area of 7240 sq m. (© ESO.)

satellite. On the other hand, the dry air and the height of Llano de Chajnantor make its atmosphere one of the most transparent on the Earth.

The project is called ALMA, acronym for Atacama Large Millimeter Array. ALMA consists of 64 telescopes — each 12 m in diameter (Figure 4.14) — that will function as if they were a single instrument with a collecting surface of 7240 sq m (the area of a hundred apartments). Its construction requires extraordinary engineering, and optical, electronic and computing capability. Just think of the signal of each single telescope being digitized and sent through an optical fiber to the control tower where a computer will execute 16 trillion operations per second to put them all together. Thirty years ago such a machine was inconceivable. The spatial resolution of ALMA for millimeter waves will be the same as for a telescope 14 km in diameter (Appendix A.5), better than that of the HST in the optical region of the spectrum.

ALMA will make extraordinary discoveries possible. It will be able to penetrate the protostellar disks of Figures 3.7 and 3.8 and perhaps allow us to understand how planets like the Earth are born. It will also detect hitherto-unobserved and relatively rare molecules, permitting the study of the complex chemistry taking place in interstellar space which, as we shall see in the next chapter, can influence the birth of life on a planet.

4.10 Molecules in Space

In 1937, two astronomers of the Mount Wilson Observatory in the United States, observing the light emitted by some luminous stars, discovered that their spectra (Section 4.9 and Appendix A.1) had a series of peaks that could not be attributed to known atoms. After some uncertainty, they realized that the peaks indicated the presence of two molecules in interstellar space, which were absorbing the light of stars on its journey toward the Earth; these were identified by the chemical symbols CN and CH.

This discovery created a great fuss, because many scientists believed that molecules could not survive in interstellar space, that if they were formed they would be destroyed by cosmic rays, or by x- and ultraviolet rays. (On the Earth, these rays are screened by its atmosphere, but they are much more intense in interstellar space.) This discovery opened a new field of research — the chemistry of the interstellar medium — and generated a whole series of observations and theoretical studies. These studies were aimed at understanding how reactions could occur in a practically empty space, with densities millions of times lower than that achieved in laboratory conditions and with temperatures ranging from the −270°C found in molecular clouds to thousands of degrees in stellar regions.

There was considerable excitement, which soon led to the work of high-ranking spectroscopists like Gerhard Herzberg, who was awarded the Nobel Prize in 1971 for his studies of molecular chemistry. Born in Germany, he became a naturalized Canadian citizen in 1935 after escaping from the Nazis. He was one of the first to discover the great importance of this chemistry for understanding both the history of the universe and the processes that generated life on the Earth. We know that organisms are made up of molecules and life is based on processes that transform molecules into other molecules; these processes, we shall see, may have even begun in interstellar space.

When they are excited, for instance by a collision with an atom (Sections 3.3 and 4.9), the majority of molecules emit radio, submillimeter or infrared waves. They are, therefore, studied primarily with radio telescopes. The great ALMA interferometer, which we discussed in the last section, was designed precisely for such studies. Its great sensitivity will allow us to see the less abundant molecules which are escaping detection today but which could play a fundamental role in many chemical reactions, including those from which life originated.

From radio studies, we know that the most abundant molecule in the universe after hydrogen gas (H_2) (composed of two atoms of hydrogen) is *carbon monoxide* (CO), which we have encountered many times in the previous pages. Not only is this molecule very abundant, it is also excited by collisions with atoms of relatively cold gases. In fact, it has the merit of being the major coolant of molecular clouds (Section 3.4).

Water (H_2O) is the third-most-abundant molecule in the universe. On the Earth, there are 2 sextillion (thousand billion) liters of water, mostly in the oceans. This is a large amount — one part per thousand of the mass of the Earth. If we could measure the amount of water trapped underground at great depths and high temperature (5 km underground the temperature is 200°C), we might find that water is even more abundant. Some estimates, based on water vapor emitted by volcanoes, prompt us to guess that there is more subterranean water than that contained in the oceans.

Water is found everywhere in the universe (even inside solar sunspots) but its greatest importance is the role it plays in making life possible. Just consider that water constitutes two-thirds of the matter of all living things, animals and vegetables. It is the only molecule on the surface of our planet that is found in all three states of solid, liquid and gas, although the liquid state is the most important for the existence of life. As we shall see in the following chapters, a planet is habitable if it has liquid water on its surface. Let us, therefore, discuss this molecule in more detail.

The spectrum of the star T Tauri, shown in Figure 4.13, represents one of the first measurements of the abundance of water on an extraterrestrial object. In fact, water cannot be observed from the Earth [except for a few lines (Appendix A.1) from extremely excited states], because the weak emission of interstellar H_2O is absorbed by the water vapor present in the atmosphere.[13] The first measurement of water in normal conditions was obtained only in the 1990s, when the *ISO (Infrared Space Observatory)* went into orbit equipped with sophisticated instruments including an infrared spectrometer.

Water is formed by chemical reactions, either on grains of dust found in interstellar space or in the hot gas surrounding stars. The latter process produces large amounts of energy: the oxy-hydrogen flame, with which one cuts and solders metals, is obtained by combining oxygen and hydrogen to form water, and is an example of how much energy this process can release.[14] Water does not form easily on the Earth; otherwise, every emission of hydrogen would result in an explosion as it combines with the oxygen in the atmosphere. Instead, the two gases must reach a temperature above $300°C$ to form H_2O, as it happens when the oxyhydrogen mixture is heated and ignited by a match.

The jets of matter expelled by newly forming stars (Section 3.7) heat the interstellar gas to temperatures of thousands of degrees, sufficient for chemical reactions to occur, as is evident from the spectra of Figure 4.13. Once formed, H_2O condenses on the grains of dust and ice that gather in giant molecular clouds. These icy grains end up in planets *via* the placental clouds and the

---

13 *Some water lines have been observed from the ground in Sun spots or in the atmospheres of some stars. They are very special lines that need gas at temperatures higher than $1500°C$ to excite or absorb them. Since the Earth's atmosphere has a much lower temperature, the lines can penetrate it without being absorbed.*

14 *The possibility of producing large amounts of energy by combining oxygen and hydrogen to form water suggests hydrogen as a clean source of energy. Do not be misled: hydrogen is not a source of energy because, unlike carbon, oil or natural gas, it does not exist in a free state in nature. To extract it from water, we need the same amount of energy that we get back when we combine it with oxygen again. (In fact, we get back less than what we put in, because of inefficiencies in the process.) Hydrogen is, therefore, only a means of accumulating and transporting energy, like the batteries of a car which give back the energy used to charge them. If the energy we use to extract hydrogen from water is produced from conventional techniques, the same process will be accompanied by the corresponding pollution.*

circumstellar disks discussed in Section 3.8. Comets, which have the same chemical composition as the matter from which the solar system was formed, all have spectra showing a great abundance of water. It follows that if the water was formed in interstellar space and arrived on the Earth *carried by* the matter that formed the solar system, it ought to be found on all the planets except those nearest to the Sun, Mercury and Venus, whose high temperature would have evaporated it.

Other molecules that are abundant in interstellar space are methane (CH_4), ammonia (NH_3), carbon dioxide (CO_2), silicon oxide (SiO) and carbon sulfide (CS). They are the simplest molecules that can be built from the atoms of H, O, C, N, S and Si, the most abundant elements in the universe (Figure 4.12). Helium is a noble gas that does not combine with any other atom and therefore does not contribute to the great zoo of reactive molecules.

Since 1937, astronomers have discovered more than 150 different molecules in space, the most complex of which are the organic molecules that have carbon in their composition. Macromolecules, with chains of 11 atoms like HC_9N, of 13 atoms like $HC_{11}N$ and of 15 atoms like $H_{13}CO$, have been observed. Larger molecules probably exist but they are rare and their emission could not be detected with existing radio telescopes.

In recent years, with the increasing sensitivity of the instruments, astronomers have begun to discover amino acids, the building blocks of the proteins that are the basis of life. It is therefore natural to ask whether *prebiotic* molecules exist in interstellar space. This question is difficult to answer, because such molecules may be too rare to be observed with existing instruments. Moreover, they could exist on interstellar dust grains, where a very complex chemistry takes place, and where they would not be seen by spectrometers that only analyze gases. Complex organic molecules like amino acids have been found on meteorites, although some of those are thought to have come from Mars. (They are rocks ejected from the planet's surface by the impact of an asteroid.) Thus, it is possible that very complex chemistry in interplanetary space may have furnished the Earth with the material from which life emerged.

It is therefore spontaneous to ask: To what extent have interstellar molecules influenced the birth and evolution of life on the Earth? From this question comes the great interest in the chemical composition of comets that contain information about the chemical composition of the Earth when life was born. vIn fact, while the Earth's atoms are those of the placental cloud, the molecules are not. From the moment of its birth, the Earth's molecules have continued to transform, producing ever-more-complex structures. They have modified the environment so much that, as we will see in Chapter 12, a distant civilization could infer the existence of life on the Earth by observing the spectra of the molecules contained in the atmosphere. To the role of molecules in the origin of life we devote the next chapter.

The Origins of Life

Chapter 5

5.1 Introduction

In 1924, Russian biochemist Alexander Oparin (1894–1980) published a book entitled *The Origins of Life*, which rapidly became famous. In the book, he proposed that life on the Earth was born in a series of processes that gradually transformed matter from an inanimate to an animate state. To support his thesis, he noted that living organisms were made up of the same atoms and molecules that formed inanimate matter. It was therefore natural to suppose that there existed a process which, over time, built ever-more-complex molecules, eventually resulting in life. Encouraged by the spectroscopic analysis of Jupiter, which showed that its atmosphere contained large quantities of methane, he suggested that the Earth's primordial atmosphere was composed mainly of methane, ammonia and water — the simplest molecules that can be built from the most abundant atoms in the universe (Figure 4.11). Lightnings, which were frequent at the time, would then have converted these into complex organic molecules, from which the first living organisms were born. As we shall see, this hypothesis was partly verified, in 1952, by Miller and Urey's experiment, which produced the famous primordial soup (Section 5.3).

Although knowledge of the early Earth was less detailed in Oparin's time than it is today, his ideas on the origin of life were substantially the same as those now held by most scientists. It is believed that the journey toward life began with the Big Bang, when the first atoms (H and He) were formed, followed by the nuclear reactions in stellar cores that created the heavy elements and the formation of the first molecules in interstellar space. The process continued with the birth of the solar system, where the proportion of heavy atoms was much greater than in the universe as a whole (Section 4.8), and where its favorable conditions enabled molecules to become ever more complex until, eventually, they were able to reproduce themselves and life was born. This final step — the origin of the first creature — is the one which scientists are trying to understand today.

The fossils tell us that this first living thing was a bacterium that appeared on the Earth 3.5 billion years ago as soon as our planet was cool enough for it to survive (Section 6.3). Evolution has produced all of today's living organisms (animals, vegetables, fungi, bacteria) from these early bacteria. Although the process has yet to be understood, the overall picture is confirmed by the recent extraordinary advances in genetics and by ample fossil evidences.

The idea that life came from inanimate matter is not new. It is found in the stories, myths and even the religions of many peoples as the simplest explanation for the origin of life. Aristotle, for example, arrived at this conclusion after having observed how insects and worms were born spontaneously in the decomposition of organic matter. His observations were, of course, erroneous: we now know that these creatures are the product of tiny eggs, which are present in the matter. Not having a microscope, the great philosopher could not have seen these eggs. Indeed, until relatively recent times, his hypothesis was generally accepted to hold for inferior species like bacteria, although not for humans. The first to cast doubt on this idea, on scientific grounds, was the Italian doctor Francesco Redi. In 1668, he showed that larvae did not appear in freshly butchered meat, sealed in an envelope to keep insects' eggs out. Nevertheless, people continued to believe in the spontaneous generation of life until 1862, when Pasteur showed that the so-called effect did not occur after sterilization.

Those erroneous beliefs arose from the lack of appropriate experimental tools. However, they show that, even in ancient times, ideas on the origins of life were not dissimilar to those of Oparin and scientists today. This is not surprising, because the only alternative to the emergence of life from inanimate matter is by supposing that the appearance of the first bacteria on the Earth, 3.5 billions of years ago, was the result of a divine intervention and not, a matter for scientific inquiry.

Moreover, we should not think that the idea of the emergence of life from inanimate matter is not, in itself, in conflict with religious belief (Section 1.4). If there is a God, responsible for the laws of nature, it is surely senseless for a believer to deny *a priori* that man, and the universe around him, have been constructed according to those laws.

Not everyone agreed with this approach. Creationists, for example, out of blind faith and against all scientific evidence, maintain that special creatures like humans can only result from a direct act of God, who bypassed His own laws to create him. Scientists, on the other hand, try to show that the laws of nature themselves — in which one may see the hand of God — enable inanimate matter to generate life because it is the only possible *scientific* hypothesis.

5.2 Life is a Chemical Process

What is life? In its simplest sense, life is the ability of an organism to reproduce itself. In a more complex definition, recognized by most scientists since Darwin's time, life is:

- An organism that contains information;
- An organism capable of reproducing itself;
- An organism capable of evolving;
- An organic (carbon-based) organism.[1]

In this definition, we use the word "*organism*" because life is, after all, a matter of the *organization* of the different components (atoms, molecules and organs) that function cooperatively in a living thing. Life, seen from this point of view, is a global property, neither reducible to a single component of an organism nor simply to the sum of its individual components. The death of an organism is the failure of this global property: after death the constituents are the same as they were an instant before, when the organism was still alive. What is missing is the "organization" that allowed them to work together.[2]

Another definition of life was given by the Nobel Prize winner Christian de Duve in a 1996 article entitled "*The Chemical Origins of Life.*"[3] He wrote, "Life is a chemical process because biological information is encoded in molecules, is replicated, decoded and expressed by means of chemical processes… thousands of chemical reactions carry out the maintenance, growth and repair of all cellular structure, as well as all other forms of biological work, with the help of chemical energy drawn from the environment. Life is chemistry, a word that has to be taken in its broadest sense to include mineral chemistry, organic, biochemistry, electrochemistry, physical chemistry and thermodynamics."

---

1 *The first three definitions could refer to a robot, capable of repairing itself and of constructing another unit, similar to itself but better adapted to the changes in the environment. Nobody yet knows how to build such robot but it may be possible some day. The fourth definition differs from the first three and cannot apply to a robot: it does not define the* functionality *of the organism but only the material from which it is built. One must logically admit, though, that it may not be an essential condition for a definition of life.*

2 *We suggest that those who want to delve deeper into this should read* Emergence of Life: From Chemical Origins to Synthetic Biology, *by Pier Luigi Luisi (Cambridge University Press, 2006).*

3 *In* Astronomical and Biochemical Origins and the Search for Life in the Universe, *C. B. Cosmovici, S. Bouyer and D. Werthimer (eds.), Editrice Compositori (1997).*

De Duve concluded that if all biological processes are chemical, *the origin of life must be explainable in chemical terms.*

The other possibilities, as we have already said, are that the first bacterium that appeared on the Earth was born by divine intervention outside the laws of nature, or by the Panspermia hypothesis (Section 5.13), which holds that life originated far from the Earth and was then transported here by some mechanism. Panspermia merely changes the location of the origin of life, but does not solve the problem of the origin itself and, as we shall see later, has not a greater probability to occur than the spontaneous birth of life on the Earth.

If we admit that life originated on the Earth, and was generated by a chemical process, we have to conclude with De Duve that because *chemistry deals with deterministic and reproducible phenomena involving trillions of molecules...also the process from which life originated should have been obligatory under the physicochemical conditions that prevailed on our planet when life appeared.* This process — which we do not yet understand — should render life "a highly probable outcome from a highly improbable premise." If this is true, life has to arise everywhere in the universe where the same physicochemical conditions occur, and therefore *life is a cosmological imperative and the search for extraterrestrial life is a scientifically valid enterprise.*

If we think of life as being the result of a chemical process, we must accept that there was a period when conditions existed that naturally favoured the creation of some complex molecules. Then, by a mechanism similar to that suggested by Oparin, these molecules combined with others to create more complex molecules and so on, at each step, increasing the complexity of the molecular structure. Eventually, the first proteins, the first enzymes, and the first cells capable of reproducing themselves and of evolving, were created. This process occurred very rapidly on the Earth, as testified by fossils containing traces of the first cells, formed over 3.5 billion years ago, when the bombardment by the infalling material that formed the Earth ended (Section 6.3), our planet cooled and solidified, and the oceans were formed. As we shall see in Chapter 6, life was created as soon as the conditions necessary for the survival of biological organisms existed. Because it was so rapid, we must conclude that the process was also extremely efficient. It should therefore be rather easy to replicate it in a good scientific laboratory, where all the environmental conditions can be controlled better than they were on the early Earth. Notwithstanding numerous attempts, however, no one has succeeded in doing so. We address this paradox in the next section.

5.3 The Primordial Soup

In 1952, Stanley Miller and Harold Urey of Chicago University, following Oparin's idea (Section 5.1) that life originated in a series of molecular

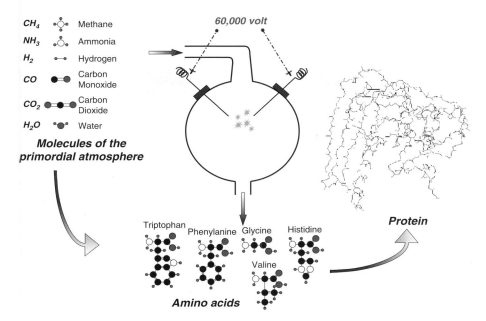

CH_4		Methane
NH_3		Ammonia
H_2		Hydrogen
CO		Carbon Monoxide
CO_2		Carbon Dioxide
H_2O		Water

Molecules of the primordial atmosphere

60,000 volt

Protein

Tryptophan Phenylanine Glycine Histidine

Valine

Amino acids

Figure 5.1: The Miller–Urey experiment. At the top-left corner the primordial Earth's simple molecules are shown. At the bottom center there are the amino acids produced by the experiment. On the right-hand side a protein molecule is shown, in which every point is an atom.

processes taking place in the Earth's primordial atmosphere, conducted an experiment (Figure 5.1) that aroused great interest. They filled a glass container with ammonia (NH_3), carbon monoxide (CO), carbon dioxide (CO_2), hydrogen (H_2) and water (H_2O) — the molecules Oparin had suggested as making up the Earth's early atmosphere. These are the simplest molecules that can be made from the most abundant atoms of Figure 4.11 but they constitute a toxic mixture, lethal to all the evolved species that inhabit the Earth today.

For several days, this mixture was subjected to a 60,000 V electrical discharge, simulating the lightning prevalent in those ancient skies. At the end of the experiment they found, at the bottom of the container, a series of amino acids, which are the constituents of proteins, the basis of every form of life (Section 5.4).

Miller and Urey's experiment is shown schematically in Figure 5.1. On the left are the molecules they put into the container to emulate the primordial atmosphere; below, the amino acids produced by subjecting the gas to electrical discharges are shown.

The news of the experiment rapidly spread around the world and raised a great clamor, particularly because radio astronomers had not yet discovered amino acids in interstellar space. For some time, people were convinced that Oparin's theory had been proven to be correct and that the origin of life on the Earth had been explained.

Many laboratories carried out every possible variation of the experiment. They found the process to be very efficient and many different amino acids should have been produced in the ancient atmosphere. Because there was no oxygen to oxidize the amino acids, nor were there any biological species to eat them, scientists concluded that they must have been produced to excess, filling the planet's oceans. The density of amino acids would have given the seas the consistency of a chicken soup; this is the origin of the name *primordial soup*, which is still used to denote any mixture from which life might emerge. After the initial experiments, the "soup" was stimulated in every possible way in the hope that the amino acids would evolve into something more complex. This did not happen and with the time passing the initial enthusiasm died down.

After 50 years of trying, we have still not been able to get beyond Miller and Urey's result. To give some idea of the difficulty in taking the next step, on the right of Figure 5.1 is a protein, a macromolecule that is fundamental to all forms of life (Section 5.4). Yet, in spite of its complexity, it is still a lot simpler than DNA (Section 5.6). Producing a protein is the very minimum that one must achieve to begin speaking of "life." It is enough to compare the structure of the protein shown in the figure, every point of which represents an atom, with that of the relatively simple amino acids produced by Miller and Urey to understand how far we are from that goal.

All the amino acids necessary to make proteins exist in the primordial soup but we have to overcome two apparently insurmountable barriers if we are to go further. First, the chemical reactions that combine amino acids into proteins are extremely slow, so slow that no proteins would be produced in a "reasonable" time. Slow reactions can be sped up, however, by employing catalysts. In living organisms, the catalysts are enzymes, which are proteins that did not exist in the soup. The problem is: How was the first protein built if another protein was needed to build it in a reasonable time?

The second problem is that proteins are not casual sequences of amino acids. As we shall see (Section 5.4), they are sequences of 20 particular amino acids put together in a precise order, according to instructions contained in the DNA. Unless this order is respected, the proteins cannot function properly inside an organism. As we shall see in Section 5.11, the probability of obtaining the right order by chance is practically zero.

At present we lack even a theoretical solution to these two problems. This leads to the paradox that we cannot produce anything more complex than Miller and Urey's amino acids, even in the controlled conditions of a laboratory, whilst life on the primitive Earth was born in such hostile prevailing conditions that the species dominating the Earth today could not even survive to them! Yet life was born as soon as the Earth became cool enough to make it possible (Section 6.3), implying that the life-generating process was easy and likely.

The usual solution of the paradox is that, although it is true that we can create any set of physical conditions in a laboratory, we do not know exactly what conditions actually existed on the Earth. And, because the environmental possibilities are practically infinite, we may never be able to reproduce these conditions correctly!

Nowadays, for example, we know that the primordial soup may have contained small quantities of unknown organic molecules resulting from a not-today-well-known chemistry happening in the interstellar space. The existence of those molecules was not even suspected by Miller and Urey. Some of them might even have reached the Earth on interplanetary dust grains or meteorites. These molecules could have played an essential role in the first steps toward the birth of life.

Another hypothesis on which some laboratories are working is that the clay formed from the ashes of volcanoes could have acted as catalyst. Moreover, this clay expanded when it absorbed water, creating cavities in their interior, where the first macromolecules would have been protected from the hostile environment of the primitive Earth. Confined in these cavities, the products of reaction would also not have dispersed in the environment, making the chain of reactions leading to the first cells more likely. The clay itself, or one of the substances it contained, could have played the role of catalyst, resolving the problem of how to accelerate reactions that would otherwise be almost impossible. If life *was* born in these cavities, we would not be saying anything new: the Bible mentioned that the first man was modeled out of clay....

The second problem is that, even with catalysts to speed up the reactions, *we still do not understand how the first proteins, and the first molecules of RNA or DNA, were born.* As we shall see, the number of possible sequences is nearly infinite and most of them do not lead to any biologically sensible product. Without the presence of another macromolecule to provide information on how to build so complex a structure, the "simple" amino acids of Miller and Urey would remain simple amino acids.

Before discussing how scientists seek to address this second problem, we shall digress to discuss a few characteristics of cells and of DNA which are necessary for understanding what follows.

5.4 The Cell

In 1680, the English scientist Robert Hooke (1635–1703), a contemporary of Newton, improved an instrument that he had seen during his travels in Europe and produced the first microscope ever constructed. With this instrument, he began to observe the fine details of objects, among them a piece of cork. He noted that its structure consisted of many little contiguous squares, covering the entire surface. Thinking that they resembled the cells of monks in a monastery, he called them cells.

Hooke had discovered the basic unit of biological organisms. Every organism consists of cells, be they monocellular like in bacteria, or multicellular like in animals, plants or fungi which are made up of an array of contiguous cells (some 100 million of them in an evolved organism). These cells are differentiated according to the functions they are responsible for. These differences are remarkable in evolved species; it is sufficient to look at the epidermis, nervous system, bone and hair cells. As we shall see, the differences are determined by the type of protein contained in the cell.

The characteristics of cells are:

- They are the components out of which the tissue of all organisms is constructed (there are large numbers of cells, which are assembled in a certain order and which recognize each other);
- They host the fundamental processes of life;
- They are born from other cells and have the ability to reproduce themselves autonomously.

Viruses, responsible for many illnesses, like influenza and AIDS, are not considered to be alive, because they cannot reproduce themselves autonomously but have to use the inner substructures of cells to do so.

The cell satisfies all the characteristics of an organism (Section 5.2), including the ability to reproduce itself. It is the simplest unit of life, so it is not surprising that bacteria, the first organisms to appear on the Earth, were monocellular (Figure 5.2). Given their simplicity, these organisms were — and still are — adapted to a wide range of environments and were able to survive all past cataclysms. They are the oldest living things on the Earth and have the best chances of surviving all future catastrophes. In them the ultimate possibility of life on the Earth resides. If it is shown someday that life is able to travel from one planet to another (Section 5.13), it will be found that the messengers of such life are bacteria; they are the only organisms that could survive in a tiny droplet of water or on a grain of dust traveling through interstellar space.

Even tough they consist of a single cell, bacteria perform all the functions of an organism: they nourish themselves, they reproduce and they move around using the cilia and the little flagella on their surfaces. If all the species populating the Earth are ultimately derived from them, they must also have the ability to evolve. Figure 5.2 is a schematic representation of a bacterium. It has an external *membrane* that protects the cell and controls the ingress and egress of the substances necessary for its metabolism. Inside, there is a gelatinous substance, the *cytoplasm*, containing the ribosomes, used for the synthesis of proteins, and the DNA, which is the genetic code of the bacterium consisting of a single large molecule occupying nearly the whole interior of the cell.

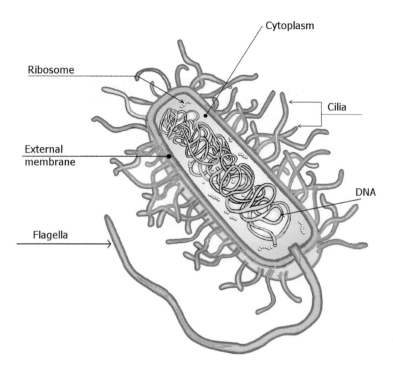

Figure 5.2: A bacterium, the simplest organism. It consists of a single cell without a nucleus. It can move, thanks to the cilia and flagella located on its surface. Inside are the ribosomes, used for the synthesis of proteins, and a single DNA molecule.

Bacteria, unlike the cells of more evolved organisms, do not have a nucleus, and for this reason they are called a *prokaryote*, from the Greek words *pro* ("before") and *karyon* ("nucleus"). Prokaryotic bacteria are between a thousandth and a ten-thousandth of a millimeter in size, 10–1000 times smaller than the cells with nucleus of the more evolved species. The first bacteria to appear on the Earth belong to the kingdom of *archeobacteria*. They do not need oxygen to live; indeed, oxygen, which did not exist in the primordial atmosphere (Section 6.2), is poisonous for most of them. A few of these bacteria have survived to the present era and belong to the group of *extremophiles* that live in the sulfurous, boiling waters of undersea volcanoes, in the freezing water of lakes below the Antarctic ice, or in the salt water of the Dead Sea, where no evolved form of life can survive. Others survivors are *methanogens*, which nourish themselves on organic matter, which they decompose to produce methane; they are found in the intestines of ruminants.

These organisms live in the strangest of environments, in conditions nearer to the primordial Earth than those inhabited by all other forms of life today. *To understand the birth of life on the Earth, we must follow the trail leading from Miller and Urey's amino acids to the archeobacteria.* Their complexity shows that this path is not straightforward and is perhaps the most difficult to follow in all the history of life on the Earth: it was the step which led from the inanimate matter to the first organisms.

For almost 2 billion years, the Earth was inhabited by archeobacteria; then came the *eubacteria*, still prokaryotic bacteria (without a nucleus), which acquired the ability to synthesize chlorophyll. To these we owe the appearance of oxygen in the atmosphere of our planet (Section 6.5). This event permitted a more efficient metabolism, leading to the birth of multicellular organisms and their evolution toward today's species. Even today, these bacteria produce a large part of the atmosphere's oxygen.

Eubacteria are very useful, because they are able to metabolize widely varied substances. Water purification plants depend on these bacteria and they are used for the elimination of hydrocarbons from the sea. Some eubacteria defend our intestines from harmful bacteria and help us to digest what we eat.

Bacteria are found practically everywhere, in organisms, in the upper atmosphere, in the depths of the oceans and beneath the surface of the Earth. A liter of pasteurized milk contains 20 million bacteria, there are 20,000 on a square centimeter of skin and every time we breathe we inhale 10,000 of them. A study by William Whitman, of the University of Georgia, estimated that there are 10^{30} (1 followed by 30 zeros) bacteria on the Earth, 92% of which are contained in the soil or in the ocean depths. In spite of their being very small and light (nearly one millionth of a gram), their contribution to the mass of the biosphere is comparable to that of the whole vegetal world or of all the fish in the sea. They also store a comparable amount of carbon.

Bacteria play a fundamental role in the survival of life on Earth, enabling many vital cycles to take place; without them, life as we know it on Earth would not exist. A small number of them are carriers of dreadful diseases such as tetanus, syphilis, tooth decay and pneumonia. We combat them with antibiotics, molds that belong to the fungus kingdom (Section 5.5), that kill *all* bacteria, including beneficial ones, and this can create other problems for an organism.

About 2 billion years after life appeared, a little before the appearance of eubacteria, a fundamental step in the evolution of life occurred: cells with nuclei were born. These are called *eukaryotic* cells, a name derived from the Greek meaning "with a well-formed nucleus." Organisms formed of these cells are called *eukaryotes* and they comprise all forms of multicellular life, including today's animals and plants. Eukaryotic cells are much more complex and much larger than prokaryotic cells; on average, they range in size from 5 to 30 thousandths of a millimeter, but those dedicated to reproduction can be much larger. Ostrich eggs, for example, can be more than 10 cm long and weigh several kg. A eukaryotic cell is represented schematically in Figure 5.3. Inside the cell we see the nucleus, which contains the chromosomes (Figure 5.4), where the DNA, containing the genetic code of the organism, is rolled up and packed (Section 5.6).

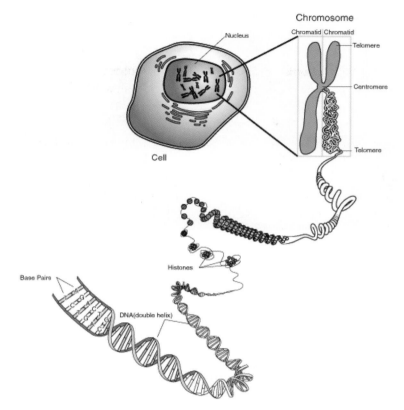

Figure 5.3: A eukaryotic cell, like those making up multicellular organisms. It has an external membrane that regulates the flow of matter in and out of the cell. Inside the membrane, is the cytoplasm, containing the nucleus, where the chromosomes reside. Inside each chromosome lies the long DNA macromolecule, which contains, rolled up and folded, the genetic code of the organism. (© Darryl Leja, National Human Genome Research Institute, National Institutes of Health, USA.)

The first eukaryotes were very simple, little different from bacteria, but some of them lived in colonies where, over time, they learned to cooperate — various cells specializing in different functions. These colonies of bacteria eventually transformed themselves into the first tiny, multicellular organisms belonging to the kingdom of *protists*. This transformation represents the transition from the world of bacteria to that of evolved organisms. The cells that make up today's animals, plants and fungi are derived from the protists.

Animal cells, like all eukaryotic cells, consist of three basic structures: the membrane, the cytoplasm and the nucleus (Figure 5.2). The first two are similar to those found in the prokaryotic cells of bacteria, although they are more complex. The membrane surrounds and protects the cell and, with the help of a few proteins disposed on its surface, controls the passage of substances into and out of the cell. It admits the glucose, which furnishes the cell with energy; the

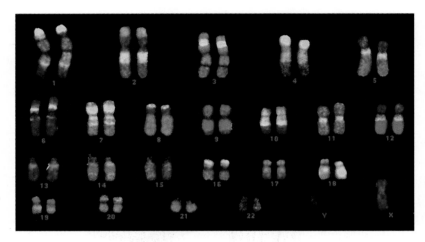

Figure 5.4: The 23 human chromosome pairs, inside which the DNA is packed (Figure 5.3). The first 22 pairs of chromosomes are similar and are called homologous; one of the chromosomes comes from the father and the other from the mother. The 23rd pair has two chromosomes, each of which is either X or Y. These define the sex of the unborn baby: if the last two chromosomes are XX the child will be female, whilst if they are XY a boy will be born. The mother always donates an X chromosome, so the father determines the sex of the baby by providing either an X or a Y chromosome. A similar process applies to all organisms that reproduce sexually, both animal and plant. (© Applied Imaging International Ltd., Newcastle-upon-Tyne, UK.)

amino acids, which are used to construct proteins; and the salts and minerals needed by the cell. On the other hand, it releases toxic material produced within the cell together with other substances, such as antibodies, that the cell produces to be used in other parts of the organism. In animal cells, the membrane has another important role, facilitating cooperation between cells: it allows similar cells to recognize and connect with one another. The rejection that sometimes follows an organ transplant, for example, results from the immune system of an organism failing to recognize the proteic structures of the membrane of a transplanted cell.

As in the prokaryotic cells, the membrane encloses the cytoplasm, which contains many small structures, called organelles, that carry out the cell's numerous activities, such as the production of proteins. Finally, in contrast to bacterial cells, the cytoplasm contains a nucleus harboring the chromosomes, in which the long DNA molecule is packed (Figure 5.3). Creatures composed of this type of cells constitute the animal kingdom; they are characterized by obtaining the vital elements they need — their food — from other living organisms.

Plant cells are similar in structure to those of animals. They have a membrane which, like that of an animal cell, protects the organelles and the nucleus, controls the passage of substances into and out of the cell, and unites and coordinates the cells among themselves. In plant cells, however, there is another envelope called the *cell wall*, composed of cellulose, which supports the plant mechanically. (In animals, this support is provided by the specialized cells of bone and cartilage.) Unlike animal membranes, these external structures are less specific to the individual plant. This makes grafting possible — a basic technique in agriculture.

Within the cytoplasm, as well as the nucleus, which contains the DNA, and the organelles, which carry out functions similar to those of animal cells (like the production of proteins), there are *chloroplasts*, which are the organelles which do the synthesis of chlorophyll, which keeps our atmosphere rich in oxygen (Section 6.5) — allowing animal species to exist. Organisms composed of this type of cells make up the plant kingdom. Apart from producing oxygen, plants use the Sun's energy to extract, from the air, soil, and water, the very rare elements (Section 4.8) needed to build their biological structures. Plants furnish the raw material for building the other organisms which are at the bottom of the alimentary chain of animals and fungi.

Fungus cells are similar to those of plants. They have more rigid outer walls than those of animals, but they have no chloroplast and are not capable of synthesizing chlorophyll, nor do they use oxygen in their metabolism. They obtain their nourishment from both living and dead animals and plants, decomposing them (often as parasites). With fungus cells are constituted both single-celled organisms, like yeasts and molds, and multicellular organisms, like fungi and lichens. Like bacteria, they play an important role in the disposal of refuse but, as parasites, they cause many animal and plant diseases. Some molds, like penicillin, are notable antibacterial agents and their discovery has been very important for medicine. Finally, yeasts are important for fermentation: to them we owe wine, beer, bread, sweets and all the products of the cheese industry.

Chromosomes (Figure 5.4) are contained within all the eukaryotic (animals, plants and fungi) cells. Each of them contains a piece of the organism's DNA, rolled up and folded around a protein that functions as a support (Figure 5.3). The number and the form of chromosomes vary with the species, and have no correlation with its state of evolution. In *Equisetum arvense*, a plant from the Carboniferous period, there are 216 chromosomes; in red fish there are 104; in chickens and dogs, 78; in cows, 60; in potatoes and chimpanzees, 48; in humans, 46 (Figure 5.4); in mice, 40; in tomatoes, 24; in maize, 20; in onions, 16; and in flies, 12.

In organisms with sexual reproduction, the number of chromosomes is always even and composed of pairs of chromosomes, called *homologous*, identical in form and dimension with the exception of those that define the sex. For each pair of homologous chromosomes, one comes from the father and the other from the mother, as we shall see in Section 5.7. They are unique to every organism, even if constructed from the genetic heritage of the preceding generations.

Proteins are manufactured within the cells of all organisms, including bacteria, according to the instructions contained in the DNA. Some proteins cross the membrane and carry out their specialized functions in other parts of the organism. They are: enzymes, which act as catalysts, facilitating chemical reactions (like the digestion of food) that would otherwise be impossible; antibodies, which defend the organism; hormones, which have regulatory functions; and hemoglobin, which transports oxygen within the red blood corpuscles.

Other proteins enable cells to carry out particular tasks. Keratin, for example, determines the characteristics of hair, horn, fingernails and birds' feathers. Collagen makes the skin elastic and is often used in beauty parlors; spidroin gives a spider's web a strength that is unequaled in natural or synthetic fibers.[4] On average, proteins represent 15% of the weight of an organism. They are made up of about 50% carbon, 7% hydrogen, 20–23% oxygen and 12–19% nitrogen (Figure 4.11). They have a very complex structure (Figure 5.1) and are formed of a long sequence of 20 different amino acids (like those produced in Miller and Urey's experiment), bound together by strong covalent chemical bonds. One hundred million different proteins exist in nature; the human body alone contains 100,000 types, each differing from the others according to the sequence in which the 20 amino acids are assembled.

All the functions of cells, of proteins and of DNA are accomplished *via* chemical reactions (Section 5.2) which, in spite of their simplicity, lead to the complex functions which are at the basis of life.

5.5   The Kingdom of the Living Things

In the old days, living things were divided into two kingdoms; plants and animals. This old division, however, did not reflect the complexity of life; it was abandoned and replaced with new subdivisions. At the end of the 1980s, the study of cells and DNA allowed a better understanding of the connection that existed between the evolution of living things and their cellular structure, as we discussed in the previous section. It was understood that the organism's evolutionary level did not depend on the size of its DNA, or on the number of chromosomes it contained (Section 5.4). Rather, it was dependent on the complexity of the organism's cells and, to some extent, on their size.

From this point of view, the Earth's living things can be divided into six kingdoms, characterized by the different kinds of cells they are made up of. These divisions trace the progress of evolution on the Earth, discussed in the previous section. The six kingdoms are shown schematically in Figure 5.5. On the left is a time scale showing the era in which those kingdoms appeared on the Earth, as determined approximately from fossil data (Section 6.1).

---

4 *Genetic engineering is trying to insert the gene responsible for spider's thread into silkworms, to obtain a species that produces large quantities of a silk (like silkworms do) as strong as a spider's web.*

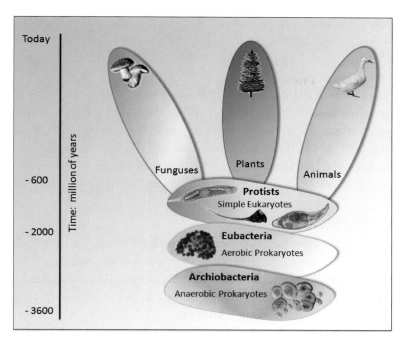

Figure 5.5: Species are divided into six kingdoms on the basis of their differing cellular structures; they are shown in the figure according to their evolutionary sequence. On the left is shown the approximate time at which the species appeared on the Earth. The oldest kingdoms are those of the prokaryotic bacteria (cells without nuclei), subdivided into archiobacteria *(anaerobic) and* eubacteria *(aerobic). Then came the* protists, *the eukaryotes (cells with nuclei) and the first multicellular organisms. From the protists, 500 million years ago, the species populating the kingdoms of* fungi, plants *and* animals *emerged.*

The first two kingdoms are inhabited by prokaryotic bacteria, consisting of a single (without a nucleus) cell. They are the *kingdom of archiobacteria*, the first inhabitants of the Earth, which are *anaerobic*, and live in extreme environments without oxygen (Sections 5.4 and 6.4); and the *kingdom of eubacteria*, which are *aerobic*, and live in an atmosphere containing oxygen. Eubacteria appeared 2 billion years after the beginning of life on the Earth (Section 6.5).

The third is the *kingdom of the protists*, composed of the more evolved eukaryotic organisms (cells with nuclei), which appeared about 300 million years after the eubacteria (Section 6.5) and represent the transition between bacteria and more complex, multicellular organisms. Some protists are single-celled, like amebae and protozoa; others live in colonies of numerous individuals; still others are multicelled, like some algae.

The last three kingdoms are composed of multicellular eukaryotic organisms, which we discussed in Section 5.4. They are: the *kingdom of plants*, which have rigid-walled cells and are capable of using solar energy for photosynthesis, obtaining nutrition from earth, air and water; the *kingdom of animals*, organisms whose cells are protected by membranes and which obtain their energy by eating other organisms; and, finally, the *kingdom of fungi*, which include single cells like yeast and molds, as well as multicelled organisms like mushrooms and lichen. They are incapable of photosynthesis and their metabolism does not involve oxygen; their sources of nourishment are other organisms, both living and dead.

The kingdoms of plants, animals and fungi are, in turn, divided into *phyla*, which we shall discuss in Section 6.6. Phyla are classified according to their body organization, under the scheme proposed by the great Swedish naturalist Linnaeus (1707–1778). They are subdivided into *class, order, family, genus,* and finally *species*, the last identifying a group of individuals with strictly similar characteristics. For example, the classification of *Homo sapiens* is:

Kingdom:	*Animals*
Phylum:	*Chordata* (all animals with a spine, among them fish, amphibians, reptiles, dinosaurs, birds, mammals)
Class:	*Mammals* (include animals ranging from whales to humans)
Order:	*Primates* (include hominids, chimpanzees, gorillas, orangutans, macaques, lemurs)
Family:	*Hominids* (include *Australopithecus* and all *Homo*)
Genus:	*Homo* (includes *Homo habilis, erectus, Neanderthalensis, sapiens,* etc.)
Species:	*Homo sapiens*

5.6 DNA: An Instruction Manual for Organisms

History. In 1886, the Swiss chemist Friedrich Miescher discovered that the nuclei of cells contained the giant molecule of *deoxyribonucleic acid,* or *DNA*; he would never have guessed that this molecule contains the genetic information that distinguishes one species from another and, within each species, one individual from another. The molecule seemed too simple and repetitive to contain complex information. The prevailing idea was that genetic information was so complex that it would require a complex structure to represent it and the protein (Figure 5.1) seemed a more suitable candidate. Scientists held this opinion for almost a century; there was not the idea, introduced later by computers, that complex information could be codified with a simple and ordered code.

It is necessary to move on to 1944, when the American biologist Oswald Theodore Avery showed experimentally that, at least for the prokaryote cells of bacteria, DNA stores the genetic information. Even then, there was much scepticism in the scientific world, which held the view that bacteria belonged to an inferior species (Figures 5.2 and 5.5), and they were so different from the world of animals and vegetables that they could not be representative of more evolved species. It was only in the 1950s that the hypothesis of DNA's being the depository of genetic information began to be taken seriously.

Finally, in 1953, the journal *Nature* published the historic article, by the American biologist James Watson and the English physicist Francis Crick, describing the structure of DNA. Employing x-rays (which are used to study the structure of crystals), they revealed the double-helical structure of DNA. This discovery opened a great new chapter in molecular biology and the two scientists were awarded the Nobel Prize in 1962.

Essentially, the DNA molecule has two functions:

- To control and program the function of every cells, and therefore the totality of an organism;
- To transmit the genetic patrimony to future generations.

To understand how this molecule works, we must understand both how the genetic information is codified and used (Section 5.4), and how this information is transmitted to future generations.

We leave a detailed exploration of this subject to biology texts; there are some excellent high school textbooks.[5] We approach the question here from the point of view of information, outlining how DNA is codified and how it reproduces itself. In the next section we shall discuss how genetic information is transmitted to future generations and we will reflect on the complexity of this molecule and its possible effects on evolution. Finally, in Section 5.11, we shall discuss how it was possible to codify a structure so complex.

Structure. The DNA molecule is contained in the chromosomes (Figure 5.3) of all cells and consists of two chains intertwined with each other in the form of a helix. The four molecules — thymine, cytosine, adenine and guanine (Figure 5.6) — are called bases and are attached to the chain in an almost infinite sequence. The two chains are made up of complementary bases: corresponding to adenine on one chain there is always thymine on the other; corresponding to cytosine there is always guanine; and vice versa. Between the bases A and T, and the bases C and G, there is a weak bond — a hydrogen bond,[6] which is easily broken. It is this bond that holds the two chains together and it is its breaking, by an enzyme, that initiates the replication of the DNA.

The diameter of the double helix is about 20 angstroms (two millionths of a millimeter), the thickness of two coupled bases. The chain is, however, very long; if you were to unfold the double helix containing the human chromosomes, you would find that it had a length of about 1.5 m. For some species, such as the rat, it can reach 10 m. There seems to be no connection between

---

5 *We suggest* The Nature Of Life, *by J. H. Postlethwait and J. L. Hopson (McGraw-Hill, 1995).*
6 *The hydrogen bond results from the polarization of a molecule, which creates a weak electrostatic attraction. It is much weaker than the covalent chemical bond in which two or more atoms are bound through sharing electrons.*

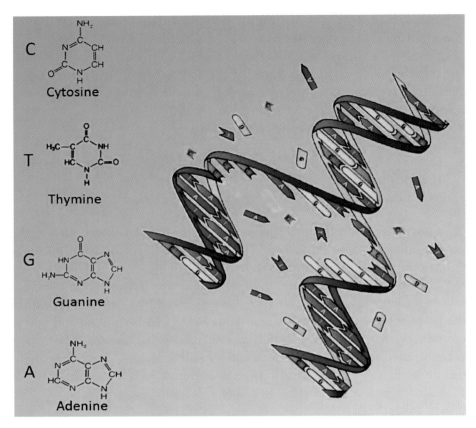

Figure 5.6: On the left are the four macromolecules C, T, A and G, called bases, which are the "characters" used to write the DNA sequence. On the right is the DNA molecule, showing double helix of complementary chains at the top and its duplication at the bottom.

the length of the DNA and the degree of evolution of species. In spite of its length, the DNA fits into each cell, rolled up and folded in on itself until it occupies less than a thousandth of a millimeter (its thickness, we have seen, is very small) and it assumes the form of the rods in Figure 5.4.

The instructions written in the DNA are used to synthesize proteins (Section 5.4) and to control the functioning of an organism, the piece of DNA used to codify a particular protein being called a gene. As we shall see, only a part of the DNA contains genes used to codify proteins; in bacteria it is 90%, whilst in humans it is less than 2%. The pieces of DNA used to codify proteins are separated by long sections (Figure 5.7) that do not codify. Until a few years ago, it was thought that these sections were completely useless, a remnant of evolution that might be eliminated in the future. Today it is thought that they could play a role in the control of an organism and that they may determine its complexity. It is suggested that the bigger the fraction of non-protein-codifying DNA, the higher the evolutionary level reached by the organism (Section 5.9).

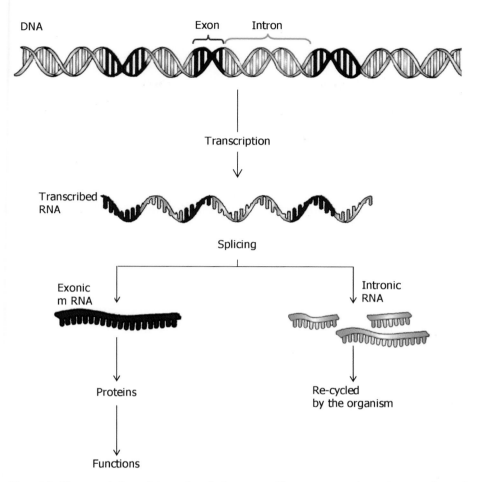

Figure 5.7: *The transcription, splicing and synthesis processes. The genes contained in DNA are made up of parts called exons (dark-colored) which codify proteins, and parts called introns which do not have this function. In the transcription process, the exons coding a protein are separated from the rest and copied into a new molecule called mRNA (messenger RNA), which will be used to produce the protein. The remaining intronic part was supposed to be recycled by the organism. Today, however, it is thought that it might contain the information necessary for controlling the organism's complexity.*

The Code. We have seen that all cells contain DNA, in which is written the information necessary for constructing a given organism and making it grow, reproduce, age and, finally, die. All cells therefore have access to the same information but, according to the functions they have to perform, they use only that part of the DNA containing the information needed for synthesizing the relevant proteins. Because each individual is different from all the other individuals, his sequence of DNA differs from those of others. Nowadays, this property is used as evidence in court to identify to whom a sample of organic material — hair, skin, blood, saliva, etc. — belongs.

How can a chemical structure contain complex information, like the genetic code, with sufficient precision, not only to differentiate every individual, be it animal or vegetable, but also to differentiate the individuals of all contemporary species from those that existed in the past and those that will exist in the future?

The technique is the same as that used for writing. When we write a book we use a code consisting of 26 letters that can be arranged in sequences to construct meaningful words and phrases. This code allows us to precisely express all concepts that can be created by the human mind.

The number of characters in a code can be any amount greater than one; computers, for example, use a binary code with only two symbols. DNA uses an alphabet of four characters — the four bases of Figure 5.6 — to write the genetic code. [Proteins are similarly codified except that they use a code of 20 characters, the 20 amino acids (Section 5.4).] If the four bases are the characters with which the text is written, each gene constitutes a "sentence" of the text, bearing a particular set of information; long sequences of genes form chromosomes that are the "chapters" of the book, characterizing an individual. Thus, as in every book, the text is written in genes and chromosomes of different lengths. Some numbers will give an idea of how long this code is. In humans the DNA contains about 3 billion bases. By comparison, a typical page of text has fewer than 3000 characters and a 1000-page novel, 3,000,000 characters, a thousandth of the characters contained in the DNA.

As we have already said, the human DNA chain is 1.5 meters long and is rolled up and packed (Figure 5.3) within the nucleus. This lengthy sequence (the book) is divided in 23 chromosomes (the chapters), each one containing between 300 and 3000 genes. Finally, each gene contains a number of characters (the bases) varying from 50,000 to 250,000.

We have already made the point that there is no connection between the number of bases in the DNA of a specie and its level of evolution; nor has any link been found between the number of chromosomes (Section 5.4) and the length of the molecule. Some vegetable species have the highest number of bases. The tulip, for example, has 30 billion bases compared with 3 billion for humans. In DNA, as for a book, it is the quality of what is written that counts, not the quantity.

Replication of DNA. Every time a cell divides, creating a copy of itself, its DNA has to be duplicated so that the new cell will have an exact copy of the mother cell's genetic code. This process is activated and controlled by numerous proteins, especially enzymes. It is initiated by the action of two proteins, the *DNA tropoisomerase* and *helicase*, which split the DNA of each chromosome (Figure 5.3) by breaking the weak bond that binds the double helix together (Figure 5.6). As soon as the two chains are separated, the bases that form the characters of the DNA begin to capture the complementary molecules, floating freely in the cytoplasm, by weak bonds. The new chain is produced rather slowly: in mammals the rate is about 50 bases per second and it would take weeks to finish the process if it progressed from one end of the DNA molecule to the other. This difficulty is resolved in multicelled organisms (eukaryotes) by enzymes that start the process contemporaneously at many points of the chain.

Once the old bases are paired with the new ones, the new bases join themselves together with a strong chemical (covalent) bond under the action of an enzyme (DNA polymerase), constructing the new chain. In this way, two double helixes are formed, each made up of an old and a new chain. Each new cell thus has an entire double helix of DNA in its nucleus, containing the complete code of the individual. The bases of each single chain are bound by a strong bond that is difficult to break, whilst the two chains are connected with a weak bond that is easy to break when duplication of the DNA is needed.

The DNA polymerase also corrects any errors that may occur in the complementary chain. In this way, the replication is very accurate, with little more that one error in every billion bases. This is an extraordinary result; it is comparable to writing 300 books, each of 1000 pages, or 3,000,000 characters, with at most a single error in the whole set. Once its genetic code has been duplicated, the cell constructs a new nucleus and splits into two new identical cells.

Execution of the instructions. In nature, there are more than 10 million different proteins — the human body has about 100,000 — that are synthesized inside cells, according to the instructions contained in the DNA. For every gene of DNA, there corresponds a specific protein that determines the cell's specific function — to become a cell of epidermis, of bone, of other animal or vegetable tissue, or to produce proteins that have functions outside the cell, as antibodies, enzymes or hormones. There is, therefore, a correspondence between the four bases[7] that constitute the DNA chain and the 20 amino acids that characterize the proteins (Section 5.4). The current "mapping" of a species' genomes is equivalent to determining this correspondence and to understanding which sequence of bases corresponds to a specific protein.

Synthesis of proteins. This occurs in three phases. The first phase is *transcription*. The part of the DNA containing the information needed for the synthesis is copied (in a manner similar to the duplication of DNA) onto another molecule — ribonucleic acid, or RNA — which is also in the form of a chain, but with only a single helix (Figure 5.7). The number of bases in RNA varies between 1000 and 100,000, depending on how many are needed to carry out the instructions. In this way, the synthesis of the protein does not directly

---

7 *The number of bases needed to codify a protein in DNA is at least three times as many as the amino acids necessary for constructing a protein. In fact, at least three bases are needed to identify one of the 20 amino acids of a protein chain (Section 5.4). Two bases is not enough because the number of combinations of four bases, taken two at a time, is only 16. With three bases the number becomes 64, which is more than enough.*

involve the DNA, which would otherwise be destroyed by the chemical processes involved in the synthesis.

After the transcription of the code onto the RNA molecule, the second phase — the *splicing* — starts (Figure 5.7). The pieces of RNA that contain the useful coding information (the exons) are separated from the introns, and are bound into a new molecule, mRNA (messenger RNA), containing only the necessary information.

At this point the third phase begins — the *synthesis of protein.* With the use of another molecule (tRNA), the amino acids needed to codify the protein are identified in the cell's cytoplasm and arranged in the order specified by the mRNA. The information contained in the DNA is thus used without involving the DNA directly and therefore destroying it.

These processes represent a logical procedure executed in sequence and in an orderly manner. We should not forget, however, that they are only chemical reactions (Section 5.2), no different from any other chemical reaction. The molecules bind and separate according to the laws of chemistry, with the help of enzymes — also proteins — produced by the same processes, resulting finally in the incredible miracle of life.

5.7 Variations in the Genetic Code

Reproduction. Children are similar to their parents and siblings, but they are never identical, in either their appearance or their behavior. This is the result of an extraordinary phenomenon, called *genetic recombination,* the process by which nature creates individuals that are always unique even though their genome is constructed entirely of the genetic patrimony of their ancestors. We shall devote a few lines to describing the most complex of these processes, namely *sexual reproduction,* leaving it to specialized texts to delve deeper.

In species employing sexual reproduction, the sexual organs produce special cells called *gametes* (the egg in the female and the spermatozoa in the male), which have only one chromosome for every pair of homologous chromosomes. In humans there are 23 chromosomes in the gamete, versus 46 in normal cells (Figure 5.4). These chromosomes are obtained by a special process called *meiosis* which, under the influence of an enzyme, randomly mixes portions of DNA from the homologous chromosomes of normal cells. As you will remember (Figure 5.4), one of these chromosomes comes from the mother and the other from the father. Each gamete therefore possesses a sequence of genes taken randomly from the DNA of the two parents. The number of sequences that can be made in this way is practically infinite, and accordingly each gamete is practically unique — different from every other gamete produced by the individual — even though it is formed using only the genes of the parents.

In fertilization, the egg fuses with the spermatozoa to form a normal cell whose DNA contains copies of the homologous chromosomes, one of which comes from the mother and the other from the father. The DNA of the new individual will thus be constructed from random combinations of the genes of its four grandparents, themselves containing information from previous generations. In this way, a unique individual is formed entirely of the genes of its ancestors, resembling its parents but is different from both of them, from all its brothers and sisters,[8] from all its ancestors and from all its progeny. After the first cell is formed, it will reproduce itself rapidly, following the instructions of its DNA, constructing a new and "unique" individual.

Mutations. The reproduction of a chain of DNA takes place with great precision (about one error for every billion bases, as we have seen). It can happen, however, that a wrong bond forms during replication. For example, base A, instead of connecting to a T, connects to a G, producing a mutation. These mutations are rare because, as we have seen in the preceding section, the DNA polymerase and other "repairing" enzymes run up and down the DNA during the reproduction of a cell, correcting the errors. Mutations are often induced by external causes such as certain chemical agents or ionizing radiation (radioactivity). When this happens a character is changed and a mutation is produced.

Normally, mutation occurs in a single cell and has no consequences for the other cells. The mutation can be mortal for the affected cell, in which case it will be replaced by another. (With the exception of those of the nervous system, the organism's cells are renewed on timescales that vary from weeks to months; only in a few cases are the times as long as a year.) At other times, the mutation may occur in a germinal cell; in this case, if the mutation does not modify the reproductive capacity of an individual, it becomes inherited and, in time, it can propagate to the entire population.

In most cases, mutations have a negative effect. As in an ordinary text, where the change of a letter or a word normally produces nonsense, the same is true with DNA. Some sequences make sense but the vast majority do not. It only needs some character or base to get out of place for the organism to cease functioning or to function badly; we then have a so-called *genetic illness*. An example is *falciform anemia*, a well-known blood disease arising from the substitution of a single DNA base with another in the sequence of genes containing the information on the shape of red blood cells.[9] Instead of being round, these cells become oblong and cannot pass through the capillaries, giving rise to serious problems in the oxygenation of tissue. As it normally kills the

---

8 *With the exception of monozygotic twins, which, coming from the same egg and the same spermatozoa, have the same genetic patrimony.*
9 *Red blood cells transport oxygen to all the parts of the body by means of a protein in the hemoglobin that binds oxygen.*

individual before he has the chance of reproducing, the illness should have rapidly disappeared. The illness, instead, survives over time because of a strange combination of circumstances.

The illness is derived from five different mutations that occurred between 500 and 1000 AD in five parts of the world. Three of them appeared in Africa (Senegal, Benin and the southern region inhabited by the Bantu tribes) and two in Asia (Arabia and India); from these regions, the disease spread to all the continents. The illness has survived because an individual who has a parent with one of the mutations (or both parents with the same mutation) does not become sick, but rather has a higher resistance to malaria. He lives longer and more healthily, and therefore reproduces more efficiently. On the other hand, an individual who has parents with two *different* mutations becomes ill and dies without reproducing. The competition between these two factors has kept the level of the disease stable for 10 centuries. Over the past few hundred years, the incidence of the illness has declined in Asian and African immigrants to the United States because malaria is not endemic there, so there is no advantage in having any resistance to it, whilst the disadvantages of the illness remain and reduce the number of carriers.

Today, we know of over a thousand genetic illnesses arising from random mutations of genes, some surviving because they have similar mechanisms to falciform anemia and do not prevent the organism from reproducing. Tumors are also considered genetic illnesses nowadays; they are caused by a modification of the gene that controls the growth or other cell functions. The mutated cells then start to develop in an uncontrolled way and to move within the organism, infecting other parts and leading to death. The positive aspect of genetic recombination and mutation is that they modify individuals, generation by generation, adapting them to new climactic conditions, to new food, and to new illnesses that may historically occur in various parts of the world. Individuals who are adapted to the new conditions live better and have a better chance of reproducing; thus the species "evolves."

The Concept of race. In humans, the genetic variations that differentiate the individuals living on the Earth today are those that occurred in the last 50,000 years, since our common ancestors left Africa (Section 6.9). Some variations — the most visible ones — characterize the various ethnicities that some people call "race," a concept with little scientific basis. In fact, if one tries to characterize a group of individuals, selected by the color of their skin (white, black or yellow), by looking for the most frequent genetic characteristics of the group — that is, trying to define a mean genetic code characterizing a "race" — one finds that the differences between ethnicities ("races") are less than the differences found between the individuals of each ethnicity (or "race"). If we must talk about race, we have therefore to accept that there are as many races as there are inhabitants of the Earth. Racists, contrary to all scientific evidence, maintain that these differences extend to other characteristics, such as intelligence. As a reaction to this view, which has had terrible consequences in recent history, some scientists have started to deny that there is *any* diversity between individuals. This, of course, flies in the face of evidence which anyone can see and is the basis of evolution. We must reject both these extremes and accept diversity as part of the richness of nature.

The genetic differences that characterize the various ethnicities of the Earth are found above all in: the color of the skin and of the hair (which depend on climatic conditions); the digestive enzymes (which adapt people to the food found in the regions where they live); and the capacity to resist certain diseases — contact between peoples that have long been isolated always gives rise to serious epidemics. As we have said, these variations, caused by the environment in which individuals live, are small compared with the differences between the individuals themselves.

For example, if we try to define a black race, we have to include Africans, all the peoples of Oceania, of the Fiji islands, of New Guinea of Australia and of southern India, who have similar skin color. From a genetic point of view this classification does not work. We shall discover that Africans and the inhabitants of Oceania are genetically more similar to Europeans and to Asiatics than they are to each other. This is because *Homo sapiens*, emerging from Africa, first colonized Europe and Asia before going to Oceania. Africans and the inhabitants of Oceania are therefore genetically the most distantly related populations. The color of the skin, black, is a secondary characteristic adapted to better protect them from the Sun. The inhabitants of Europe and Asia lost this characteristic when they left Africa, and regained it in Oceania (or in southern India) when they again found themselves in a region with strong solar radiation.

Changes induced by the environment are therefore secondary and are small compared with the differences observed between individuals of a given ethnicity. The differences that one sees today among the peoples of the Earth represent less than one part in 10,000 of the genetic code. It is a predictable result, because the 50,000 years that separate us from our common African ancestors amount to only 2000 generations. More surprising is the discovery that the chimpanzee is closer to us than we thought: what differentiates it from us is about 1% of our DNA. According to some scientists, this means that we should include chimpanzees in the "genus" *Homo* as *Homo chimpanzee*, along with *Homo neanderthalensis, Homo erectus, Homo habilis* and *Homo sapiens* (ourselves). This would be consistent with the classification we are applying to other species when such small genetic differences are found, that we group these species in the same "genus."

Induced genetic variations. In the recent history of the Earth, many of the genetic variations in animals and vegetables have been induced by humans, who have favored the useful species and eliminated those that were dangerous. These changes were produced by a process of artificial selection, which gave rise to varieties that would not otherwise have existed and which would quickly cease to exist if we did not continue to take care of them. They have been made by selecting particular species and by crossing (or grafting) different animal and vegetable species; nowadays, they can also be produced in the laboratory by genetic engineering. This "guided evolution" led to the abundance of food that now nourishes the billions of people who live on the Earth. Just compare this with the primitive tribes that live by gathering natural products; each of those individuals needs a piece of land that, with modern agricultural methods, could feed hundreds, if not thousands, of people.

5.8 Darwin's Theory of Evolution

The genetic variations (mutations and genetic recombinations) that we discussed in the previous section are the foundation of all theories of evolution. The most prominent is that which Charles Darwin (1809–1882) presented in his famous essay *On the Origin of Species by Means of Natural Selection*, a book that revolutionized the way scientists viewed nature.

Before Darwin, the prevailing idea was that the variety of species seen in nature did not change in time; they were the same as when the world was created (a doctrine called "fixism"). This idea was held by Aristotle and also by the great Swedish naturalist Linnaeus (who lived 50 years earlier than Darwin), whose classification of creatures is still used today (Section 5.5). Such a view is understandable, because the timescale on which evolution takes place is very long compared to that of human life. Even over the course of many generations, we do not notice any changes. To see how remote the idea of evolution was, 50 years before Darwin's time, we need only read what is written in *Natural Systems*, Linnaeus' principal work: "*...we count so many species, how many different forms were created by the Infinite Being.*" In his book, he gives an accurate description of nature but does not seek any explanation for the differences and similarities of forms that were the basis of his classification.

What led Darwin to his theory was the examination of the data he had collected during a journey around the world when he was very young. In December of 1831, at the age of 22, he embarked on the barque *HMS Beagle* of the Royal Navy. During a five-year voyage, he visited South America, Australia, New Zealand, South Africa and numerous islands (among them the Galapagos, the Canaries, Cape Verde, Tasmania, Mauritius and the Keeling Islands). In each of these places, he studied the species he found and collected a large number of samples, which he cataloged with great care: even after 20 years, when some of the cases containing his material were opened, their contents could be analyzed without difficulty. Darwin concluded his journey in 1836 and then studied his collection for many years before finally publishing his book in 1859, 23 years after the end of the voyage. *The Origin of Species* had been a long-awaited book and was a great success. The first edition sold out in a day and successive editions followed in 1860, 1861, 1866 and 1869 before the definitive version in 1872.

Darwin's theory starts with the observation that individuals in each species differ from one another because of small changes.

Darwin he asked himself whether these small changes, given enough time, could lead to the observed differences between species. He observed that if some mechanism could select certain of these variations in preference to others, then with succeeding generations this mechanism would be able to produce the evolution we observe. He identified the environment as the selection mechanism.

Darwin thus introduced the idea of a randomly progressing evolution, producing a large number of individuals, each slightly different from the others, which were then selected by the environment. Advantageous variations increased individuals' probability of survival and of transmitting their advantages to their progeny; disadvantageous variations reduced the probability of survival and of reproduction, leading to the extinction of such individuals with time. The Earth's living species were therefore the result of a series of fortunate variations, occurring over hundreds of millions of years, with the selection carried out by the environment.

Darwin's idea is therefore based on three fundamental concepts:

- A common ancestor of many different species (or perhaps a single ancestor for all existing species);
- An evolution proceeding by small random variations and therefore gradual in time;
- A selection by the environment, permitting the best-adapted species to survive.

Darwin's theory provoked much enthusiasm and gave a great impulse to biology, which until then was limited to the cataloging of species. Biology then became a dynamic science that sought rational explanations for what was observed.

The theory also elicited fierce opposition; the idea that humanity originated from animals was found to be especially distasteful. Even today, in spite of all the confirmatory fossil and genetic evidence, there are movements that use pseudoscience to fight against an idea of evolution they believe to be an affront to human dignity.

Today, the mechanism proposed by Darwin is commonly accepted in the world of science and explains very well the small differences and the similarities found in nature. For relatively recent species, the hypothesis of a gradual evolution is used by geneticists to determine the age of a common ancestor on the basis of the present diversity of the genetic code. Today, we have succeeded in extracting the DNA of bacteria and vegetables that lived up to 500,000 years ago, and 200,000 years for animals (Section 6.9)[10]; the changes seen in this DNA have confirmed the evolutionary process.

In spite of the great success of Darwin's theory in explaining the processes that led to differentiation among the species, not everything observed in nature can be explained by this theory. Implicit in Darwinism is the idea that evolution has to proceed gradually. However, what we observe on large time-scales (Section 7.1) is that large differences between species occur not gradually but in sudden jumps. As we shall see in Section 5.11, we cannot explain the origin of a complex code like DNA with the random changes envisaged in Darwin's theory. As we shall discuss in the next section, new hypotheses are being studied that, in parallel with Darwin's theory, try to explain how evolution could take place, from the first bacteria to the living things of today. It should be emphasized, though, that it is not Darwin's idea of evolution itself that is being questioned but the notion of gradualism.

5.9 Evolution and Complexity

As we mentioned in the previous section, many scientists have noted that it is difficult to relate the evolutionary sequence represented by fossils directly to Darwin's thesis because there are many missing links. The intermediate species that should trace the path of evolution down to present-day organisms have not, on the whole, been found. Yet, according to Darwin's theory of *gradual* evolution from a common ancestor, they should exist.

In fact, to evolve from species A to species Z, an organism has to pass through many intermediate species, B, C, D,..., until it arrives at Z. According to Darwin, each of these intermediate species is little different from those preceding and following it. Each intermediate species also needs to embrace a large number of individuals, first to avoid the risk of its dying out and secondly to enhance the probability of a mutation's leading to the next step; within a very small population, the probability of a mutation is close to zero.

Because this is required at every step until we arrive at Z, a large number of individuals of the intermediate species should have existed on the Earth and should have left their imprints in the fossil record. However, when we search for the fossils of these species, we do not find them. They are the famous missing links and there are enough of them to form entire chains! Looking at the

---

10 *With the progress of technology, DNA is being extracted from the ever-smaller bits of tissue found in ever-older samples, allowing us to go further back in time. In 2007, a collagen protein, extracted from the femur of a* Tyrannosaurus rex *which lived 85 million years ago in Montana, USA, was sequenced — an achievement considered impossible a few years ago.*

fossils in any museum, we may find some intermediate species but nothing like the number we should expect. Open an encyclopedia and look under the word "evolution": you find various beautiful schemes, with arrows linking one species to another, but these are only speculative because there are no fossils to fill the gaps.

Another fact to emerge from the fossil record is that evolution is not gradual but proceeds by jumps: there are periods during which evolution accelerates (especially after mass extinctions; Section 7.4) and others, lasting more than 100 million years, during which the number of species is constant (Figure 7.3). All this evidence stands in sharp contrast to the Darwinian concept of evolution, according to which the differentiation of species should proceed smoothly with time. Darwin's defenders claim that the jumps correspond to epochs when resources were particularly abundant, but the explanation seems unconvincing.

Finally, perhaps the strongest argument against purely random evolution is that, as we shall see in Section 5.11, an evolution of this kind is highly improbable: the universe is simply too young to have created structures as complex as the DNA of existing species by random mutations.

For all of these reasons, many researchers, whilst not doubting that evolution took place, find that the Darwinian picture does not entirely explain the evolution we observe, and needs to be complemented in some way. One such complementary theory considers the level of complexity that a cell can control.

One of the great surprises found in the study of DNA is that only a small part of it is used to codify proteins. In humans, this *exonic* DNA amounts to less than 2% of the total; the remainder seems to consist of sequences that have no apparent role (Figure 5.8). As we said above, it was assumed for many years that the noncodifying part of the DNA was the residue of failed attempts at evolution. This "useless" part was called "rubbish DNA", because, once separated from the exons, it was destined to be recycled by the organism, just as we recycle rubbish. Surprisingly, the proportion of the useless (*intronic*) DNA is very low in simple organisms like prokaryotic bacteria; it is larger in fungi and invertebrates, and reaches its maximum in humans (Figure 5.8). The fraction of useless DNA, the "evolutionary residue," increases with the complexity of the species until it reaches 98% in the human genome.

Some of the new genetic theories attribute a different significance to this part of the DNA: instead of containing the code needed to construct new proteins, it contains information on how the organism should be managed and how its functions should be regulated. Naturally, therefore, it increases with the complexity of the species. This part of the DNA controls, for example, the individual's growth, a complex process involving the whole organism, whose parts must follow different but well-coordinated paths if the organism is to

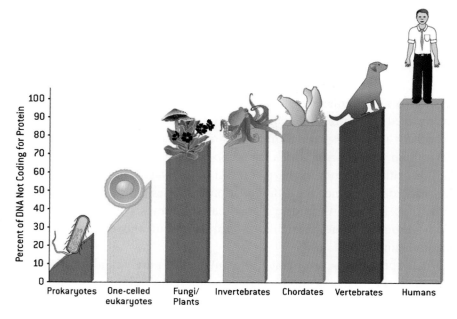

Figure 5.8: *Percentage of DNA that is not used to codify proteins. It increases with the complexity of species until it reaches 98% in the human genome. (From: John Mattick,* "The hidden genetic program of complex organisms," *Scientific American, October 2004.)*

survive. In simple organisms, like bacteria, a small percentage of the DNA is enough to control this process, whereas in more complex creatures like humans it needs practically all of it (Figure 5.8).

Therefore, what distinguishes an evolved from a less-evolved organism, other than the material from which it is constructed,[11] is the purpose for which the material is used. The genes that codify the proteins in humans are practically the same as those of monkeys, mice, pigs and chickens[12]; 70% of human genes correspond to those of nematode worms or of fruit flies. What makes a real difference is the method by which they are regulated, assembled and turned on or off in an organism. In other words, the difference lies in the "program" that may be contained in the noncodifying part of the DNA; it is this "program" that differentiates man from the other animals. Complex organisms need more information for their operation than for building themselves! Should this hypothesis be confirmed, it would mean that the level of "DNA rubbish" would become an index of how evolved a species is.

This interpretation of how the organism makes use of the intronic DNA could lead to a different picture of evolution from that we used to have. As we have seen, according to Darwinism, evolution is conditioned by casual mutations and

---

11 *The cells do not differ much between species.*
12 *That is the reason these species share several common diseases.*

by the environment. According to this theory, evolution is determined by the ability of an organism to control an increased complexity in its makeup. Species obviously remain subject to natural selection because they must be able to survive in their environment, but the environment does not define the mode in which evolution takes place, nor does it define how fast a species evolves.

This new view explains the jumps that occur in evolution. Evolution remains static until it finds, possibly by chance, a mechanism capable of controlling a greater complexity. When it succeeds, all the species permitted by the new level of control are rapidly born, and this appears as a jump in the evolution, a sudden diversity among the fossils.

As we shall see in Section 7.4, this hypothesis explains how, after every mass extinction, the variety of existing species quickly matches that before the extinction. The information needed to control the level of complexity reached before the extinction is already present in the species that survived.

This new approach to evolution explains why, once life had been born, it underwent a very modest evolution for about 3000 million years, and the only complexity they could control was that of single-celled organisms (Figure 5.5). Then, as soon as evolution permitted the control of multicelled organisms, it was possible to produce rapidly a variety of species[13] and evolution underwent a jump. In this way, one can explain all the evolutionary jumps found in the fossil record, including the explosion of life that took place in the Cambrian, 500 million years ago, when the ancestors of all the Earth's present-day species suddenly appeared (Section 6.6). The evolutionary jumps all correspond to an increase in the complexity that an organism can control.

From this perspective, the main problem that evolution has to solve is not the building of complex organisms (the hardware of the system) but how to control them in order to allow them to function (the software). This is very similar to the computer systems of today.

The Future. Fifty years after Watson and Crick's discovery, the mapping of the human genome, and of several other species, has been completed. An atlas of genes has been produced, associating the hundreds of thousands of DNA genes with the proteins they manufacture and the functions they perform. In the near future, we expect to map the proteins in detail and to identify their functions better — a task of great importance for all the living things.

True genetic engineering is carried out today in the laboratories; specific pieces of DNA are cut out and united with pieces of diverse origin, creating partly artificial sequences. Humans could soon become the first living things able to modify their biological destiny. In 2009, Craig Venter and Hamilton

---

13 *This theory does not put in doubt the Darwinian idea of evolution; nor does it the idea that the existing species share a common ancestor, as the "creationist" ideas claim. It is enough to see the similarities among various mammals, whether they are whales, antelopes or humans, to realize that these species should have close origins.*

Smith produced the first living thing with an artificial DNA designed at the computer. They introduced this DNA into the cell of bacteria from which its genome had been taken away, building the first artificial organism able to reproduce itself. With OGM new species are built using part of the DNA of existing species, and here the DNA is totally artificial; the first step toward a synthetic biology is taken.

This field is witnessing extraordinary developments that can influence the future of humanity; naturally, this gives rise to a great deal of apprehension. Used correctly, these technologies can cure diseases, improve agriculture, reduce the use of fertilizers and pesticides, and construct species that would otherwise not exist. If misused, they could have terrible consequences.

The future for humanity will be difficult but we cannot just reject these new technologies, as some people would like to and as often happened with new ideas: think of the secular opposition to Copernicus' theory. These techniques are part of our cultural patrimony, and history shows that such progress can never be "undiscovered," at least not without a collapse of the society. Their proper use will be a great opportunity to show that evolution has finally rendered our species wise.

Before continuing our discussion on the origin of life, we shall talk briefly about another kind of evolution of great importance for humans: cultural evolution.

5.10   The Cultural Evolution

Biologists and ethnologists recognize a second type of evolution of the Earth's species, called *cultural evolution*. This occurs in all species that *learned to use their intelligence to improve their quality of life, to invent useful things, to acquire experience and to transfer these skills to other members of the species.*

In this case, evolution occurs via *invention*, which plays the same role as *mutation* does in biological evolution. Cultural evolution differs from biological evolution in that it propagates in a vertical, Lamarckian[14] manner, transmitting everything it has acquired to the next generation. Sometimes it also works in a horizontal mode, between individuals of the same generation, if they possess a sufficiently high level of communication, as it is among humans today. In cultural evolution, selection occurs by means of *choice*, i.e. whether an individual accepts a given change or not. It is a conscious process, very different from that of biological evolution.

---

14 *The evolution theory of the naturalist J. B. Monet de Lamarck (1744–1829) is based on the hypothesis that the changes which occur during the life of an organism are passed to its progeny. The evidence does not support the theory: a man who becomes strong through exercises will not pass his strength to his children, who have to do the same exercises in order to be also strong. It works for cultural evolution, however, because the acquired experiences are passed to the following generation.*

Cultural evolution is observed in many animals, particularly those having a social life; it even existed among dinosaurs. This type of evolution has its maximum expression in the genus *Homo*. According to ethnologists, the privileged role of humans in the animal world is derived from its extraordinary cultural evolution. As a result, our biologically weaker and comparatively defenseless species has come to dominate the world.

The manner in which cultural evolution is transmitted to other individuals is based on the ability to communicate. It exhibits jumps, just as does biological evolution. The jumps occur whenever an innovative idea is spread within a species, and is accelerated with the growing capacity for communication.

The first jump, which differentiated our species from the other animals, occurred 2.5 million years ago, when our ancient ancestor (*Homo habilis*) constructed the first tools from pieces of stone. This completely new development became possible because of previous biological evolution. If, for example, our ancient ancestors had hooves instead of prehensile fingers, they would not have been able to construct the first tools, and we would not have developed the complex technology that dominates our society and that, itself, is conditioning our biological evolution.

Another great jump came with the *invention of language*. Although we are not sure when that happened, we estimate that it was between 45,000 and 35,000 years ago, when our species had already existed for more than 100,000 years. Language is an essential step in cultural evolution: it not only allows individuals to communicate with one another but has also forced *Homo sapiens* to order his thoughts so that they can be expressed with articulate sounds. With this ordering, names and categories were assigned to objects and experiences, and reasoning was transferred to other individuals; it was the beginning of abstract symbolic thought.[15] Even language was only made possible by a particular genetic evolution of the vocal apparatus, which occurred in *Homo sapiens* at least 70,000 years before he learned to speak — A mutation that probably never occurred with the Neanderthals, who likely were never able to articulate complex sounds.

With the birth of abstract symbolic thought, statues, pictures and complex instruments appeared and *Homo sapiens* achieved the superiority that allowed him to dominate the whole animal kingdom. According to some researchers, the first to pay the price of this conquest were the Neanderthals, who, after having coexisted geographically with *Homo sapiens* for 50,000 years, were

---

15 *It is a common experience that an idea which seems robust to us collapses when we try to convey it to someone else. It does not survive to the order imposed by language.*

"extinguished" by their physically weaker cousins when these acquired the overwhelming advantage of articulate language.[16]

Another great jump in cultural evolution occurred with the invention of writing, which allowed our ancestors to leave a concrete trace of their ideas and experiences. Writing marks the end of prehistory and the beginning of history. Since then the sequence of events can be deduced not only from fossils and archeological remains but also from the written documents left by the protagonists of history. Modern society begins: agreements between nations are written down and relations between individuals are regulated by law.

The next jump occurred with the arrival of the press, marking the beginning of mass culture: information and innovation were rapidly transferred to an ever-greater number of people.

Cultural evolution is now accelerating following the invention of computers, with their capacity for managing great quantities of information and making these universally available through the Internet.

Biological evolution and cultural evolution are closely intertwined and influence each other. Cultural evolution depends on biological evolution: we have already noted that a *Homo sapiens* with hooves, instead of hands, would never have developed our technology and that, without the vocal apparatus, would never have learned to speak. But, vice versa, biological evolution depends on cultural evolution because it is influenced by the environment which is modified from the human activity resulting from the cultural evolution. An example is that the brain of *Homo sapiens* developed remarkably after the birth of language and of symbolic thought.

In concluding this discussion of DNA and evolution, we can say that we now understand how genetic information is codified and how it is passed on to the future generations; we may also have some insight into the factors that have guided the evolution of species.

What is unclear is how the first cell and the first molecule of DNA came about. This first cell was the model for subsequent cells and, via evolution, all of the Earth's species were derived from it. We do not understand how such a first complex structure could arise without a model to follow. It seems impossible that such an event could have occurred completely randomly. We devote the next section to this problem.

---

16 Not all anthropologists agree that our ancestors were responsible for the Neanderthal's extinction. The hypothesis is based on the fact that the two species lived in the same areas but did not interbreed (a fact shown from genetics) and they were therefore in competition for the available resources. Moreover, the extinction of the Neanderthals coincided with the appearance of statues (Figure 6.13), graffiti and pictures among sapiens, that they had acquired the symbolic thought and therefore the advantage of language. On the other hand, there is no better explanation of the Neanderthals' extinction except for a hypothetical epidemic, for which there is no evidence. In fact, the Neanderthals were stronger and better adapted to the prevailing cold climate of that time than were the sapiens.

5.11   Does Life Come by Chance?

Here, we take up the discussion that we interrupted at the end of Section 5.3. We must face the second of the two problems in understanding the origins of life: how was it possible for the first sequence of DNA to be written, choosing the correct sequence from among an almost-infinite number of possibilities, most of which are meaningless? It is a problem that cannot be solved at the present time, even from a theoretical point of view.

It is difficult to go beyond Urey and Miller's experiment in Section 5.3 and produce a specific macromolecule of protein or of DNA in the laboratory by randomly combining the amino acids or bases of Figure 5.6 — a process that should have occurred spontaneously in the primordial soup. The difficulty is that the number of combinations is extremely large and most of them seem to have no biological sense. It is therefore practically impossible to get the right combination by chance. For example, in the case of a protein, the number of possible sequences obtainable by a random combination of all its constituent elements is 1 followed by 50 zeros whilst in nature "only" 10,000,000 proteins have ever existed (1 followed by "only" 7 zeros) and in the human body there are only 100,000 (1 followed by 5 zeros). Therefore, the probability of randomly generating a useful protein combination seems to be ridiculously low. For DNA, the number of possible sequences is 1 followed by 100 zeros. To understand just how big this number is, compare it with the number of atoms in the universe: 1 followed by 80 zeros. To make thing more difficult, we must remember that it is not sufficient to build a single cell containing a DNA macromolecule; a relatively large number of them are needed, otherwise life will not emerge. Even today, species whose population falls below a certain level become extinct.

To give an idea of how small the probability is of creating a given macromolecule by chance, consider 5.5×10^{21} litres of water, enough to cover the Earth with an ocean 10 km deep. (One-third of the Earth is actually covered by oceans with an average depth of 3.8 km.) Suppose that, for 10 billion years (and the Earth is only 4 billion years old), a billion chains of DNA were produced per second in every liter of water. During all this time, only 10^{48} chains would have been produced in this ocean. Since there are 10^{100} possible sequences of DNA, only an infinitesimal fraction of the possible cases (1 in 10^{52}) would have been explored.

All the amino acids necessary for creating enzymes, proteins and DNA could very well have existed in the primordial soup but it is very difficult to suppose that the first protein, or the first molecule of DNA, was formed by assembling them randomly. In the 13.7 billion years of life of the universe, as we have seen, only a small fraction of the possibilities could have been tried, and still fewer in the 4 billion years that the Earth has existed. However, the event that led to the emergence of life seems to have been highly probable because, as we shall see in Chapter 6, life was born almost immediately on the Earth, as soon as the prevailing conditions allowed it to survive.

To overcome this problem, some scientists have proposed the existence of a *deterministic* chemistry in the process that built the first macromolecules. They suppose that once a certain molecule had been created, another was necessarily created, and so on, until the more complex molecules that are at the basis of life were built. They seek the existence of a series of deterministic phenomena, involving millions of molecules, to show that the number of possible cases is well below the number with 100 zeros we mentioned earlier. This would render possible the undeniable fact that life exists.

If the answer to the question of the origins of life is to be found in a deterministic chemistry (helped by catalysts that accelerated the reaction; Section 5.3), we arrive at the conclusion that life must exist at every place in our galaxy, and in all the other galaxies, where physical conditions are the same as they were on the Earth when life was born — a conclusion already suggested in Section 5.2, when we discussed the chemical origin of life.[17]

Since it seems impossible to think that a planet like the Earth is unique among the billions of Sun-like stars populating our galaxy (to say nothing of the billions of galaxies that populate the universe; Section 10.5), we must conclude that those conditions which existed on the Earth when life was born must also exist "elsewhere." It therefore seems impossible that life exists only on the Earth, even if Earth-like planets are so rare that there are none close to us. Only the observations discussed in the final chapter will be able to tell us how rare the Earth is and whether habitable planets exist near us.

5.12 Why Carbon?

Organic chemistry is based on the carbon atom and on its unique characteristics (Figure 5.9). Thanks to these characteristics, carbon together with other atoms, especially H, O and N, (the most abundant elements in the universe; Figure 4.11), gives rise to an extremely large variety of molecules. These molecules are called organic, because they are at the basis of all biological organisms. Today, we know of more than 500,000 different organic molecules, a very large number that is growing from year to year, not only because of the new

---

17   *There are two other solutions that might explain the origin of life. One is to assume that the number of useful combinations of molecules is similar to the number of possible ones; namely that the number of protein chains and of DNA chains that make biological sense are enormously greater than the few used on the Earth. In other words, most of the possible combinations produce something that works. For DNA, at least, this seems unlikely. We have already noted that the change of base in the chain generally produces something that does not work. Moreover, if the number of "useful" proteins is so large, why has life on the Earth used only a tiny fraction of them? Studies of synthetic proteins, not found in nature, are continuing and may one day provide an answer. This hypothesis also favors the existence of life elsewhere in the universe.*

The other possibility is that we are a unique outcome of a highly improbable event. In that case, we simply exist and have no reason to discuss why. This hypothesis is often put forward by those who believe in intelligent design: our almost-impossible existence is rendered possible by a superior intelligence.

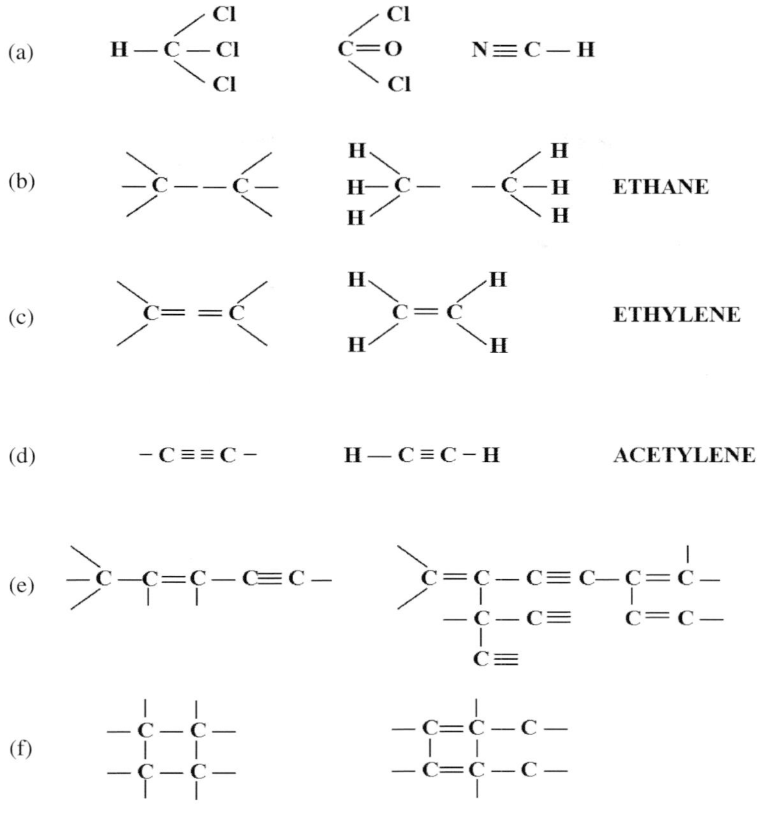

Figure 5.9: (a) Carbon (C) is tetravalent, i.e.-it tends to bind with other atoms using four bonds. These bonds can be saturated by other atoms or with other carbon atoms. Various cases are therefore possible: (a) one carbon atom forms bonds with four other atoms; (b) two carbon atoms bind with six other atoms; (c) two carbon atoms bind with four other atoms; (d) two carbon atoms bind with two other atoms. What makes carbon unique is that it binds with other carbon atoms to generate: (e) contiguous chains of atoms; (f) branched chains. These chains are the basis of organic molecules.

molecules produced by evolution but also because of those we are continuously inventing and producing, which are essential for our society.[18]

As we saw at the beginning of Section 5.2, organic chemistry is part of the definition of life itself. Even so, some science fiction, written by authors with good scientific background, is envisaging alien cultures based on atoms other than carbon. We may therefore ask how likely it is that there are forms of life, based on these different atoms which are able to construct complex structures, similar to those of organic chemistry, that can reproduce themselves and are capable of storing the vast amount of information contained in a genetic code.

---

18 A few examples of synthetic organic molecules in the field of medicine are aspirin, antipirina and Veronal. In the industrial field: artificial colorants such as aniline; explosives such as nitroglycerine, TNT.; plastic materials such as bakelite, nylon, PVC, formica.

GROUPS

PERIODS	1	2	3	4	5	6	7	8	9	10	11	12	13	14	15	16	17	18
	IA	IIA	IIIB	IVB	VB	VIB	VIIB		VIII		IB	IIB	IIIA	IVA	VA	VIA	VIIA	VIIIA
1	1.008 H 1																	4.003 He 2
2	6.941 Li 3	9.012 Be 4											10.811 B 5	12.011 C 6	14.007 N 7	15.999 O 8	18.998 F 9	20.180 Ne 10
3	22.900 Na 11	24.305 Mg 12											26.982 Al 13	28.085 Si 14	30.974 P 15	32.065 S 16	35.453 Cl 17	39.948 Ar 18
4	39.083 K 19	40.078 Ca 20	44.956 Sc 21	47.88 Ti 22	50.941 V 23	51.996 Cr 24	54.938 Mn 25	55.847 Fe 26	58.933 Co 27	58.69 Ni 28	63.546 Cu 29	65.39 Zn 30	69.723 Ga 31	72.61 Ge 32	74.922 As 33	78.96 Se 34	79.904 Br 35	83.80 Kr 36
5	85.468 Rb 37	87.62 Sr 38	88.096 Y 39	91.224 Zr 40	92.906 Nb 41	95.94 Mo 42	98 Tc 43	101.07 Ru 44	102.91 Rh 45	106.42 Pd 46	107.87 Ag 47	112.41 Cd 48	114.82 In 49	118.71 Sn 50	121.76 Sb 51	127.60 Te 52	126.90 I 53	131.29 Xe 54
6	132.91 Cs 55	137.33 Ba 56	La-Lu 57 - 71	178.49 Hf 72	180.95 Ta 73	183.84 W 74	186.21 Re 75	190.23 Os 76	192.22 Ir 77	195.08 Pt 78	196.97 Au 79	200.59 Hg 80	204.38 Tl 81	207.2 Pb 82	208.98 Bi 83	(209) Po 84	(210) At 85	(222) Rn 86
7	(223) Fr 87	(226) Ra 88	Ac-Lr 89-103	(261) Rf 104	(262) Db 105	(263) Sg 106	(264) Bh 107	(265) Hs 108	(266) Mt 109									

Figure 5.10: The periodic table. The elements are ordered horizontally according to increasing atomic weight, written above the atomic symbol in each cell, indicating the mass of the atom relative to that of hydrogen. Below the symbol is the atomic number, which is the number of protons in the nucleus (and, therefore, the total *number of electrons in their neutral form). In the columns of the table (also called groups) are listed the atoms that have the same number of outer electrons. This means that even though they have different atomic weights and numbers, the atoms of the same column have similar chemical characteristics and make similar molecules. For instance, the elements of the group 13 are oxidized by binding with an oxygen atom to produce BO, AlO, GaO, etc. Those of group 14 bind with two oxygen atoms to produce CO_2, SiO_2 and PbO_2. Those of group 16 bind with three oxygen atoms to produce SO_3, SeO_3; etc. Silicon, closest to carbon in its group, is the best candidate for having a chemical behavior similar to that of carbon.*

The alternative atom, often proposed in science fiction, is silicon, one of the most abundant atoms in the Earth's crust (where it is 100 times more abundant than carbon; Table 4.1) and the principal constituent of rocks. However, the choice of silicon as an alternative to carbon is not because of its abundance but because it is *homologous* to carbon: it is, below carbon, in the same column (group) of the periodic table. The columns of the periodic table contain elements that have similar compounds because they have the same number of electrons in their outermost electronic orbits and therefore the same chemical affinity (see the caption of Figure 5.10). The similarity between the elements gradually diminishes, however, as one moves down a column: the mass of the atom increases and its chemical compounds get heavier. The *total* number of electrons also increases along the column, and with their number there increases the tendency to become a metal.

Silicon, being the closest element to carbon in its column, has the most similar chemical compounds and is therefore the best candidate for having a chemistry capable of storing a genetic code; this is the reason why it has been used in science fiction. As an example, silicon is one of the few inorganic elements that can polymerize (note Section 5.8) even if its bonds differ from those of carbon (Figure 5.9).[19]

19 Unlike carbon in organic polymers (Figure 5.12), a silicon atom cannot bind directly with another silicon atom but only via *oxygen atoms.*

In spite of its affinity to carbon, silicon cannot build complex molecules like DNA. The differences between the two atoms are considerable, starting with the atomic mass (12 for carbon and 28 for silicon), which significantly influences the compounds they produce. Many carbon compounds are gases at room temperature and are soluble in water; those of silicon are not.

For example, carbon dioxide (CO_2) is a soluble gas that plays an important role in the biological cycle (Section 6.5). It is produced by animals when they burn oxygen in their life processes, and hence it is emitted into the atmosphere through respiration. It is then absorbed by vegetation, which extracts the carbon by means of photosynthesis and restores the oxygen to the atmosphere. Silicon dioxide (SiO_2) cannot behave like this, because it is a hard and insoluble crystal (quartz), much heavier than CO_2, and cannot participate in any biological cycle.

An amusing aside is that we do owe a particular type of intelligence to silicon, the artificial intelligence of computers, on which a considerable part of our life now depends: most electronic components are based on silicon technology. There is therefore the possibility of one day encountering forms of life based on silicon in the form of robots that have replaced the creatures who created them, just like the computer in Kubrick's film which refuses to be deactivated.... In any case, it is interesting to note that two such very different forms of intelligence, biological and artificial, depend on similar atoms!

The complexity needed for biological structures therefore seems, as said in Section 5.2, obliged to use the carbon chemistry. If some day we come into contact with creatures from another world, they might be similar to amebas, reptiles, squids, etc., or they might differ completely from the creatures that exist on the Earth. Nevertheless, they will be organisms based on organic chemistry — the chemistry of carbon — the only one with the complexity needed to codify life and permit the functioning of complex machines like the biological ones.

5.13   Suppose Life Came from Space

We have peered into the origin of atoms and molecules. We can all claim with some pride to have been in the center of a star at least once, because the atoms which make up our bodies — carbon, oxygen and the other heavy elements — were made there. Stars have therefore contributed to life by providing all the elements needed for its development, and to building the environment we live in (Figure 4.9).

Is this the only contribution of stars? Some people say, "No." This is the suggestive hypothesis of *panspermia*. This word encompasses all theories which are speculating about the existence of the seeds of life everywhere in the universe, seeds that not only have the capacity of bringing forth life but also of determining its evolution. These theses have been supported by a few scientists, among them Fred Hoyle, a talented theorist and cosmologist we have already encountered in this book. To him we owe the strenuous battle against the Big Bang, the merit of having named it and of having found the first proof of its reality with his calculation of the abundance of primordial helium. Hoyle had a great ability to think beyond the generally accepted scheme of things. Together with other scientists, among them Wickramasinghe, he maintained for over 30 years that the placental cloud of Figure 3.1 was placental in more ways than was commonly thought. He believed that the cloud contained much more than the simple atoms from which the solar system was built, that among the molecules within the cloud there were also macromolecules: those that were the origin of life and that we are not able to construct from the primordial soup. These molecules arrived on the Earth along with all the other materials that fall on our planet every year. Here, they adapted to the environment and, under the influence of other materials coming from space, reproduced themselves and evolved into all the species inhabiting the Earth today. In other words, the process that gave rise to the first molecules capable of reproducing and developing into ever-more-complex forms was guided at each step by an "intelligent universe." This would explain the jumps found in the evolution (Figure 7.3): they were just the result of more information received from space.

For Darwinists, the explanation for the discontinuities found in evolution lies in the environment that stimulates evolution more at certain times than others. For some modern evolutionary theorists, they result from changes in the control of complexity. For Hoyle, who died before these later theories were put forward, the explanation lay in the increase in the genetic materials falling on the Earth. The problem of the birth of life and of its evolution, which some scientists try to solve by invoking a deterministic chemistry, was solved by Hoyle via an "external" intervention. It is not difficult to imagine the skepticism with which this thesis was received in the scientific world! Hoyle's intelligent universe was rather similar to God (even though he was a convinced atheist).

In his revolt against Darwinism, Hoyle used all the arguments we raised in previous sections, such as the many missing links in the evolutionary chain. Bacteria fossils as old as 3.6 billion years have been found (when life appeared on the Earth), but there are very few traces of species that existed in the years that followed (Section 5.8). Is it conceivable that all of these have been missed or overlooked?

Everything becomes easier to understand if we assume that changes are induced by some external agent and not through the more or less continuous evolution envisaged by Darwin. In Section 6.6 we shall see than an explosion of life occurred in the Cambrian, when, in a relatively short time, nearly all the ancestors of today's species were born. Such an explosion is almost impossible to explain by Darwinism but easily explained by an insemination from "outside."

There was also the failure of producing anything useful from the primordial soup (Section 5.3). The fact that, after decades of trying, nothing significant had been produced in a controlled laboratory environment, seemed to Hoyle a proof that life originated beyond the Earth.

The theory of panspermia also maintained that, among the matter falling to the Earth from space, there was something that told biological structures how to evolve. Darwin claimed that mutations occur by chance and that only organisms better adapted to their surroundings survive. Hoyle did not think it possible that random modifications of the genetic code could result in progressive evolution because the number of possibilities is practically infinite, as we have noted in Section 5.11.

Another phenomenon that is difficult to explain with Darwinism is isomorphism, which is commonly found in nature. It occurs when an animal (usually an insect) becomes similar in appearance to the vegetation on which it lives. This enables it either to easily capture its prey or to camouflage itself against predators. Figure 5.11 shows an insect that imitates the leaves which it inhabits. How was it possible for two separate evolutionary threads to produce such similar vegetable and animal species? If one examines their genetic patrimony, one finds that the parts of their DNA containing information on shape and color are similar in the leaf and the insect. Hoyle's explanation is simple: the leaf and the insect were near to each other when the mutation was induced and therefore underwent the same mutation.

Figure 5.11: An example of isomorphism in nature. (© Prof. Giuseppe L. Pesce, Aquila University.)

Hoyle proposed therefore that mutations occur under the external influence of germs, bacteria or other things arriving every year from space. Because they are microscopic, these entities can fall to the Earth without being burned up in the atmosphere. Experiments conducted in laboratories have shown that the hypothesis is tenable; bacteria can survive in much more hostile conditions than penetrating our atmosphere. The ever-increasing traces of biological matter found in meteorites give robust support to this thesis, as would the discovery of biological structures by the probes sent to Mars. These results would, however, equally support the hypothesis of a chemical origin of life: if, in the past, Mars had experienced the same conditions that existed on the Earth when life was born here — a particularly effective greenhouse effect which allowed liquid water to exist on the planet — the same process leading to life on the Earth would have taken place there.[20]

Other factors in favor of Hoyle's theory have been found in the study of the epidemics that have struck humanity from time to time, starting with the flu. Looking at studies of the first half of the 19th century, he found that these epidemics often struck almost simultaneously peoples living so far apart geographically that it would take months to travel between them; there were no airplanes in those days. Hoyle found it difficult to explain these outbreaks unless they were the result of germs that had fallen from space. Yet more evidence in Hoyle's favor lies in the fact that materials taken into space by astronauts are often found to be contaminated with germs and bacteria on their return to the Earth. NASA's interpretation is that they were contaminated on the Earth by mistake; Hoyle believed that they were contaminated in space.

He was also a great writer — he wrote the famous science fiction novel *The Black Cloud* — and he put forward his theory in many popular books. They are easy to read and accessible to everybody.

This vision of the origin of life explains in part Hoyle's antipathy to the Big Bang: a universe barely three times older than the Earth has little more chance of developing life than does the Earth itself, a fact that severely compromises his theory. The panspermia theory is incompatible with a universe only 13.7 billion years old; moreover, it simply moves the problem of the origin of life to another part of the universe without solving it, making the solution more difficult.

Even if this theory does not explain the origin of life, it creates a new issue to be explored — the possibility that one planet can biologically infect another and that organic matter from space could have affected the birth and evolution of life on the Earth. Traces of bacteria have been found on meteorites; it is difficult to estimate today how much they contributed to the story of life on our planet.

---

20 Even if it is possible that life (bacteria) born on a planet can be transferred to another via the material ejected after the impact of a large meteorite on the soil of a planet.

Another possibility, which lies entirely in the realm of science fiction, and for which there is no evidence, is that travelers from another planet contributed to the birth of life here. It is a possibility tied to the hope that interstellar travel is possible (Chapter 11) — A hope that some people like to nurture so as to give a future to humanity, rather than foresee that all that has happened on our planet will die with the Sun.

We stop here. Let us recall that panspermia is only a theory, and will remain so until it is proven or disproven; we must be prudent in considering any theory, and in this case we must be very prudent. Why, then, have we raised this idea at all? Have we forgotten Galileo and the importance of measurement? No, we have not. What exists in space and what falls to the Earth from space are measurable and therefore the contribution of this material to the origin and evolution of life will be established one day.

5.14 Conclusion

Many — even highly educated — people find the idea that our species may have evolved from animal ancestors, not to mention bacteria, to be distasteful. Benedetto Croce, a great 20th century Italian philosopher, considered this possibility "discouraging and dangerous." He came to define the study of evolution as a "pseudoscience," which denied the intrinsic nature of humanity. Who knows what he might have said of our sharing 99% of our genes with those of the chimpanzee and 70% with those of the nematode worm? Even Stalin did not like this hypothesis; he considered evolutionary science a useless expression of bourgeois culture.

Others found the possibility of humans having an animal origin to be in conflict with their religious sentiments.[21] Humans, they say, are the result of a particular expression of God's will and cannot be reduced to a simple act of nature. Today, especially in the United States, there are active movements that promote the teaching of creationism in schools, a doctrine based on pseudoscience which denies that any form of evolution took place. Others, whilst accepting that evolution took place, believe that all creatures are the result of "intelligent design" by God and, therefore, that humans have no connection with other species on the Earth.

It is difficult to share such views: accepting them is equivalent to admitting, on the one hand, that the dignity of humans depends on their origins and not on what they have accomplished and, on the other hand, that the magnificence of creation is not in itself sufficient evidence for the existence of God. These

---

21 *Seen properly, the origin of man described in the first few lines of "Genesis" in the Bible is not very different from a description of the modern theory of evolution: God first separated light from the darkness (the Big Bang and the birth of the first stars), then land emerged from the waters and He populated the seas with fish, then creatures arrived on land and, finally, on the sixth day, man came.*

"theories" are searching for a framework that attests to a noble origin for humanity, something they believe is denied by evolution. It demonstrates a poverty of thought similar to that displayed by the Inquisition in persecuting people of faith, like Galileo and Bruno (Section 3.1). In those days, the fear of losing the privileged position of the Earth, and of humanity, in the universe pushed the Inquisition into defending the geocentric concept against all the evidence. Today, the battle for the central position of the Earth having been lost, a similar fear forces some people to hold an anthropocentric vision of humanity, separate from the rest of nature.

Such attitudes have produced equally irrational reactions in *defence* of evolution, to the point of pushing Hoyle, a convinced atheist, to accuse the scientific world of uncritically accepting all Darwinian ideas (Section 5.9), simply because that theory had been condemned by certain religions. An uncritical approach like this is completely at odds with the scientific method we discussed in Section 1.2.

An extraordinary difference between our species and other animals is our cultural evolution (Section 5.10), which has accelerated over the past few thousand years. People who are aware of this value are so proud of it that they do not need "a certificate of origin." They regard the universe with humility, realizing that it was not constructed just for them. They know that humans have existed on the Earth for only a few million years — "only" 150,000 years for *Homo sapiens* — a very little amount of time compared with the 3.8 billion years separating us from the origin of life on the Earth and the 135 million years during which the planet was dominated by dinosaurs. They know that, should our species disappear, the history of the universe would not be affected in any way.

If they have a religious faith, they are conscious that it would be very strange to conceive of a God not asking us to use our intelligence for understanding the world He has created. They are conditioned neither by superstition nor by fear of scientific results, in which they see only the manifestation of their beliefs. They have no difficulty in admitting that if a Creator exists, He decided that evolution should determine the course of nature.

We conclude the first part of the book, devoted to the genesis of all that surrounds us, with a chapter in which the questions we posed are probably more than the answers we gave. This should not come as a surprise, because biology is taking its first steps in this field. The question of how life originated had not been properly addressed when Miller and Urey devised their experiment; even today, we may not be posing it correctly. Miller and Urey had only vague notions about the conditions that existed on the Earth at the time when life appeared. Today, the situation is different because not only do we know about these conditions better, but biology and genetics have entered a rich era that should bring great results.

We can draw some conclusions from what we have discussed in this chapter. The hypotheses of the chemical origin of life, and of panspermia, both predict that life exists outside the solar system. The first hypothesis, based on the objectivity of the laws of chemistry and physics, suggests that life must exist everywhere in the universe wherever the environmental conditions of the primordial Earth had existed. On the other hand, the panspermia hypothesis predicts that the seeds of life exist everywhere in an "intelligent universe" that knows how we should evolve. Therefore, if we reject the possibility of a divine intervention that created an Earth different from the rest of the universe, we must conclude the first part of the book by saying that there is a high probability of life elsewhere.

How common is life in the universe? What is the probability of our finding it? That depends on how often we may find environmental conditions similar to those that existed on the primordial Earth, conditions in which life was born. In other words, how common is a planet like the Earth in the universe? We shall devote the second part of the book to this question, albeit without finding a clear answer. The third part of the book will be devoted to the search for life outside the solar system.

Part II

The Case of the Earth

History of the Earth

Chapter 6

6.1 Fossils and the Age of the Rocks

For many years to come, if not forever, our civilization will not be able to explore planetary systems other than our own because even the closest stars are too far away to be reached in a reasonable time. If a day should come when a few bold humans will start such a trip, they will know that they will never be able to return to the Earth and that only their progeny may one day reach their destination (Section 11.5). The Earth is the only planet in our Solar System hosting a form of evolved life[1] and so is the only case we can study to understand the origin of life, the only testing ground for the theories of the preceding chapter. Let us begin the second part of this book, therefore, with the story of our planet from its beginning until today. Before starting the tale, we will speak briefly of the *fossils*, which are the traces ancient species have left, testifying their existence on the Earth, and of the *dating of the rocks* containing these fossils, which allows us to order events chronologically. Without this dating, we cannot speak of "history".

The Fossils. The biological history of the Earth can be reconstructed through the study of fossils, the petrified remains of plants and animals that lived in the past. When a plant or an animal dies, its body is normally destroyed

---

1 *Some primitive organisms, like bacteria, might exist deep under the surface of Mars or on some of the satellites of the giant planets such as Europa where, under the icy surface, there is a big lake of liquid water heated by the tides induced by Jupiter. But there cannot be any form of evolved life, comparable to that existing on the Earth.*

by microorganisms and by atmospheric agents. In few rare cases, when death occurs in water, the body can be covered by mud and sediment which build a protective shell around it, preventing decomposition. With the passage of time, water penetrates this shell, carrying minerals that gradually replace the organic structure whilst preserving its form. Fossils are therefore composed of minerals that have assumed the form of the organism. Sometimes they also contain a small amount of the original organic matter that, one day, might permit us to reconstruct the DNA of these ancient living creatures, as was done in the science fiction film *Jurassic Park.*[2]

Fossilization is a slow process and there is, therefore, a high probability that the body will be destroyed before its completion. It is estimated that fewer than one in every thousands of the species ever existed have left a trace in the fossils. Organisms without bones, carapace or cartilage have greater difficulty in leaving such a trace because they decompose very quickly, before there is time for them to be covered by mud. Decomposition of bones and cartilage is slower and their fossilization is therefore easier; they therefore constitute nearly all the fossils. Fossilization is therefore a process that occurs rarely. This explains, according to biologists, the numerous gaps observed in the evolutionary sequence traced by fossils. Entire generations of organisms might have existed without leaving trace of their existence.

Determining *the ages of rocks,* and of the fossils they contain, is fundamental in reconstructing the history of the planet. The dating method uses the radioactive atoms which, in small quantities, are contained in nearly all Solar System rocks. They are heavy atoms produced in the explosion of one or more supernovae (Section 4.5) which occurred in the vicinity of our placental cloud (Figure 3.1), before the birth of the Solar System. The radioactive atoms are therefore contained, in similar proportions, in all matter found in our zone of the galaxy. They were present in the material constituting our placental clouds (Section 3.2) and are therefore present, in similar proportions, in all Solar System rocks, including every meteorite that falls to the Earth.

These radioactive elements are truly natural clocks because, after a time, they decay, emitting heat and expelling an electron or an alpha particle (which weighs as much as four atoms of hydrogen). In this way, the atom produces energy and becomes lighter. If the new atom is itself unstable, it decays into another, still-lighter atom, always emitting an alpha particle in the process. This decay will continue until a stable atom is reached. In this way, a sequence of atoms is created, each is four hydrogen atoms lighter than its predecessor. The sequence is called a "family"; the original atom, from which the others are generated, is called the "parent" whilst the others are called "daughters."

---

2 *As we shall see in Section 6.12, the oldest living things for which DNA has been reconstructed are vegetables and bacteria 500,000 years old.*

The process is probabilistic, which means that not all the atoms of the same element decay at the same time — some decay earlier, others later. To characterize this property we use a quantity known as the *half-life*, that is the amount of time in which half of the "parent" atoms are transformed into "daughters." For example, half the ^{238}U atoms,[3] which are the "parents" of the uranium family, decay in 4.47 billion years; namely, after 4.47 billion years, a kilogram of ^{238}U is reduced to half a kilogram. On the other hand, the number of ^{232}Th atoms, the "parents" of the thorium family, is halved in 14 billion years. We are talking of a very long period of time here, comparable to the age of the universe in the case of thorium, and they are therefore suitable for measuring the age of ancient rocks, like those making up the Solar System. Moreover, these half-lives can be measured with great precision in the laboratory.

It is intuitively easy to understand how these clocks work. When a rock forms, solidifying from magma, small quantities of radioactive elements are sealed inside. With the passage of time, these elements decay, increasing the numbers of "daughter" elements and decreasing those of the "parent." The ratio of daughter to parent increases with time and measuring this ratio provides the clock used to establish the age of rocks.

There are also radioactive atoms, lighter than those we have cited and with shorter half-lives, very short in some cases; for example ^{8}Be, produced by the collision of two He atoms, decays in a millionth of a billionth of a second (Section 4.1). Such atoms are obviously not a useful clock.

An element with a moderate half-life is ^{14}C, with 5730 years. It cannot be used to date geological eras but is perfect for dating archeological finds. With a half-life of a few thousand years, the ^{14}C, now found in the Earth, cannot have been produced in supernovae explosions occurred five billion years ago (before the birth of solar system).

It arises from the collision of cosmic rays with nitrogen atoms in the atmosphere, the resulting nuclear reaction transforming ^{14}N into ^{14}C. It is a reaction taking place continuously because, to a first approximation, the bombardment by cosmic rays is rather constant as it is the quantity of nitrogen in the atmosphere on a scale of a few thousand years. We can therefore assume that the rate of production of ^{14}C is constant in time, as is its concentration in the atmosphere. The ^{14}C then enters the carbon cycle through photosynthesis (Section 6.5); it is absorbed by plants and propagated to other living creatures through the food chain. Its concentration is therefore similar in all living things, the amount varying only slightly from species to species. When an organism dies, it stops absorbing carbon, and so the amount of ^{14}C it contains, begins to decrease at a rate determined by the half-life. By measuring

---

3 *Remember that the number before the U means that it weighs 238 times more than a hydrogen atom.*

the ^{14}C abundance in a sample of organic matter, one can therefore establish its age. Even a small sample, such as the one contained in pottery clay, is enough.

The atoms that are used to date rocks millions or billions of years old have half-lives much longer than that of ^{14}C. Such massive atoms are formed only in supernovae (Section 4.5). They are not only important as clocks. As we will see in Chapter 9, the heat they emit when they decay heats the interior of the Earth and supplies the energy necessary for continental drift, volcanic eruptions and earthquakes.

6.2   The Primordial Earth

No living organism could have existed on the early Earth. The planet was a great ball of molten magma,[4] heated by planetesimals (Section 3.8) that fell on its surface. The heat released by the impact of those bodies (Section 10.3) was so high that they melted, liberating the gases contained in their interior — principally hydrogen, helium, water, carbon dioxide and ammonia. The hydrogen and helium, being lighter than the other gases, rose rapidly into the upper atmosphere and were dispersed into space; the heavier gases were trapped by the planet's gravity. The proto-Earth, when it was about the size of the Moon, was already surrounded by a dense layer of gas that, at the end of its accretion phase, contained as much water as there is in the oceans today. This primordial atmosphere was, therefore, very different from today's. Its temperature exceeded 1000°C and its pressure was more than 200 times that of the present-day atmosphere, the same pressure found at a depth of 2000 m in the oceans. These dense gases created an enhanced greenhouse effect, similar to that found today on Venus (Section 8.5), trapping the heat arriving from the Sun. The temperature of our planet then varied from 1000°C in the atmosphere, through 1300–1500°C on the surface, up to 5000–10,000°C in the interior, this in spite of the fact that the Sun was less luminous than it is today. At these temperatures, all the rocky material on the Earth was in liquid form. The planetesimals melted and their substance differentiated according to weight: the gases rose into the atmosphere while the heavy materials, including the metals, sank toward the center of the planet, forming the Earth's metallic core (Section 9.2) and giving rise to our planet's magnetism. In their descent toward the center of the Earth, these materials released gravitational energy, increasing the internal temperature of the planet. It is estimated that this gravitational energy alone was enough to keep the Earth in its molten state for many centuries.

---

4 *For deeper study, we suggest the book by Jones and Newson,* Origin of the Earth, *Oxford University Press.*

When the accretion of matter from the disk began to wane, and the impact of large planetesimals became infrequent,[5] the Earth cooled and the first solid rocks formed. Zircons, small hard stones that can be found in several places on the Earth, are 4.4 billion years old and are the oldest known solids. (These stones are very resistant to atmospheric agents. They make up 0.03% of the Earth's sands, so we have all walked on them.)

The oldest rock known today, firmly anchored to the Earth's crust, is the Gneiss of Acosta; it was discovered in Canada's Northwest Territory and is dated at 4 billion years. So far, no older rocks have been found, therefore it is assumed that the surface of the Earth started to solidify and to form the crust between 4 and 4.4 billion years ago. Along with that primitive crust, the first volcanoes were born, rapidly achieving enormous size because of the great quantities of heat they had to dissipate[6] and the lava they ejected.

With the formation of this first layer of solid rock, a barrier between the surface and the interior of the Earth was erected. The atmosphere began to cool, water vapor began to condense into an even denser and colder clouds, until the first droplets were formed, and it began to rain. At first, the rain fell on an incandescent Earth, but then the soil started to cool gradually. The rainfall lasted a long time, perhaps hundreds of years, and inundated the Earth, forming rivers, lakes and, finally, oceans covering the entire planet. At this stage, the Earth was a world of water, with a few volcanoes emerging above the seas to form the first dry land (Section 9.1).

In those first years, the Earth must have looked terrifying; Figure 6.1 shows an artist's impression of what it would have been like. The sky was dark, covered by a thick layer of cloud that allowed little light to filter through. Under that leaden sky, there lay a sea shattered by the eruption of volcanoes and the impact of meteorites and comets. Today, there remains little to trace what the Earth's solid surface was like, giving us no clue to reconstruct how it was; erosion has destroyed everything (Section 9.1). If we look at the Moon, however, we do get an idea of what the Earth must have been like. Because our satellite has no atmosphere and liquid water, no erosion took place and therefore its surface has conserved its original aspect, covered with craters caused by the impacts of meteorites. The only significant difference between the primordial Earth and the Moon was the large quantity of water that covered the planet's surface. Nevertheless, it was the most important difference because water created the conditions necessary for the beginning of life. It was exactly in these conditions that life was born on the Earth.

---

5 The impact of the Mars-sized asteroid which gave birth to the Moon occurred after the heavy materials had sunk to the Earth's core (Section 3.8). This explains why the Moon's density and composition is similar to that of the Earth's crust.
6 The heat produced by radioactive nuclei was then 4–6 times that of today (Figure 9.11).

Figure 6.1: An artist's impression of what the Earth looked like when life was born. Volcanoes are emerging from a stormy sea whilst the dark sky is crossed by meteorites and lit up by lightning storms. (© J. M. Poissenot Enciclopedie des jeunes Larousse, 1997.)

6.3   The Birth of Life

Through the use of radioactivity, it has been possible to establish the age of the rocks containing the oldest fossils. Among the most famous examples are the *stromatolites*, found in the semisubmerged reefs of Sharks Bay at Warrawoona in the north of Western Australia. These rocks have an age of 3.5 billion years, they are very close to the time when the surface of the Earth solidified. Inside these rocks, filamentary blue structures have been found. They are the fossil remains of chains of bacteria similar to the *cyanobacteria* (Figure 6.2), still found on Earth today.

The stromatolites are rocks made by the fossilization of great colonies of bacteria which, when they are alive, transform calcium bicarbonate, dissolved in water, into calcium carbonate. This precipitates and forms rocks, consisting of layers of calcium carbonate and fossilized cyanobacteria. Such rocks are shown in Figure 6.2; they are similar to corals, which also consist of small organisms and which, via a different process, produce structures so large that they can become islands.

Stromatolites 3.4 billion years old have been found in Australia at Pilbar Block, and others, 3.5 billion years old, at Masvingo in Zimbabwe. At Barberton in Mpumalanga, a province created from the northeastern part of the Erstwhile Transvaal, 3.5 billion-year-old carbonaceous microorganisms were found; others have been discovered in Greenland. The hunt for these ancient rocks began only recently and the results lead us to believe that we shall soon find many more. There is little doubt, therefore, that life existed on Earth at least 3.5 billion years ago.

Figure 6.2: Sharks Bay at Warrawoona, in the north of Western Australia. On the left are rocks 3500 million years old. In these rocks, remains of chains of bacteria were found (right), with filamentary blue structures like those of cyanobacteria (Figure 5.3), which still exist on Earth today. (Left: © Department of Geological Sciences, University of South Carolina. Right: J. William Schopf, University of California, Los Angeles.)

How long after the Earth cooled were these bacteria born? The answer lies in the samples collected by astronauts from the great lunar lakes of solidified lava. That lava was produced by the same rain of meteorites that was then falling on to the Earth. Measurements of radioactive materials in these samples show that those lakes solidified 3.8 billion years ago. This date is therefore taken to indicate the end of the meteoritic bombardment of the Moon and, therefore, of the Earth. Before 3.8 billion years ago, the temperature of our planet was too high to allow life to exist even if, as we have seen, the oldest rocks are 4 billion years old. This is also the age of the first marine sediments on our planet, confirming that the crust had cooled and that there was liquid water on the Earth's surface. This date establishes the moment when life became possible on the planet: if it had been born earlier, it would have been destroyed by the hostile environment.

If we compare the two dates, minus 3.8 billion years for the end of the bombardment by meteorites and minus 3.5 billion years for the age of the first fossils, we see that the difference is only 300 million years. This is a very small difference and the real interval for the appearance of life is likely to be even smaller because the Earth was probably uninhabitable well after the 3.8 billion years established by the lunar rocks. The bombardment by meteorites and comets continued for a long time and, although it was with a lower intensity, it would have had terrible consequences for any form of life that might have existed on the planet. In Section 6.11, we shall discuss what happened 65 million years ago when an asteroid 10 km in diameter fell on the Earth and caused the extinction of the dinosaurs. In its early years, our planet was struck by even larger asteroids, with diameters of hundreds of kilometers. The heat released by these impacts (Section 10.3) was sufficient to boil and evaporate all the Earth's oceans. It is even possible that the Earth's oceans appeared and evaporated more than once during those years and that, life itself followed a similar cycle of birth and destruction. We estimate that, for at least another 100 million years, the Earth was still uninhabitable, covered with a fiery atmosphere as a result of the impact of large asteroids.

If we consider that life on Earth was born before the date established by the first fossils (because this date merely establishes the time at which life was able to leave observable traces, rather than the moment when life was actually born), we must therefore conclude that the interval between the moment when the Earth became habitable and the moment when life actually arose was, indeed, brief. Taking into account the uncertainties involved in this dating — of the order of 50 million years — the two moments might actually have coincided. We can therefore conclude that *life arose on Earth as soon as the environmental conditions made it possible*. This result, established in the case of the Earth, suggests that the birth of life is a highly probable event even though it has not yet been possible to reproduce it in the laboratory.

6.4   The First Bacteria

In the beginning, the terrestrial atmosphere had no oxygen; it contained only water (H_2O), carbon dioxide (CO_2) methane (CH_4), and ammonium (NH_3), the simplest molecules that can be constructed from the most abundant elements in the universe (Figure 4.10). These are also the most abundant elements found today in comets, which retain the original composition of the material that formed the Solar System.

The explanation for the lack of atomic oxygen in the primordial atmosphere is that, as we shall see in Section 6.5, oxygen needs to be created continuously. The first bacterium that inhabited the Earth, unable to make use of oxygen, found its nutrients in chemical reactions involving sulfur from volcanoes and methane from the atmosphere. Even today, we find bacteria of this type in the solfataras (such as the one in Pozzuoli, Italy or that in Yellowstone Park, United States) and in the depths of the seas, where there is no air for them to breathe. We are speaking of *anaerobic* prokaryotic archiobacteria (Section 5.4 and Figure 5.5), which live without oxygen and are killed by its presence, and *thermophilia* ("lovers of heat"), which normally live at temperatures above 80°C (185°F), conditions that are impossibly hot for most of the organisms we know.

To understand the range of temperature occupied by the majority of the organisms that inhabit the Earth today, in Figure 6.3 we show the growth curve (the rate at which they reproduce) of a very common prokaryotic

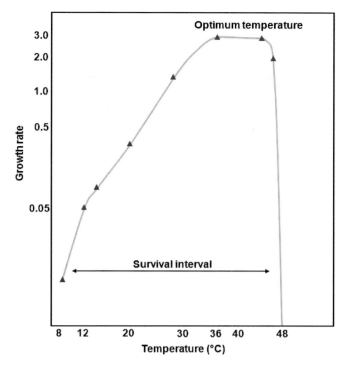

Figure 6.3: The growth curve of the bacterium Escherichia coli. *Most categories of organisms on Earth live within this temperature range; outside, they die.*

bacterium: the *Escherichia coli*, a useful inhabitant of the human digestive system. Its growth curve can be taken as an indicator of the range of temperature within which the majority of species that inhabit the Earth can live. Outside this range of temperature, the equilibrium of many chemical reactions that form the basis of life is changed, proteins are denatured and life we know cannot survive.

At temperatures below 8°C and above 45°C the bacteria do not reproduce; if the temperature is higher they die, as would most of the organisms living on our planet. We have confirmation of that in our everyday experience. Refrigerators, for example, conserve food at a temperature a few degrees above zero because, at that temperature, the bacteria responsible for the decomposition of food do not reproduce or do so very slowly.[7] Warm-blooded animals, including humans, can survive at temperatures below 8°C because they produce the heat necessary for keeping their body temperature within the correct interval. To do this, they expend energy, and therefore need more food, than cold-blooded animals. The latter do not consume energy to heat themselves because their bodies adopt the temperature of their environment but, as a

---

7 *They can actually survive at very low temperatures, even at 70 K, which is used for storing many of them.*

compensation, they live only in warm regions to remain within the curve of Figure 6.3. As an example crocodiles, when they are not moving, consume so little energy that, unlike warm-blooded animals, they can go for months without eating, waiting for the prey to arrive.

At temperatures above 50°C, the majority of the Earth's organisms are unable to survive because there is no way for them to lower their temperature efficiently. Sterilization is performed by raising the temperature for the time needed to kill bacteria; the higher the temperature, the shorter the time needed.

The survival interval shown in Figure 6.3, as we have seen, does not apply to all the bacteria that exist on Earth. It certainly does not apply to certain prokaryotic archeobacteria, the thermophiles ones. Recently, these bacteria have been discovered living at temperatures above 120°C, and reproducing at 112°C. In the hot water surrounding very deep submarine volcanoes, where the Sun's light never penetrates. Research on materials extracted during the drilling of petroleum wells seems to indicate that, at 4 km below the Earth's surface, microorganisms exist at even higher temperatures — temperatures at which the majority of proteins would be destroyed through the process of "cooking", which we use to make them more digestible for humans.

The study of the ability of these thermophilic organisms, which belong to the family of *extremophiles*, of living at extreme temperatures and pressures, began only a few years ago and the results are still uncertain. Nevertheless, biologists think that we may find organisms able to survive at up to 200°C without being cooked, hidden deep within the Earth!

Other organisms, also belonging to the group of extremophiles, are the *psychrophiles* which live at low temperatures. In the Antarctica, at 77° S latitude, there is a small lake under the ice where the concentration of salt, and the pressure of the ice, are high enough to maintain water in its liquid state at − 20°C. To the great astonishment of biologists, various types of bacteria, fungi and algae have been found living and reproducing in these cold waters without any difficulty.

Are these organisms, hiding in some of the most hostile parts of the Earth, the descendants of the first organisms that populated the Earth? They are certainly adapted to living in conditions very different from those experienced by most of the Earth's present-day inhabitants. For the extremophiles, the constraints shown in Figure 6.3 clearly do not apply.

These exotic organisms seem to demonstrate that, as long as the water within the cells remains liquid, the cell survives even if, for thermophilic bacteria, the enormous pressure at great ocean depths is needed to stop the water vaporizing; for those living at low temperature, pressure helped by the high concentration of salt plays a similar role in stopping the water freezing.[8] In any case, the extremophiles retain a characteristic of other terrestrial organisms: they need water to live. One can therefore say that *life can exist without oxygen but not without liquid water!*

The range of temperature in which life is possible therefore seems to be extended beyond that shown in Figure 6.3 to the less clearly defined range in which water can exist in the liquid state. Nevertheless, scientists are still trying to understand the mechanisms that allow life to exist under such extreme conditions.

The discovery of organisms living at very low temperatures increases the probability of finding life on Mars or on Europa, Jupiter's moon, which has underground lakes of liquid water. The existence of such organisms also leads us to suppose that life on a planet is not easily destroyed, and that it could survive the worst of cataclysms, as long as water remains and the environmental conditions permit it to be liquid....

6.5   The First Three Billion Years

Figure 6.4 shows a timeline of the history of the Earth and of the evolution of its organisms; it is the final part of a path of continuously growing in complexity started with the Big Bang. At the beginning, the matter in the universe was composed of only two simple atoms, hydrogen and helium; the story we are telling is of how, the today's world, including its living organisms, has developed from these two atoms.

The first level of complexity was achieved via nuclear reactions within stars, which produced all the atoms heavier than helium (Chapter 4). This is the only process that can be considered completed in the sense that all the possible stable atoms allowed by the laws of physics have been already built in the universe. What will change in future is the relative abundance of these atoms: the proportion of the light elements will decrease, whilst that of the heavier elements will increase. This change in relative abundance will continue for a very long time because only 2.3% of the initial hydrogen and helium has been burned (Figure 4.12).

The second step in complexity involves molecules: atoms ejected from stars at the end of their lives were bound together to form molecules. This process began in interstellar space and continued on planets with a continuous increasing in complexity. This complexity will possibly continue to increase, for the duration of the Earth's life if not for the life of the whole universe, because the number of possible combinations of atoms that can be bound into molecules is practically infinite — think of proteins or DNA.

---

8 *We recall that water, unlike other liquids, increases its volume when it solidifies; (ice floats). Therefore an increase in pressure will oppose to an increase in volume allowing water to remain liquid at temperatures lower than 0°C.*

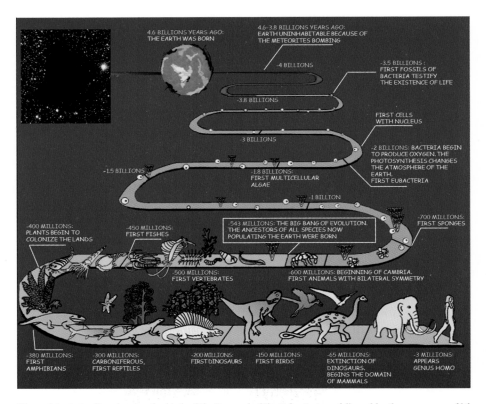

Figure 6.4: A timeline showing the birth of the Sun and of the Solar System, followed by the emergence of life and its evolution. The times range from 4.6 billion years ago (upper-left) to today (lower-right).

With molecules came the third step (Section 5.9): they allowed life to be born and to evolve with a complexity increasing with time (together with that of the DNA molecules where the progress of evolution are recorded).

There is a last step, the one reached by the association of individuals exchanging their experience and knowledge that is expressed through cultural evolution (Section 5.10).

Figure 6.4 tells the story of the organisms of our planet. It starts with the placental cloud, in the upper left-hand corner, and continues with the formation of the Solar System and the Earth. It ends at our present era in the lower right-hand corner. Let us follow the principal events shown in the figure, indicating with a minus sign the time at which they occurred, measured backwards from today. Starting from the beginning, we have:

- *13.7 billion years*: Our universe was born.
- *12 billion years*: Our galaxy was born.
- *5 billion years* (Figure 6.4, upper-left): The placental cloud (Figure 3.1) was formed, containing all the matter from which the Sun and the Solar System are built.

- *4.6 billion years*: The Solar System and the Earth were born (Section 3.8).
- *4.5 billion years*: A meteorite, the size of Mars, hit the Earth, expelling the material that formed the Moon.
- *4 billion years*: The first solid rocks appeared.
- *3.8 billion years*: After about 800 million years, the bombardment of meteorites ended. From this epoch, the first traces of marine sediments are found (Section 6.2).
- *3.5 billion years*: Traces of the first bacteria show that life existed on Earth at this time.

From the placental cloud of Figure 3.1 to the birth of the first terrestrial organisms it took less than a billion years, whilst about 3.5 billion more years elapsed before we humans arrived. We can therefore say that life has inhabited our planet for three quarters of its existence, even if it has consisted of very simple organisms for most of that time. As we shall see, for more than 3000 million years and up to *500 million years ago*, our planet was inhabited by bacteria and simple algae which had limited evolutionary ability. On the other hand, they were able to adapt to different environmental conditions, to such an extent that a few of them have survived to the present day.[9]

The first three billion years saw the occurrence of three very important events:

(a) the *photosynthesis: the emergence of bacteria capable of producing oxygen* and thus able to modify the Earth's atmosphere;
(b) the emergence of cells with a *nucleus*;
(c) the emergence of *multicellular* organisms.

The next three sections are devoted to these events.

6.6 Photosynthesis

Two billion years ago: in the geological layers of the period, we find the first evidence of highly oxidized rocks. What is often called the *oxygen revolution* had begun — an occurrence that changed the biological history of our planet. A few bacteria, swimming close to the surface of the sea, succeeded in changing their diet. They used the energy of the Sun, the water and the atmosphere's carbon dioxide (CO_2) to create sugars, on which they fed, and expelled into the atmosphere the oxygen in excess. There was an evolution in the organization of cells and *eubacteria* (Figure 5.5), which constitute most of today's bacteria, were born.

---

9 *With a few exceptions, today's animals have existed for only a few million years. Sharks are examples of long-lived species, having existed for 300 million years.*

The eubacteria brought *photosynthesis*, the mechanism on which the biology of today's plants is based (Section 5.5) and on which all forms of multicellular life depend for their food chain. By means of photosynthesis, the chlorophyll of plankton, algae and leaves use the light of the Sun to break up molecules of water into their constituent atoms, oxygen and hydrogen. These two atoms are then combined with the air's carbon dioxide (CO_2) to produce glucose molecules, which plants, algae and plankton use to grow and to repair any damaged or sick parts of their structure. At the end of this process, some oxygen remains and this is released into the atmosphere, where it will be breathed in by animals and "burned" to produce the energy they need to function. Finally, the animals breathe out the CO_2 produced in this process, to be again used by plants in a new cycle of photosynthesis.

Photosynthesis, therefore, stores the Sun's energy in chemical form within plankton and plants. When we burn wood, we exploit just this stored energy. All the atoms of the wood are still present in the smoke and ash; what provides the heat is the release of the binding energy that the Sun stored in the plant when water was split into H and O by photosynthesis.

When we eat it is always this energy we use. When we ingest meat for example, we ingest material containing energy that a herbivore had obtained from plants, which in turn had stored by means of photosynthesis. All forms of vegetable life are based on photosynthesis and, consequently, so are all the multicelled organisms, belonging to the animal and fungi kingdoms (Section 5.5), which feed on plants or of animals that themselves have fed on plants.

All the living creatures, therefore, use the Sun's energy, produced by the nuclear processes taking place in its core. With these processes, stars not only manufacture the atoms needed to create biological species (Section 4.8) but also furnish the energy needed to nourish them. Atoms, even if incredibly small. contain the source of energy that sustains everything in the universe, from stars to biological organisms — a source of energy that has barely diminished in 13.7 billion years (Figure 4.11).

A second benefit of photosynthesis is that of maintaining a high concentration of oxygen in the atmosphere. Oxygen, apart from being consumed by all animal species, is also "burned" rapidly in nature. It is, in fact, one of the most reactive atoms, it combines with nearly all the elements to oxidize them. If it were not for the photosynthesis that continually produces it, atomic oxygen would disappear quickly from the atmosphere and, with it, the majority of living species.

Photosynthesis represented an enormous jump in the evolution and laid the basis for new forms of life, those that have dominated our planet since. The organisms using oxygen in their metabolism are much more efficient in producing energy than the anaerobic bacteria. Bacteria making use of photosynthesis therefore proliferated, generating ever more oxygen and, in

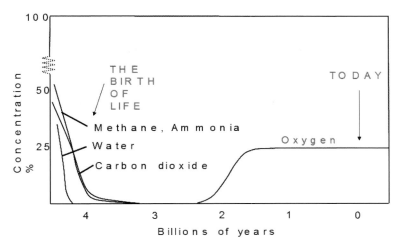

Figure 6.5: Composition of the Earth's atmosphere. At the time when life was born, the abundant gases were methane, ammonia, water and carbon dioxide. Two billion years ago, bacteria started producing oxygen via photosynthesis, modifying the atmospheric composition. Anaerobic bacteria, which had inhabited the Earth up to then, became in large part extinct and aerobic bacteria began to develop. (From the homepage of the Darwin mission's website.)

time, producing the concentration of oxygen found in today's atmosphere (Figure 6.5).

If, on the one hand, the oxygen revolution created the basis for new forms of animal life, it also caused the extinction of the first inhabitants of our planet. Anaerobic bacteria could not survive in an atmosphere rich in oxygen, which was poisonous to them. A few of them survived in airless places deep in the ocean or below ground. Their descendants are the extremophilic archeobacteria which we spoke of in a previous section. The oxygen revolution provoked the first known extinction in the history of our planet which, like all other extinctions, resulted from a drastic change in environmental conditions.

At the same time as the oxygen revolution took place, another very important event occurred in the story of evolution: *the first cells containing a nucleus* — the *eukaryotic* cells — were born (Section 5.4). These cells are the basis of all multicellular organisms. They are more complex than prokaryotic cells: they protect the DNA inside the nucleus; and they have specialized organelles in the cytoplasm

6.7 From Cells to Multicellular Organisms

It is not easy to trace the evolution from single-celled to multi-celled organisms via the fossil records, because very few of them actually left a record. The organisms through which this transformation took place belong to the kingdom of protists (Section 5.5 and Figure 5.5). Some of them were single-celled organisms, like amoebae and protozoa, whilst others were living in colonies of

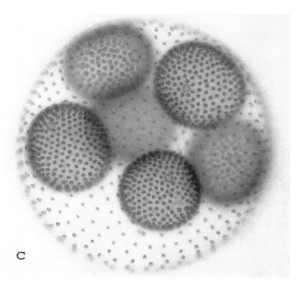

c

Figure 6.6: *Volvox is an example of colonies of individual cells linked together by filaments, making up struc-tures larger than a millimeter and visible to the naked eye. This was the level of complexity achieved by life 1.5 billion years after its birth. (© Stephen Durr, http://www.btinternet.com/~stephen.durr.)*

numerous individual cells. Yet others were multi-celled organisms that are difficult to classify among the kingdoms of animals, plants and fungi.

After cells with a nucleus appeared, some bacteria began to organize them-selves in colonies like those we still find today, such as *Volvox*, which we shall return to shortly. As time passed, the singles cells in these colonies started to carry out specialized tasks, allowing the whole colony to use the available energy more efficiently. As an example, some colonies became divided into an "anterior" and a "posterior" part, with the anterior cells taking on the role of steering the colony toward light or nutrients, whilst the posterior cells are con-centrated on the reproduction. In these colonies, the functions carried out by single-celled bacteria are shared between the member of the colony. As time went on, these tasks became more and more specialized; the first step toward multicellular organisms had been taken.

Examples of colonies of cells are still found today belonging to the family of Volvox (Figure 6.6). They are groups of similar, monocellular organisms form-ing beautifully colored spheres that move about in water by means of flagella.[10]

Certain Volvox have developed a sexual capability; they build colonies com-prising only male or female cells. The former release microgametes that swim toward a female Volvox and fertilize its cells. If this differentiation also

---

10 *Volvox was discovered in 1600, when drops of pond water were examined with the first microscopes. Carl Linnaeus gave them this name, being captivated by the gentle revolving motion of these colonies. Those colonies are employing sexual reproduction, it is thought that these movements help the diffusion of the micro-gametes into the environment to fertilize the female cells.*

occurred in ancient colonies, it means that these organisms had already discovered *sexual reproduction*, even though they were not yet multicellular. This was a very important jump in evolution. By mixing genes of different origins, sexual reproduction increases the differentiation among organisms (Section 5.6) and accelerates evolution. Such a mechanism is quite different from that used by bacteria, which reproduce themselves by duplicating their cells, thus creating a twin identical to the parent. That may explain why bacteria evolved so slowly, despite their very large numbers.

Let's return to the chronology:

- *1.8 billion years (approximately)*: About 200 million years after the appearance of oxygen the third fundamental event in the evolution of life occurred — *multicellular organisms* appeared. The oldest known multi-celled fossil is the *Grypania*, a strange alga about a millimeter in diameter and a meter long. It probably resulted from the evolution of colonies of bacteria that had increased their capability of dividing functions among themselves and had finally acquired the ability to control a greater level of complexity (Section 5.7). With the passage of time, the cells developed external membranes, which bound them stably together and formed the first organic tissues. Then they increased the differentiation of their cells, which performed more and more different functions. The variety of algae also increased, and included the first green algae. However, for another billion years, the Earth continued to be inhabited by very simple organisms — bacteria, algae and plankton.

- *650 million years*: *Sponges* — the porifers — appeared. Although they are classified as belonging to the kingdom of animals, they had an organization not very different from that of the first colonies of protists, having no nervous tissue, no internal organs and no mouth, no anus. They fed by filtering water through the pores filling nearly the whole body.

  Their tissues were little differentiated and were composed of cells that functioned almost entirely independently of one another. Nevertheless, they represent a step forward with respect to the protists; the cells are linked together to form a true tissue, even though they lack the differentiation we find in the tissues of developed organisms. Like the more evolved bacteria colonies, they also employ sexual reproduction. They represent an evolutionary dead end and, since then, they did not evolve into anything more complex. Nevertheless, they give us an idea of what the first multi-celled structures were like.

- *600 million years (approximately)*: Another important step in the Earth's biological evolution occurrs: the first organism with *bilateral symmetry* appear in the fossil records. It was *Vernanimalcula*, discovered in the eastern provinces of China in the 1990s and classified in the kingdom of animals. Although tiny, a few tenths of a millimeter in size, it represents an important step in evolution. Apart from its left — right symmetry, it has longitudinal differentiation. For the first time, we have an organism with a mouth through which food enters, an internal cavity in which nutrients can be stored, an intestine in which these are digested and an anus through which the refuses are expelled. This marked a step forward with respect to sponges. The entire animal kingdom, from worms to mammals, exhibits bilateral symmetry and has a similar body organization. These animals gradually increased in size until, at –555 million years, the Kimberella — which some consider to be the ancestor of modern mollusks — reached 10 cm in length.

These animals were the precursors of the great jump that occurred a little later, at the beginning of the Cambrian era when, at — 543 million years occurred an event unique in the history of biological evolution — the so-called "Big Bang" of animal evolution that was the origin of all the today's species. This event marked the end of the *Precambrian* era, the first of the five great eras into which the history of the Earth is divided, an era that spans from the origin of the Earth up to the explosion of life at the beginning of the Cambrian era.

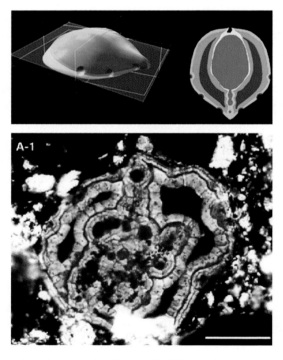

Figure 6.7: A Vernanimalcula *fossil, found in China in the 1990s. At the top is a reconstruction of how this animal might have looked. Whilst it is of microscopic dimensions — only a few tenths of a millimetre and barely visible to the naked eye — its fossils are the first to have a symmetrical structure and a differentiation in the longitudinal direction, with a mouth, intestine and anus: the typical structure of all animals. This was the level of evolution achieved 3 billion years after the birth of life, just before the Cambrian explosion. (© Photo by* Junyuan Chen, Science, July 2004, **305** (5681), 218.)

6.8 The "Big Bang" of Evolution: The Origin of Species

–543 million years: The second era of the history of the Earth begins. It is called the *Paleozoic* or *Primary Era*. It started with the Cambrian, a period in which occurred a single extraordinary event in the biological history of the Earth, an event for which there is no explanation even today. Nature, which until then had been sleeping, suddenly exploded. In a while, thousands of new species appeared in the oceans; in them, the progenitors of all of today's species can be recognized. It is as if Nature had reflected on what to do to our planet for 3 billion years and once she had made up her mind, did it in one fell swoop. Before the Cambrian, the only big developments were the cells containing nuclei, the oxygen revolution, the first multicellular organisms (sponges) and the first organisms with bilateral symmetry like the *Vernanimalcula*. Then, suddenly, at 543 million years, Nature produced the ancestors of all the animals, plants and funguses that populate the Earth today (Figure 5.5).

The first animals were invertebrates, like their predecessors, but they quickly transformed themselves into creatures with shells, teeth, skeletons and jaws. Our planet's seas were rapidly populated by an incredible variety of species that, in a relatively short time, emerged to colonize the dry land, which until then had been a lifeless desert. The first to emerge were plants, followed by insects, amphibians, reptiles, mammals, dinosaurs and, finally, birds — which descended from dinosaurs.

The classification that we find in scientific books is based on what happened in this period: before then, no species existed! At that time, the various *phyla* were born — the biological categories that correspond to well-defined types of body organization and structure and into which are divided the species that populate the different kingdoms (Figure 5.5). Some phyla are shown in Figure 6.8. Among the best known, we find: the *chordates*, which comprise all the animals with an internal skeleton (fish, amphibians, reptiles, dinosaurs, birds, mammals); the *arthropods,* characterized by an exoskeleton that covers the entire body with a series of rigid plates, permitting the structure to articulate like medieval armor (crustaceans, crabs, scorpions, ants and insects in general); the *mollusks,* characterized by a single rigid skeleton (bivalve molluscs, octopus, snails and ammonites (which became extinct together with the dinosaurs), and of which *Vernanimalcula*, discussed at the end of the previous section, is perhaps an ancestor; the *celenterati,* characterized by a radial symmetry (medusa, chorals, sea anemones); the *porifers* (the sponges of which we spoke earlier, one of the few phyla that existed before the Cambrian), which have a feeding system lacking a mouth but with pores in which nutrient-bearing water flows.

Other phyla constitute the vegetable kingdom: *Pinophyta* which includes the conifers, araucaria and the cypresses (550 species); *Tracheophyta*, with the

CHORDATES: FISHES, ANPHIBIANS, REPTILES, DINOSAURS, BIRDS, MAMMALS

ARTHROPODS: INSECTS, CRUSTACEANS, SCORPIONS

MOLLUSCS: BIVALVES, OCTOPUSES, SNAILS

CELENTERATES: CHORALS, MEDUSA, SEA, ANEMONES

Figure 6.8: Examples of phyla, corresponding to precise body organization. Today, 40 phyla are recognized, all of them dating back to the Cambrian and none born thereafter.

cactus and the Indian fig; *Magnoliophyta*, characterized by large flowers and including the magnolia and angiosperm (250,000 species).

The actual classification of phyla was done by Darwin in 1859, starting from Linnaeus' classification of 1735. Within every phylum, living creatures are subdivided into classes, orders and groups, arriving finally at individual species (Section 5.5).

As we said above, the puzzle is that *the progenitors of all the phyla that exist on Earth today were born in the Cambrian* with a few exceptions, like the *porifers*, which were born earlier. Nowadays, we recognize about 40 phyla but not a single one of these was born after the Cambrian. In fact, a few phyla disappeared in the various mass extinctions occurring between the Cambrian and our epoch. (One of them is the *amomalocaris* of Figure 6.9.)

We attribute the origin of all existing species to the explosion of life that took place then. For this reason, the period is also known as the "Big Bang" of species. What is difficult to explain is why all those species emerged in a relatively short time, surely less than the 60 million years postulated by Darwin; recent evidence points to a period well under 10 million years.

The Cambrian explosion has always posed a difficulty for Darwinists. Darwin's defenders have claimed that the explosion was only apparent, a trick of the fossil record. Darwin attempted to explain the explosion by supposing that organisms were already divided into different phyla before the Cambrian

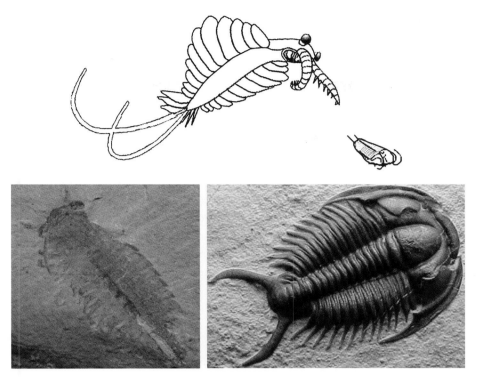

Figure 6.9: Prey and predator. At the top, an anomalocaris *is shown attacking a* trilobite. *Below are photographs of fossils of the two animals. With a length of 60 cm,* anomalocaris *was one of the biggest creatures of that period. Its fossils are found in Australia, Canada, China, Greenland and the USA. It had no teeth and probably fed by sucking its prey, immobilized with its tentacles. Its phylum has not been identified. On the right is one of the numerous fossils of* trilobites, *probably the victim of the* anomalocaris; *the name derives from the fact that the body is divided into three lobes — one central and two lateral. (© Virtual Fossil Museum, www. fossilmuseum.net.)*

but, for various reasons, had not left a trace in the fossil record: invertebrates, for example, do not fossilize easily. Or perhaps all the fossils tracing their evolution have been lost. These explanations are no longer acceptable today. The dating of fossils is more precise and the period during which the explosion occurred is now known to be shorter. Moreover, the fossils bearing witness to the explosion are found in many places — China, Namibia, America, Siberia and Antarctica among others. All these fossils indicate that the explosion began 543 million years ago. Previously, only simple multi-celled organisms had existed on those continents. Then, after a few million years, the ancestors of all the species inhabiting the Earth today had appeared.

It is understandable that this phenomenon, to some extent at least, results in a crisis for the Darwinian idea of evolution. A planet that drowses for 3 billion years, and then creates the whole of the animal and vegetable kingdoms in less than 10 million years, is difficult to explain with a theory that envisages gradual change. It is difficult to believe that this happened — in every part of the world and at the same time — without leaving traces of a preceding phase in the fossil records.

There have been many hypotheses for the possible cause of the explosion, all based on the ambient conditions prevailing at the time. These include high values of oxygen or temperature or the sudden appearance of large quantities of food. However, such hypotheses are all difficult to verify, nor do we understand why the same condition did not appear in either earlier or later historical periods.

The idea of panspermia that life was seeded from outside by a universe that had decided how to make us evolve, is very well suited to explaining the, Cambrian explosion. In fact, Hoyle put forward his theory (Section 5.13) precisely because of the difficulty that evolutionary theory had in explaining evolutionary jumps, of which the Cambrian is the major example. But we should not forget that the idea of an extraterrestrial insemination might explain the origin of life on Earth but does not explain the origin of life itself. It does not explain how a universe, only three times older than the Earth, could produce the seeds of all animal and vegetable species and bring them to the Earth.

The new genetic theories, those we discussed in Section 5.9, work better; they explain the jumps in evolution as resulting from an increased capacity of nature to control higher levels of complexity. Once this capacity is acquired, all the species whose complexity can be controlled are rapidly produced and, suddenly, a great variety in the fossil record appear.

Another point, even more difficult to explain, is why another explosion, like that of the Cambrian, did not occur later. One hypothesis is that all possible phyla were those born in the Cambrian; thereafter, any new phyla would have to be so similar to those already existing that they would not be recognized as different. This argument is questionable, though; some phyla became extinct in the course of evolution and never reborn.

6.9   From the Origin of the Species to the Dinosaurs

With the Cambrian explosion, 543 million years ago, the oceans became populated by thousands of species. Initially, these were invertebrates. Some had the protection of a carapace, which also provided a rigid structure to which muscles could be attached. Signs of injury are found on some of these carapaces showing that, even then, nature was divided into predator and prey. One of the most feared predators in those early times was *anomalocaris*, which reached a length of 60 cm, a large size for these first animals.

The upper part of Figure 6.9 shows an *anomalocaris* attacking a *trilobite,* whilst photographs of fossils of the two species are shown below. The *trilobites* are among the most common fossils of that period: we know of more than

Figure 6.10: An ammonite, one of the more common fossils. Some specimens can be more than a meter in diameter. Ammonites belong to the phylum of molluscs and to the class of cephalopods. They lived between 400 and 65 million years ago and disappeared along with the dinosaurs. (© Virtual Fossil Museum, www.fossilmuseum.net.)

2000 species. Their average length is about a centimeter and the biggest ever found is 10 cm long. They resemble millipedes protected by a robust carapace, and belong to the phylum of the arthropods. Another fossil, frequently found in the geological strata of that era, is that of the *ammonite* (Figure 6.10); ammonites resemble large snails and could be over a meter in diameter. They belong to the phylum of mollusks and became extinct along with the dinosaurs.

These animals did not remain for long the only inhabitants of the sea. In a relatively short time (tens of millions of years), *vertebrates* began to appear; their fossils are characterized by a spinal cord, which made their bodies robust and flexible and furnished a robust structure to which the muscles were attached.

Life, which had existed only in the sea until then, began to colonize the dry land. The first inhabitants were mosses, lichens and plants. Then came the animals — first insects and then amphibians, reptiles and mammals. Meanwhile, lichens transformed themselves into shrubs, with roots that enabled them to search for nutrients over a relatively large area. They then evolved into trees and made up the great carboniferous forests.

The abundance of food and the robust skeletons, capable of supporting large weight, allowed to grow ever more the species of the Earth and, in "only" 300 million years from the "origin of the species", nature produced the largest animals ever existed: the dinosaurs. We will run through the principal events of those years, as traced by the fossil record.

- *500 million years*: From this period, we find fossils of the *first vertebrates*. These were small and lived at the bottom of the sea; they moved with difficulty and had no jaws, taking in food through a sucker, like their invertebrate predecessors. The modern lamprey may be an example of these first animals.

- *450 million years*: From this period, we find the fossils of the *first fish*, with teeth, jaws and fins. They quickly became very large and the great predators were born, like the today's sharks, a species hundreds of millions of years old.

- *440 million years*: about 100 million years after the Cambrian explosion, *the colonization of dry land by plants began*. Fossils of the spores of fungi and lichens are found.[11] The dry land ceased to be desert.

- *410 million years*: This was the age of the first *shrub*. It was 10 cm high and had no roots. Shrubs were a well-developed species of lichens. In this period the first *gymnosperms*[12] were born; they reproduced by means of seeds rather than spores.

- *400 million years*: From this period, about 140 million years after the Cambrian explosion, fossils of the first *insects* are found. They were the first animals to abandon the sea. They were covered by a carapace, like the trilobites, which fossilized easily. (Worms may also have existed but, having no bones, they could have not left fossils.) Insects were the first inhabitants of the Earth and today they constitute the largest number of existing species, about 800,000 against the 4600 species of mammals.

- *380 million years*: This is the age of the first *amphibian* fossils, the *ichthyostega*, the first specimen of which was found in Greenland in 1931. Ichthyostega was a meter long, with a fish's tail, paws and nostrils. It lived on the land, but its eggs were without shells, similar to those of fish, so they had to be deposited in water, where they hatched and where the initial development of the animal took place. This is the case today with frogs, which begin life as tadpoles, more similar to a fish than to an amphibian. Fossil footprints of older amphibians, estimated 390 million years old, have also been found.

- *370 million years*: This is the age in which the first fossil tree (20 meters high) is found. It had roots that could seek out nutrients far from the plant itself.

- 359 to 300 *million years*: This period is called carboniferous because the Earth was covered by luxuriant forests that produced large part of the great deposits of coal (and probably oil).

- *300 million years*: From this period, the first fossils of land-adapted reptiles are found. They laid eggs protected by a shell, which could therefore be

---

11 Unlike their spores, fungi and lichens decompose quickly and therefore leave very few, if any, fossil traces.
12 "Gymnospermae" comes from the Greek "gymnos" ("naked") and "sperma" ("seed"). These plants were characterized by seeds arranged on scales, as in pine cones. They differed from the more evolved angiosperms (plants with flowers), whose seeds grow inside a fruit. Conifers and ginko are examples of gymnospermae.

laid on land. This allowed the animals to roam into terrains far away from wetlands and therefore to colonize the entire planet.

- *280 million years*: This is the age in which the first *conifer* fossil is found. Conifers spread rapidly over all the dry land, contributing to the carboniferous deposits.

- *251 million years*: The second-greatest era of the Earth, the *Paleozoic*, ended with the worst extinction in our biological history; 90% of marine species and 60% of land species disappeared (Section 7.2). At that time, the dry land, which were united in one big continent, the *Pangea* (Figure 9.4), began to split up into what would become the present-day continents.

   At the end of this great extinction, there began the third geological era of the Earth, the *Mesozoic* or *Secondary* Era, characterized by the supremacy of the dinosaurs.

- *220 million years*: From this period, 320 million years after the Cambrian explosion, we find the first mammalian fossils (Section 6.12). The animals were about the size of mice, laid eggs and suckled their young like the platypus does today. For over 150 million years, while the dinosaurs dominated the planet, the mammals remained in their own niche, evolving very little. Later, when the dinosaurs disappeared, they multiplied and nowadays, they dominate the entire planet.

6.10   The Dinosaurs

- *200 million years*: From this period, 340 million years after the Cambrian explosion, we begin to find fossils of the largest animals ever existed on the Earth — the dinosaurs.[13] They ranged in size from that of a chicken to that of Brachiosaurus, which weighed 70 tons. Among the giants there were the Sauroposeiden, 18 m high (comparable with a five-story building), and Supersaurus, 40 m long. Since dinosaurs became extinct, no other creature has attained such dimensions. The biggest dinosaur lived for more than 100 years and, according to some scientists, might even have exceeded two centuries.[14]

   The dinosaurs were more evolved than reptiles. They had straight legs, like mammals, and walked well above the ground, whereas most reptiles have bent legs and stomachs that lie on the ground.

   Many dinosaurs were bipeds with an erect posture, a characteristic retained by their descendants, the birds. In contrast to reptiles, there were

---

13 The name "dinosaurs" was suggested in 1842 by Richard Owen, the English scientist, after seeing the large dimensions of these old animals. The name comes from the Greek word "deinos" (terrific) and "sauros" (lizard).

14 The lifespan of an animal is approximately proportional to its size. The average life of a mouse is 3 years, of a bear 25, and of an elephant 70; the average life of a blue whale — the largest living animal — is 100 years. For humans, we should expect it to be around 30 years as, indeed, it was before the beginning of civilization (and of medicine) and as it is for most primates.

no flying dinosaurs (pterosaurs were reptiles), nor did any of them have a marine habit. And, even if some had a semiaquatic life, they differed from amphibians in laying their eggs in terrestrial nests.

Another peculiar characteristic the dinosaurs shared with the mammals is that of living (in the case of herbivores) or hunting (in the case of carnivores) in packs, with well-defined hierarchies. They had therefore developed strong social behavior and remarkable intelligence. Many fossilized nests have been found in which dinosaurs deposited their eggs. Studies have shown that they took great care of these nests and repaired them annually. In contrast to most reptiles and fish, which abandon their eggs as soon as they are laid, dinosaurs looked after their offspring, increasing the probability of their survival — behavior still found in birds.

Dinosaurs were warm blooded animals, as is proven by the large quantity of food they consumed[15] and by the feathers or fur covering their skin, a characteristic found only in warm-blooded animals. They lived at all latitudes: remains have been found of "polar dinosaurs", living above the Arctic Circle and in Antarctica, with eyes bigger than normal, adapted to see in near-darkness. Even if the Earth's overall temperature was then higher than it is today, winters at those latitudes were long, dark and icy, as confirmed by studies of pollen.[16] It is not clear how animals of this size — the polar dinosaurs reached 1.5 tons — could feed themselves during the arctic winters. Scientists suggest either that they migrated southward or northward (no nests have been found in the polar regions) or that they hibernated, as bears do today.

Carnivorous dinosaurs were fearful predators, intelligent, fast and capable of adapting to any situation and any climate. The gaps in the chain of development of other species, including mammals, are attributed to their undisputed supremacy. They dominated the world for 135 million years until 65 million years ago, when a meteorite, 10 km in diameter, fell to Earth, causing a catastrophic change of climate and a disastrous reduction in food supply, leading to the dinosaurs' extinction. Let us start the chronology again.

- *150 million years*: This is the age of the fossils of the first bird, *Archaeopterix*, of which six specimens have been discovered. The first fossil was discovered in 1862, in Bavaria. It was the size of a blackbird, was covered with feathers, and had teeth and clawed feet. Its structure was very similar to

---

15 *The dinosaurs' need for food was similar to, if not greater than, that of mammals. This is one of the pieces of evidence leading us to suggest that they were warm-blooded creatures. Warm-blooded animals need more food than cold-blooded ones, because they need more energy to keep warm. A lion eats more than a crocodile, which can fast for months waiting for the prey's arrival; a lion cannot stay without food for long, lacking the energy needed to stay warm, will die.*

16 *The study of the fossil of pollen and wood shows that the plants living in these regions were conifers, ferns and cycas, which live in sub-arctic areas where the average temperature ranges from 3 to 13°C.*

that of the *miniraptor*, a small dinosaur. The discovery of *Archaeopteryx* is considered the proof that *birds descended from dinosaurs*, which are not, therefore, completely extinct. More evidence in favor of this assertion lies in their respiratory systems. Relative to their weight, birds are more efficient at oxygenating their blood than other animals and dinosaurs had a similar characteristic. The study of their bones also shows that dinosaurs grew faster than mammals. A Tyrannosaurus Rex was fully grown in 15 years, whilst an elephant takes 30. This characteristic has been inherited by birds (which grow even faster) and distinguishes them from reptiles, which are the slowest-growing animals in nature, even slower than mammals. This is another argument favoring the view that dinosaurs were warm-blooded, like the birds which have descended from them.

–   *135 million years*: The *angiosperms* were born. These were plants that produced flowers, in which seeds developed in an ovary before becoming fruit. They were the precursors to most of today's plants and to nearly all those used in agriculture.

6.11   The End of the Dinosaurs

Sixty-five million years ago, a meteorite 10 km in diameter fell into the Gulf of Mexico, near the Yucatan peninsula, with catastrophic consequences for all the Earth's species. The impact occurred at a velocity of over 10 km per second, and the energy released was estimated to be the equivalent of 15 trillion tons of TNT. It was like the simultaneous explosion of a billion nuclear bombs similar to that detonated over Hiroshima. The catastrophe caused the extinction of 45% of the species existing at that time, including all the dinosaurs of large size. It was the last great extinction on our planet and, being the nearest in time, it is also the best-studied.

The extinction of the dinosaurs marked the end of the *Cretaceous* and of the *Mesozoic* or *Secondary Era*, and the beginning of the *Cenozoic* or *Tertiary Era*, which covers the period from –65 million years to –1.8 million years, when our era, the *Neozoic* or *Quaternary*, began.

In 1980, the Physics Nobel laureate, Luis Alvarez, investigating the geological layers of that period in the "Gola del Battacchione" near Gubbio in Italy, discovered a stratum with an anomalous abundance of iridium. It was then found that such layer existed everywhere around the world (Figure 6.11). It was called the *K–T layer* and it is used today to identify the geological boundary between the Cretaceous (which in German starts with a K) and Tertiary eras.

Because no dinosaur fossils are found above it, the K–T layer has been associated with the event that provoked the extinction of the dinosaurs (called the *K–T event*). The K–T layer varies in thickness according to the location, ranging from a few centimeters to over a meter. Within the layer, apart from an anomalous abundance of iridium, quartz powder and ash particles have been

Figure 6.11: A geological stratum, called the K–T boundary layer, is found on all the continents and is contemporary with the extinction of the dinosaurs. This layer contains iridium, together with powder and ash of vegetable origin which are attributed to massive fires. At the top-left the layer seen at Gubbio, Italy is shown; at the top-right, the same layer at Ramonal in the state of Quintana, Mexico; at the bottom, that at Carvaca in Spain. The layer is a few centimeters thick in Italy and Spain but more than a meter in Mexico close to the impact zone. (Top: © Museum of Natural History Berlin; bottom: © USGS.)

found. The ash particles have the same composition as those from a forest fire, and the total quantity contained in the layer is estimated to be about 70 billion tonnes. As we shall see, the presence of these substances allows us to reconstruct what happened.

Iridium is very rare in sedimentary rocks but is abundant in meteorites and is sometimes found in the ejecta of volcanoes. Accordingly, for many years, the most reasonable hypotheses for explaining the extinction of the dinosaurs were two: a very intense volcanic eruption and the impact of a meteorite. Both could have provoked the drastic climatic changes that caused the extinction, and probably both occurred.

The volcanic eruption hypothesis was plausible because, at that time, a violent series of volcanic eruptions occurred in India — which was still an island not connected to the Asian continent (Figure 9.4) — forming the great Deccan Plateau with a million cubic kilometres of lava. Recent studies have shown, however, that these eruptions could not be the principal cause of the extinction because they occurred a million years earlier. Moreover, the dinosaurs disappeared very rapidly, whilst the eruptions lasted more than 500,000 years. Some paleontologists, however, have noted, before the K–T event, a progressive

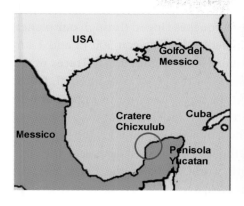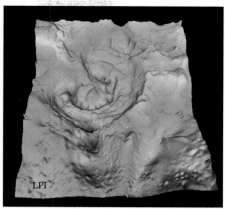

Figure 6.12: Left: The impact point of the meteorite that caused the dinosaurs' extinction. Right: The image of the 180 km diameter crater as seen after 65 million years. It shows strong signs of erosion, mostly in the north, where part of the crater is missing. Today, the crater is submerged under a thick layer of rubble that prevents its being seen directly. The image of the crater was obtained using accurate measurements of the local gravitational field to reconstruct the structure that caused it. (This technology is employed to find oil deposits under the sea-bed.) The image of the crater is very clear because it has a higher density than the sediments covering it. (© LPI.)

decline of the existing species, which would be compatible with the climate changes brought by these terrible eruptions.

If the volcanic hypothesis is excluded, only a meteoritic impact is left to explain the extinction. All doubts vanished when, in 1990, Alan R. Hildebrand of the University of Calgary, Canada, using gravimetric measurement, found a crater 180 km in diameter at Chicxulub, in the Gulf of Mexico, caused by the meteorite's impact. The crater was not visible on the seabed, because it was covered by a great quantity of sediment (Figure 6.12). From the diameter of the crater, the energy released in the impact has been evaluated.

The crater, and the composition of the K–T layer (Figure 6.11), have helped researchers to reconstruct what happened in that dramatic event.

When the asteroid crashed into the Earth, it threw thousands of tons of rocks into the sky. It is estimated that part of the material reached 100,000 km above the Earth before falling back again and that 10% of it escaped into space. Most of the matter falling back to Earth traveled through the atmosphere at several thousands of kilometers per hour, becoming small, fiery meteorites that heated the atmosphere. The heaviest of them reached the ground, causing widespread catastrophic fires. It is estimated that nearly half of the forests covering the Earth were destroyed, producing the 70 billion tonnes of ash found in the K–T layer, as well as a huge quantity of greenhouse gases: at least 10 trillion tons of carbon dioxide and 100 billion tons of methane — the equivalent of 500 years of fossil fuel burning at today's rate. The dust thrown up into the atmosphere by the impact constitutes the quartz powder found in the K–T layer. These dust particles blanketed the Earth with a thick mantle that blocked the sunlight for at least one year.

What occurred must have been terrible for every living creature on the Earth. At the moment of impact, vast earthquakes ripped through our planet, followed by tsunamis with waves estimated to be between 500 m and 1 km high, sufficient to inundate most of the dry land (Figure 9.1).[17] Traces of the material carried by these enormous waves have been found in Haiti, Cuba, Mexico and Texas. They have helped to identify the point of impact and to find the crater. The K–T layer in these regions is much thicker than elsewhere and contains the detritus brought by the tsunami (Figure 6.11, top-right).

Those species that survived the earthquakes, the seaquakes and the fires found themselves in a completely dark world. Initially, the Earth was hot because of the enormous energy released in the impact and the subsequent fires; this alone would have been enough to kill many species. Then the temperature began to fall and, after some time, our planet was covered with ice. Without sunlight, the plants and the plankton in seas died and, with them, the basis of the food chain for all living species (Section 6.5). The large animals that escaped the initial cataclysm had no chance of surviving because they depended on large quantities of food that no longer existed.

After a few years, when the sky began to clear again, the greenhouse gases, produced by the fires, brought a significant rise in temperature. Because of this, the surviving species had to contend with a torrid climate, but the dinosaurs, in spite of their great adaptability to the different climates, were already gone, along with the flying and marine reptiles and the ammonites. A few small dinosaurs survived and became birds.

It is hard to imagine how some species survived such a cataclysm. Computer simulations have shown that some northern parts of America, Europe and Asia might not have suffered fires. Pollen studies have also shown, that the number of plant species was dramatically reduced, but plants did not disappear completely. When sunlight returned, seeds germinated and life started all over again but it took a long time to return to normality. Studies of sediments have shown that it took around 130,000 years for plankton and plants to return the carbon cycle to its previous level, and 3 million years for the organic material carried by rivers to the ocean to reach its previous value.

- 65–50 *million years ago*: When the dinosaurs disappeared, one of their descendants, the nonflying bird *Diatryma,* became the most ferocious carnivore around. It was 2 m tall and, because it had no competitors, it dominated the Earth for 15 million years.

---

17 Seaquakes or tsunamis (Section 10.6) appear as a sudden rise in the sea level, sometimes lasting hours. During all this time, the water swamps the hinterland, submerging everything lower than the height of the wave. The waves generated by this meteorite are estimated to have been higher than the average height of the dry land (Figure 9.1).

6.12 Mammals

The disappearance of terrible predators like the dinosaurs led to an abrupt change in the animal species living on the Earth: mammals, which had appeared before the dinosaurs but had been confined in an evolutionary niche, developed quickly and dominated the planet. In the beginning, they laid eggs like the mammalian reptiles from which they descended but, with the disappearance of dinosaurs, they adapted to the new conditions and evolved into horses, elephants, rhinoceroses, wolves and monkeys. Their evolutionary story has left traces in the animals we find on Earth today, which are classified into three large groups:

- the *monotremata*, the earliest, laying eggs and suckling their young, they are represented by three living species, of which the *platypus*, living in Australia and New Guinea, is an evolutionary remnant;
- the *marsupials*, whose youngs are born alive (not from eggs) and conclude their development in an external pocket of the maternal body, where they are suckled and carried about; the kangaroo, the koala and the opossum are examples; they are represented by 270 species today, of which 60 live in America and 210 in Australia and New Guinea;
- the *placentals*, the most evolved and most recent, they have a placenta connecting the mother to the foetus and they number about 4400 species, including humans, and live in all continents.

In contrast with most reptiles and fish, which abandon their eggs as soon as they are laid, mammals increase the chances of their offspring's survival by defending and taking care of them after they are born, just like dinosaurs and birds. The story of the mammals begins at:

- *270 million years*: the *mammalian reptiles*. We believe that these were the ancestors of mammals because, although they had reptilian bodies, they had the teeth and crania of mammals. These predecessors disappeared with the appearance of dinosaurs.
- *220 Million years*: the date of the first mammalian fossils, among which is the *megazostrodon*, with the size of a mouse. They were warm–blooded vertebrates, they laid eggs like a reptile but they had mammary glands to nurse their young, like the platypus today.
- *120 million years ago*: during the reign of the dinosaurs, the first viviparous mammals appear. They were *marsupials*, like the kangaroo and the opossum.
- *100 million years*: the fossils of the first placental mammals appeared. Their embryos developed entirely in the mother's womb and were born completely formed.
- *67 million years*: the ancestors of the primates appeared. They had, in addition to other animals, prehensile digits, namely the ability to grasp objects

with their four paws by opposing their thumb and hallux to the other digits. This is one of the characteristics which lead to the extraordinary development of our species. The technological development and cultural evolution (Section 5.10) characterising the genera *Homo* would have been impossible without these prehensile hands.

- *40 million years*: the age of the first apes which rapidly propagated on all the continents.

The rest of the story is the final part of Figure 6.4, a relatively recent tale that began in the forests of Africa, whence our species went on to the conquest of the Earth. We can write this tale with considerable accuracy nowadays, thanks to *molecular anthropology*, a science based on the study of ancient and modern DNA.

In the case of ancient DNA, we can now extract readable parts from bacteria and plant up to 500,000 old whilst, for the ancestors of our species, it is possible to go back up to 150,000 years. New techniques should soon allow us to read the DNA contained in microscopic organic remains of species that lived several million years ago. It will then be possible to trace our evolution from apes by measuring how this DNA changed with time.

The study of the DNA of the existing species, allows us to estimate how long ago similar species existing today diverged from a common ancestor, by measuring the differences in their DNA on the assumption that these changes occurred gradually.

The study of the DNA of the existing species, on the other hand, allows us to estimate how long ago, similar species diverged from a common ancestor. It is done by measuring the differences in their DNA assuming that these changes occurred gradually: the Darwin's assumption which certainly holds true for at least the last million years. These techniques have been applied to the recently-sequenced DNA of several apes, including macaques and chimpanzees. By studying the differences between them, and by comparing their DNA with that of humans, we can even better dtaw the evolution of our species.

Let us continue the chronology of events we can now consider as "recent":

- *25 million years*: the separation of the evolutionary line leading to the Rhesus macaque (*Macaca mulatta*) — little monkeys living in India, China and Afghanistan – from that of *Homo* and *Chimpanzees*. The date is determined by looking at the differences between the macaque and human genomes, amounting to less than 7%.[18]

---

18 *The genome of a female macaque, living in a research center in San Antonio, Texas, was sequenced in 2007 by an international consortium of 170 scientists and 35 institutions; this was the third genome to be sequenced after those of humans and chimpanzees. Of the 3 billion base–pairs of its DNA, 93.5% are identical to the corresponding ones found in the human genome, as expected for a species that had separated from the human line only 25 million years ago.*

- *8 million years*: approximately the time of the divergence of the gorillas' evolutionary line from that of humans.
- *6 million years*: the approximate date of the divergence of the *chimpanzees*'[19] evolutionary line from that of humans. The differences in DNA between the two species is about 1.3%.
- *3.5 million years*: the age of Lucy, one of our most famous ancestors. She was an *Australopithecus afarensis* whose skeleton was discovered in 1974 in Ethiopia by Donald Johanson and Tom Gray. Lucy was possibly killed at the age of 25 and her body dumped into a marsh, where her skeleton was preserved. Being 107 cm tall and weighing less than 30 kg, she was not very different from the primates from which she evolved. She was probably spending the whole day searching for food on the ground and at night she went to sleep safely in a tree.
- *2.5 million years*: the beginning of the Paleolithic. This is the age of, the first remains of *Homo habilis* found. He evolved from Australopithecus in eastern and central Africa. He walked upright, like us, and knew how to fashion stone, signing the beginning of man's technological development. He disappeared 1.3 million years ago.
- *2 million years*: *Homo erectus* was born in the Rift Valley, having probably evolved from *Homo habilis*. He was the first hominid to travel widely and diffused rapidly throughout Europe and Asia. [Peking Man[20] is an *Homo erectus*.] He was the first *Homo* capable of building shelters, of clothing himself in fur, and of living in the coolest climates. He was also the first hominid to live in small groups of hunter-gatherers (as still happens today in some African and Amazonian forests). He became extinct 150,000 years ago.
- *1.8 million years*: our era, the *Neozoic* or *Quaternary* era began, it is characterized by the rapid emergence of the genus *Homo*.
- *250,000 years*: in Europe and Asia, the *Neanderthal Man* appeared. Evolved from *Homo erectus*, he was strong with a massive, shortish body, a large mouth and a receding forehead. He was able to live in the severe ice-age conditions, taking refuge in caves. He knew how to make fire and manufacture tools and he wore clothes made from animal skins. He lived in family groups and buried his dead. It is supposed that his vocal equipment

---

19 *The genome of chimpanzees was sequenced in 2005, it is the second to be mapped after that of humans, sequenced in 2001. It allowed us to make an unprecedented gene-to-gene comparison between the species, and it showed that only 40 million (1.3% of the total) molecular changes had occurred over a period of 3 billion years. And, of these, no more than 300,000 seem to be responsible for the differences between the two species today. By means of this comparison, it was established that chimpanzees and humans diverged from a common ancestor about 6 million years ago.*
20 *The fossil of Peking Man disappeared in 1941, during World War II, when it was shipped to the United States to prevent its possible destruction.*

did not allow him to develop an evolved language and that this limited his development. In 2011 the 70% of its DNA was sequenced.

- *150,000 years*: Homo sapiens — our own species — emerged in Africa. He built the first shelters from the branches of trees and had vocal equipment that enabled him to speak.
- *110,000 years*: the Earth began to cool, leading to the last glaciation, which put America in communication with Asia. The land bridge across the Bering Strait was more than 50 meters above sea level (Section 9.7).
- *50,000 years*: Homo sapiens left Africa and began the conquest of the continents. The first attempts at sailing in coastal waters took place during this epoch. These ancient voyagers were the progenitors of all the human races. It may seem far back in time but was only 2000 generations ago.

It was about this time that *Homo* sapiens encountered the Neanderthals. Studies of these two species' DNA shows that they never interbred, in spite of having a common ancestor. Born in different parts of the world, they remained separate after they met.

- *40,000 years*: humans arrived in Australia.
- *32,000 years*: the age of the Lion-Man, established by Carbon 14 dating. This is a remarkably sophisticated ivory statuette, representing a man with the face of a lion, and it shows that the *Homo sapiens* was already capable of abstract and symbolic thought. According to anthropologists, this also indicates that sapiens possessed an articulate language (Section 5.10).
- *30,000 years*: *Neanderthals disappeared,* probably exterminated by *Homo sapiens* who, whilst physically weaker had the advantage of communicating with an evolved language.
- *25,000 years*: this is the age of the fossil footsteps found on the banks of the Valsequillo lake in Puebla, Mexico, showing that humans existed in America at that time. Mankind had therefore completed his conquest of continents.
- *18,000 years*: the last glaciation reached its maximum extension (see the minimum temperature shown in Figure 7.9, on the left). The arctic ice cap extended to the south of Berlin and New York. Glaciers then created the valleys and lakes of the alpine arc.
- *16,000 years*: humans began *to domesticate animals*, the first being dogs used for hunting. A study of the mitochondrial DNA of more than 500 dogs and 50 wolves, completed in 2009, indicates that all present-day dogs derive from a wolf domesticated 16,000 years ago in the Yangtze River region in China.
- *14,000 years*: the end of the last glacial epoch. It was also the end of the long period during which, Siberia and Alaska were joined by a land bridge,

Figure 6.13: The Lion-Man, found in 1939 in the Stadel cave, in the Hohlenstein mountains of Germany. The statue was carved out of mammoth's tusk using a stone knife. It was broken into 70 fragments that were not re-assembled until 1969. The sculpture has a lion's head and a body combining human and animal anatomy. 32,000 years old, it is one of the oldest animorphic sculptures known. The significance of the statue is obscure but the lion probably played an important role in the mythology of the early Paleolithic. It clearly shows the existence of symbolic thought. The sculpture is 29.6 cm (11.7 inches) tall and 5.6 cm (2.2 inches) wide.

because of the low ocean levels, and of human migration from Asia to America.

– *12,000 years*: the temperature of the Earth, after a rapid rising, reached its maximum in the interglacial period (Figure 7.9). The seas rose by 150–200 metres. Mammoths disappeared, probably exterminated by *Homo* sapiens.

– *12,000–10,000 years*: clay vases appeared. They were the first durable objects made by man from artificial materials, not carved from stones or animal bones. The oldest ones are found in the Amazon, the Sahara and Japan. This marked the end of the Paleolithic age and the beginning of the Neolithic.

– *10,000 years*: the beginning of agriculture and farming. *Homo sapiens* began to modify his environment, using fire to clear forest from relatively large areas of land, reducing nature's ability to recycle the greenhouse gases. *The first climate change caused by man* is attributed to this period. This period also saw the beginning of an alliance between man and cats, to keep huts clear of small harmful creatures like mice.[21] It also saw the *breeding of animals for food*, first sheep and goats and then chicken, cows and pigs.

- *6000 years*: in the far east, large areas were cleared of trees in order to cultivate rice, resulting in large quantities of methane, whose greenhouse effect is about 23 times greater than that of carbon dioxide. According to some researchers, this is the beginning *of a substantial change in the climates* caused by humans (that is increasing nowadays, Section 7.4).
- *5,500 years ago*: the date at which the *domestication of the horse* is estimated to have occurred. It probably took place in Kazakhstan, where the remains of horse teeth damaged by bridles have been found. In the 2000 years that followed, horse culture spread from Asia to Europe, where the first traces of domesticated horses go back to about 3500 years ago.
- *5300 years*: the date of the earliest-known *written texts*. They employ cuneiform characters and were found in Mesopotamia. During the following centuries, writing appeared in Egypt and China as well. The invention of writing marks the end of prehistory and the beginning of history, when events are mainly reconstructed with the help of documents written by the protagonists.
- *4500 years*: the Egyptians constructed the *pyramid of Cheop* that remained the tallest man-made structure in the world for over 3800 years In Iran, the first copper tools appeared: it is the beginning of the *bronze age*.

1439 AD: Gutenberg invented *the printing press* and mass culture began and *Homo sapiens* cultural evolution (Section 5.7) accelerated.

1784 AD: the Scottish James Watt built the first *steam engine*. The industrial age began, considerably increasing the resources available to mankind. It was the beginning of the Anthropocene (Section 7.4), characterized by a dizzying increase in the world's population, accompanied by the extinction of many species and the emission of great quantities of greenhouse gases into the atmosphere. *The Earth's climate is ever more determined by humans.*

- 1942, December 2: in Chicago, Enrico Fermi implemented the first nuclear reactor capable of producing energy; it was the beginning of the nuclear era.
- 1953: the discovery of DNA.

---

21 *DNA studies show that all the world's cats descend from a single race, tamed in Mesopotamia about 10,000 years ago. The alliance between man and cats is associated with the beginning of agriculture. In this area, the oldest-known cereals repository have been found within mouse excrement. It is likely that cats first approached human habitation to hunt mice and found themselves rewarded by man, resulting in permanent collaboration.*

1969, July 20: Neil Armstrong landed on the Moon. For the first time in history, a man leaves an imprint on a celestial body other than the Earth.

The story leading to our species is a difficult one and one that leaves many open questions. Suppose the Cambrian Explosion had not happened? What would have occurred if dinosaurs had not become extinct? Would we humans be here today? As we have seen, the dinosaurs were extraordinary animals, capable of adapting to many different environments. They had no enemies outside their own ranks. It is hard to believe that they would have become extinct spontaneously. Is it possible that our species exists because of the impact of a meteorite? What will happen in the future? We must not forget that the dinosaurs dominated the planet for 135 million years, whilst the mammals have seized it for less than 60 million and *Homo* sapiens has existed for "only" 150,000 years. The Earth will survive for billions of years. Will another species dominate our planet after us?

To give an idea, of how insignificant our species is in the Earth's history, an analogy using a metre rule is often used. If the time separating us from the beginning of life on Earth (3.5 billion years) is equated to 100 cm, then only 57 cm separates us from the oxygen revolution whilst there is only 15 cm between us and the Cambrian age, the origin of species. The extinction of the dinosaurs occurred 1.8 cm ago, 10 mm separates us from Lucy, whilst Neanderthals became extinct and *Homo sapiens* became the dominant species only 8.6 thousandths of mm ago...

Extinctions

Chapter 7

7.1 Extinctions of Species in Biological Evolution

Extinctions have played an important role in the biological history of the Earth because they eliminated some species and allowed others to develop. As an example, the dinosaurs appeared after the greatest extinction of the Earth history, that of the Permian, and only when they became extinct was it possible for the mammals, appeared before them (Section 6.9), to develop. Extinctions have been a fundamental ingredient in biological evolution. They occurred in the past and will also occur in the future. They are the index of the continuous change happening to living things. Just remember a few numbers: 99% of the species displayed by fossils have disappeared, and the number of species extinct from the Cambrian (Section 6.6) till today is about 2 billion (of which only a small fraction left traces in the fossil record), while there are "only" 10 million species alive today on the Earth.

The majority of extinctions are connected with the competition between species to gain control over the available resources, and this usually happens between similar species; when nature produces a species that is more efficient (and therefore better suited to the environment) at hoarding the available resources, the weaker species disappears. The extinction does not happen (except in very rare cases) because of predators, whose survival depends on the availability of prey and therefore finds an equilibrium between them. Evolution cannot occur without extinctions because species tend to increase until they use all the available resources so that new species are possible only if the old one disappears. The rule is valid also for humans, and it explains the large rise in population after 1800, when industrialization increased the resources available. It may also explain the disappearance of the Neanderthals, in competition with *Homo sapiens* for resources (Section 5.10), and the massacres in recent times when a stronger ethnic group came into contact with a weaker one.[1] These extinctions, linked with the fight for survival, have resulted in the disappearance, every century in the number of species ranging between one to ten thousand and one to hundred thousand species.

Other extinctions, not related to the competition between species, are caused by physical phenomena that modify the climate of the Earth and eliminate species not adapted to these changes. Some of these extinctions have been particularly intense and, when they cause a loss of more than 40% of the existing species, are called "*mass extinctions.*"

When we consider mass extinctions, we often think of a traumatic event that happens quickly, but fossils show that they mostly occur slowly over thousands, or even millions of years. The fastest extinctions are due to climate change caused by an asteroid impact but, as we shall see, they are rare events; the KT event is the only great extinction for which there exists a strong evidence (Section 6.8). For all the others the evidence shows that they were caused by climate change provoked by slow phenomena and they lasted as long as it was necessary for the ecosystem to reach a new equilibrium. The longest extinctions are probably those caused by glaciations. The glaciers reflect back the heat coming from the Sun and keep the Earth from heating; the further the glaciers extend, the longer the extinction will last, because it will be more difficult for the planet to return to its initial conditions. It is hypothesized that a total glaciation involving all the oceans, turning the Earth into a giant snowball, must be irreversible, after which our planet will be unable to reheat itself.[2]

An example of irreversible climate change in the opposite sense exists in the Solar System: the case of Venus where a particularly severe greenhouse effect raised the temperature to 480°C for a time long enough to evaporate all the existing water, which should have arrived initially on Venus like on the Earth (Section 6.2). Having lost the water, the rains that made the Earth habitable were no longer possible, and Venus remained a hot world, inaccessible to life.

The Earth's past is divided into *eras*, which are in turn divided into *periods*. The classification is made on the basis of fossils found in the geological layers. The passage from one era to another and from one period to another is

---

1 *The idea of an animal origin of man came to Darwin when he saw the wild behavior in which the Gauchos killed the natives of Tierra del Fuego today nearly extinct. Ethnic cleansing, still happening on our planet are explained as the killing of weaker populations to hoard their resources.*

2 *As we will see in Section 8.6, the temperature of the Earth, given its distance from the Sun, should be −18°C. The average Earth temperature is instead +15°C because of the greenhouse effect of the atmosphere. Once the whole planet is frozen and the greenhouse gases reduced (they decrease with temperature), it is difficult to find a mechanism that can restore the initial conditions and make the temperature rise. Perhaps a particularly intense volcanic eruption which increases the greenhouse gases might cause the change.*

characterized by a change in the quantity and variety of fossils. There are four eras, each starting or finishing with a mass extinction or an explosion of life in the biological history of our planet. They are:

- *The Archeozoic or Precambrian era.* This is the oldest and longest era. It lasted from the birth of the Earth, which happened around 3.8 — 3.6 billion years ago (Section 6.3), to the Cambrian period, 540 million years ago at the origin of the species (Section 6.6). This era is subdivided into three periods.
- *The Paleozoic or Primary era.* This began with the Cambrian and concluded with the most devastating mass extinction, 250 million years ago, in the Permian. During this period animal and vegetable species colonized the dry land. Toward its end the continents merged into one: the Pangaea (Figure 9.4). This era is subdivided into six periods (Figure 7.1).
- *The Mesozoic or Secondary era.* This began 250 million years ago with the Permian extinction and the fragmentation of Pangaea into the present continents. It was characterized by the dinosaurs and their dominion on the Earth. It terminated 65 million years ago with their extinction. This era is subdivided into three periods (Figure 7.1).
- *The Cenozoic or Tertiary era.* This era started with the extinction of the dinosaurs, witnessed the rise of the mammals, and concluded 1.8 million year ago. It is divided into five periods.
- *The Neozoic or Quartenary era.* This is the present era, which started 1.8 million years ago and has not yet concluded. It is characterized by the rise of the genus *Homo* dominance and is divided into two periods, the Pleistocene and the Holocene. To these, we can add a third period, the Anthropocene, as suggested by the Nobel laureate Paul Crutzen (Section 7.5), is characterized by the great capability of the species *Homo* to destroy the environment in which he lives.

7.2 Causes of Extinctions

To study extinctions we need to estimate the quantity and variety of organisms that existed in the past. The valuations must be precise, otherwise it is not possible to compare results from different epochs. The work is enormous and requires classifying and arranging in chronological order great quantities of fossils, so as to have a sample that is statistically significant — work that paleontologists have done for at least two centuries. Among the recent collections of data, one of the most accurate is from the American paleontologist J. John Sepkoski (1948–1999), who constructed a database with 3700 different genera.[3] From his work come Figures 7.1 and 7.3, which synthesize what we can say today about the phenomenon of extinction.

---

3 The genera is one of *the taxonomical ranks of the classification of biological organisms (end of Section 5.5).*

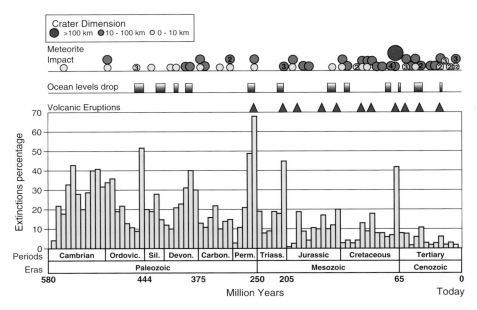

Figure 7.1: Extinctions occurred in the last 600 million years (adapted from a figure by J. J. Sepkoski Jr., 1994, Extinction and the fossil record, Geotimes, March 15–17). The horizontal axis shows in chronological order the 83 geological layers in which the last 600 millions years, since the Cambrian, are divided. In the vertical axis the percentage of extinct species found in each layer is shown. Below, the geological periods, eras and the dates of the principal extinctions are listed. The Neozoic, or Quaternary era (the one in which we are living), started 1.8 million years ago is not shown because it is too small to be represented by the scale. The upper part of the figure shows the events that may have caused extinctions. They are:

*— **Impacting meteors** (from Greve et al., 1996, The record of terrestrial impact cratering, GSA today 5:193–195), divided into three groups according to the craters' dimensions;*

*— **Intense volcanic eruptions** (from Courtillot et al., 1996, The influence of continental flood basalts on mass extinction, GSA (special paper) 307: 513–525);*

*— **Large-scale receding of sea levels** (from A. Hallam, 1992, Phanerozoic Sea-level Changes, Columbia University Press, New York) which indicate intense glaciations.*

In Figure 7.1, on the vertical axis is labeled the percentage of species extinct in the last 600 million years, since the origin of species in the Cambrian. On the horizontal axis is labeled in chronological order the 83 geological layers into which the period is divided. Because the layers do not have equal duration the timescale is approximated. In the lower part of the graph are listed the periods, the geological eras and the dates of the principal extinctions.

Observing the figure, we note that the extinctions are diminishing with time. From the Cambrian till today, there have been 55 layers (out of 83) presenting the extinction of more than 10% of the species, and of these layers only one is in the 65 million years that separate us from the KT event, while during the 70 million years of the Cambrian, a period equally long, nearly all layers show extinction rates of more than 20% of the existing species.

With the exception of the KT event (Section 6.8), we do not have sufficient data to determine, with certainty, the events that caused the extinctions. There are hypotheses on events that could have rendered the Earth too hot, too cold or too toxic. For understanding the statistical incidence of hypothetical causes of extinctions, in the upper part of Figure 7.1 are recorded: volcanic eruptions, meteorite impacts and variations in sea level that happened in that period, traces of which exist in the geological record.[4]

Intense volcanic eruptions

The extinction is caused by the greenhouse gases emitted into the atmosphere by volcanoes that cause the temperature of the Earth to raise (an effect that is most dramatically seen on Venus; Section 8.5). For volcanic eruptions the only data available are from the last 250 million years after the fragmentation of Pangaea (the earlier ones are not considered statistically reliable). The observed correlations are very strong: *there is always an extinction corresponding to an eruption*. The worst of the extinctions which happened at the end of the Permian correlates with the great eruption happened in that period in Siberia, which produced 10 million cubic km of basaltic rock. The extinction of the dinosaurs is also associated with an eruption — the one that created the great plateau of Deccan in India with a million cubic km of lava (Section 6.8).

Large drop in sea levels

These are the result of mass glaciations. On land the extinction is caused by the advance of ice masses and the consequent reduction in food and habitat. In the seas the extinction is caused by the lowering of the water level (150–200 meters below the present) which, reduces the habitat, limits the circulation of the water (some basins are left isolated), diminishes the oxygenation of waters and increases its salinity, creating conditions unfavorable for life. Figure 7.1 shows that half of the events registering strong reductions in the sea level are in correlation with extinctions.

Meteorite impacts

This is the third phenomenon which scientists consider. The impact results in extinctions by provoking earthquakes, seaquakes (or tsunami)

---

4 *The explosion of a supernova (Section 4.5) near the Earth is another possible cause of mass extinction. It produces an intense gamma emission that can sterilize a nearby planet. It is an event difficult to prove, yet could not be disproved with the existing data.*

and disastrous climactic effects (Sections 6.8 and 10.3). The destructive effects of asteroids increases with their mass (Figure 10.5), which can be evaluated from the dimensions of the crater they leave. In Figure 7.1 the power of impacts is divided into three groups, according to the diameter of their craters. The figure shows that the KT event, which caused the extinction of the dinosaurs, has been the most serious in the last 600 million years. Of the remaining impacts, all are smaller, and some happened concurrently with extinction, while others of similar size did not have serious effect. Because it is unthinkable that similar impacts produce different effects, it is concluded that there is no association between these meteorite impacts and extinctions. The only exception to this is the KT event, which 65 million years ago produced the crater of Chicxulub (Figure 6.12), causing a serious extinction that is only the fifth in gravity among those represented in Figure 7.1. The data are scarce, however; many large craters could have been destroyed by erosion, while others could have been covered with detritus, as happened to the Chicxulub crater (Figure 6.12), and may be discovered in the future. Before Hildebrand's discovery of the crater in 1990, the extinction of the dinosaurs was attributed to various causes; the meteorite impact was only one of the hypotheses and the most probable cause was supposed again to be a volcanic eruption. Therefore we cannot exclude the possibility that other great impacts, whose traces have been deleted by erosion, may have contributed to mass extinctions of the past.

In any case, these are the data we can use; the events are numerous enough to give statistical significance to the causes of extinctions. Figure 7.1 shows that volcanic eruptions are the events correlating most strongly with extinctions, with 11 large eruptions, each being associated with an extinction. For the receding of sea levels, 12 episodes are registered, and 7 of these are associated with extinctions. For meteorite impacts there are 20 episodes that have produced craters larger than 10 km and of these only one is associated with extinction: the KT event (the only one with a crater greater than 100 km). This does not mean that impacts are not dangerous (Figure 10.5), but that to cause a mass extinction meteorites leaving craters larger than 100 km are needed Chicxulub is the only crater of this size found till now.

Even with uncertainty in the data, we can affirm that *climate change induced by volcanoes is the principal cause of extinctions of the past*. For the future it is difficult to predict how real is the risk of large eruptions.

On the other hand, humans are showing the capability to be substitutes for volcanoes with emissions 100 times larger (Section 7.5). If we continue to emit greenhouse gases into the atmosphere without restraint, we have a good probability of becoming responsible for the next mass extinction.

7.3   Mass Extinctions

Figure 7.1 shows that certain extinctions are worse than others. Those that have eliminated more than 40% of the existing species are called *mass extinctions*. For the whole biological history of the Earth nine episodes of this type have been identified. Two happened before the Cambrian when the organisms were less sensitive to ambient catastrophes, and the other seven in the 540 million years that separate us from the origin of the species (they can be seen in Figure 7.1). We will now look more closely at these nine mass extinctions.

- *2 billion years.* Oxygen appeared in the atmosphere (Section 6.5). Anaerobic species disappeared with the possible exception of those that had found refuge in the depths of the sea and of the Earth where there is no oxygen. They were the extremophiles, of which we have spoken in Section 6.4.
- *700 million years.* During the Precambrian era a strong reduction in algae and plankton is observed. It is an extinction well documented by the fossils, but for which it is difficult to know the cause. A hypothesis is of an anomalous increase in atmospheric oxygen and a strong decrease in CO_2 that could have blocked (or stopped) the development of algae.
- *540–510 million years.* There were two episodes, the first of which came immediately after the Cambrian explosion. Each of them eliminated about 45% of the marine species. They were possibly caused by a very intense glaciation. These extinctions are considered a balancing phase of nature after the great Cambrian explosion, so they are not considered real extinctions which posed threats to life on the Earth.
- *444 million years.* At the end of the Ordovician period, more than 50% of the marine species disappeared. This is considered the second largest extinction by intensity. It happened in a very short while (Figure 7.1), therefore it could have been caused by the fall of an asteroid, but there is no evidence for such an event. More probable is a volcanic eruption (but there are no data on eruptions so far in time). The extinction is correlated with a reduction in the sea level, so a fast and intense glaciation is supposed too. This possibility is also confirmed by the remains of glaciers of the epoch that are found in the Sahara (rocks smoothened by the flowing ice). This extinction is considered the first mass extinction after the Cambrian.
- *375 million years.* At the end of the Devonian more than 40% of the marine species vanished. The extinction was relatively slow, lasting 3 million years. This excludes the meteorite impact and renders a volcanic eruption unlikely. The proposed cause is, once again, an intense glaciation. The study of marine sediment and glacial remains from the period, found in

Brazil, confirms this hypothesis. This is considered the second mass extinction after the Cambrian.

- *251 million years.* In the Permian the most catastrophic extinction in the history of the Earth happened (Figure 7.1). About 90% of the marine species and 60% of the terrestrial ones disappeared, among them numerous vertebrates. The extinction was relatively rapid. Craters due to the impact of large asteroids have not been found, so it is assumed that the most probable cause of the extinction must have been the intense volcanic activity occurred at that time in the region of Norilsk in Siberia. During that eruption 10 million cubic km of lava were emitted, covering an area of 2.5 million square km with a depth of 4 km of rock.

  To justify such a large extinction, it is supposed that the gas erupted from volcanoes caused a strong greenhouse effect and an increase in temperature, which in turn liberated the CO_2 contained in the oceans (Section 9.6) and increased the water vapor in the atmosphere (water vapor is also a greenhouse gas). The CO_2 would have killed the marine species directly, asphyxiating them, and indirectly the terrestrial species, provoking, along with water vapor, a further increase in the greenhouse effect that brought the temperature (Figure 6.3) above the level of survival for large areas of the planet. According to some researchers, this hypothesis is supported by having found an increase in the limestone in this period due to the large quantity of CO_2 contained in the atmosphere (Section 9.6). This extinction is considered the third mass extinction after the Cambrian.

- *202 million years.* At the end of the Triassic, 45% of the species disappeared. This extinction has been attributed to the fall of a meteorite (smaller than the one that caused the extinction of the dinosaurs) that caused the Manicouagan crater (Figure 7.2) in the province of Quebec, Canada. There is doubt over the association with that extinction, since the two dates do not coincide and moreover on other occasions similar impacts have not caused a similar extinction. The more probable cause is, once again, an intense volcanic eruption that happened at this time in North America. This is considered the fourth mass extinction after the Cambrian.

- *65 million years.* At the end of the Cretaceous period the fall of a meteorite caused the extinction of 45% of the species on the Earth, among them the dinosaurs (Section 6.8). This is the fifth and last mass extinction.

7.4 Extinctions and the Species on the Earth

Figure 7.3 shows the abundance of genera (we have seen that they are groups of similar species, Section 5.5) observed among fossils during the last 600 million years. Below the curve various periods and eras are indicated. The mass

Figure 7.2: *The Manicouagan impact crater in Quebec, Canada, photographed in 1983 from the space shuttle* Columbia. *Formed around 200 million years ago, it is one of the oldest impact craters known on the Earth. It is 70 km in diameter and contains a ring-shaped lake feeding a hydroelectric plant (at the bottom of the picture). (© NASA & STS-9 crew.) The crater can easily be found with Google Earth.*

extinctions appear here as negative peaks that correspond to sudden reductions in the genera present on the Earth. Just after the origin of species, the number of genera increased throughout the Cambrian, and it remained stable for 200 million years, until the great mass extinction of the Permian. Then it began a second phase of growth (interrupted only by the KT event), which has not finished yet. The figure shows that *The number of species present on the Earth today is therefore the highest in the entire biological history.* From the figure we also see that after every extinction nature has recovered within a relatively short period the number of preexisting species, continuing the trend that existed before the extinction.

In Figure 7.3 the colors identify four groups of species which originated in the Cambrian and had a similar story: a period of rapid growth followed by a period in which the number of species remained constant and then a phase of decline that began with a mass extinction.

The first group is the *Cambrian fauna*, completely disappeared today. Its growth began with the origin of the species and its decline happened at the end of the Ordovician period. The *trilobites* and the *anomalocaris* in Figure 6.9 belonged to this group.

The second group is the *fauna of the Paleozoic*. It began its growth after the Ordovician extinction and declined after the great extinction of the Permian. A few of these species have survived until our times; among them are marine stars and cephalopods (the octopus, considered the most intelligent among invertebrates).

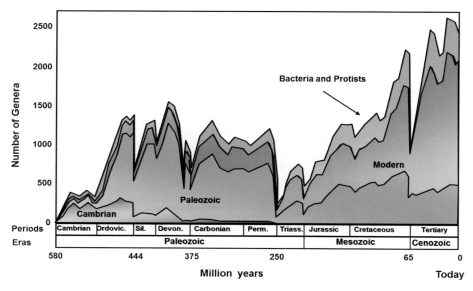

Figure 7.3: *The species that have inhabited the Earth from the Cambrian period until our times. The strong reductions in their numbers, in parallel with the great extinctions (Figure 7.1), are highly identifiable. Today the number of species is the highest in the history of the planet: around 10 million if we account only for vegetables and animals. Figure obtained from Sepkosky's database.*

The third group is the *modern fauna,* of which the vertebrates are a part (including man). It began its ascent with the dinosaurs and is still growing today.

The fourth group consists of the bacteria and the protists (Section 5.5), which have existed since the beginning and will probably continue to exist as long as the Solar System does.

It is difficult to see a general criterion in these trends. We can only observe that the diversification of the species occurred in successive waves, each beginning with an extinction and starting to decline with another extinction. We can only observe that the manner in which the diversification of the species increases is not continuous in time and is in agreement with an evolution that proceeds by jumps, an idea explainable by the theory of complexity (Section 5.9). With the dawn of the Cambrian, nature acquired the capacity to control a greater complexity and the number of species increased rapidly, then it stopped until it acquired the capacity of controlling a higher complexity and a new explosion of life began. Also the fact that after each extinction nature recovers quickly the number of species lost (no more no less) can be seen as an element in favor of the theory: the information necessary for managing that level of complexity existed in the surviving species, so that they quickly fill the space left by the lost varieties.

In conclusion, we can say that the mass extinctions were caused by climate change; some were provoked by geological events like volcanic eruptions, others by cyclic events like glaciations (Section 9.7), and still others by astronomical events like meteorite impacts.[5] To these natural events we can add the capacity of humans to modify the climate (and that of destroying themselves by the weapons they have constructed…).

What will happen in the future? There are reasons for being optimistic and others for being pessimistic. The reasons for optimism are suggested in Figures 7.1 and 7.3. The former shows that natural extinctions have been diminishing with time; the latter that life on the Earth is in an extremely favorable moment, that we never had as many species as in this period and we are far from cataloging all the living species. As an example, for insects it is thought that only 25% of them are known, with 3 million left to be cataloged. The peak and decline (if you want to give a general sense to the behavior of the figure) will happen for vertebrates in a very distant future. One can also add that technology may soon find remedies for some of the catastrophic events of the past. In Section 10.4, we will see how we can in the future deviate a danger-ous meteorite and in Section 7.6, how one may control the temperature of the Earth.

The reason for pessimism comes from the effects our civilization is bringing to the climate. If it is true that the number of species alive today is the highest in the history of the Earth, it is also true that the number of extinctions is one of the highest ever registered.

For birds and mammals it is estimated that the number of extinctions in the last century is on the order of 1/100 of the existing species, more than enough to invert the tendency shown in Figure 7.3. A value this high was set in the past only for the great mass extinctions (we have seen that in normal times extinc-tions per century are between 1/10,000 and 1/100,000 of the existing species, which is also the value registered on the Earth until 1800). Estimates for the future are even more pessimistic. Some researchers predict the loss of 10% of the species existing in a period of a few decades, it is the disappearance of a million animal and vegetable species in a short while.

We maintain that the cause of this large number of extinctions is the changes humans have caused on the environment and the climate, to which is added the inclination we have to destroy all species in particular animal and vegetable (by means of agriculture) we consider not useful. In the last century, with the demographic explosion and the growth of industry, this phenomenon has assumed disastrous proportions and there are by now many who feel that we are entering a new period of mass extinction, the sixth after the Cambrian. To this extinction, which we may be leaving today, we devote the next sections.

---

5 *We have already mentioned supernova explosions as a possible cause — an event not easy to be traced.*

7.5 The Modern Era Extinction

The possibility that our planet is entering a new phase of mass extinction is a theme debated in the scientific world. Ever-more-frequent meetings are held in which biologists, physicists, geologists and experts on the climate and atmosphere discuss this argument. If we simply consider the anthropic emission of CO_2 they evaluated in 30 billion tons per year, 100 times larger than the natural ones (including all the Earth's volcanoes) evaluated only in 300 million tons.[6] This means that the capacity of the atmosphere of regulating the soil temperature is completely compromised (Section 9.6)

The report of the Intergovernmental Panel on Climate Change (IPCC), the group of scientists charged by the UN to present a report every five years on the status of the planet, is ever-more-alarming. They speak of apocalyptic scenarios by predicting the emigration of populations in millions, and also of lands becoming hotter and more desert-like. The data we have offer hardly relief.

The world's population is increasing exponentially (Figure 7.4); we passed from a billion inhabitants in 1800 to 7 billion in 2011, and we are projected to exceed 9 billion before 2050. The increase in population is occurring especially in the poorest countries those that consume the least today and will for sure increase their consumption in the future. An example is Uganda: which in 1962, at the point of independence, had 7 million inhabitants; in 2006 it had 30 million, and in 2050 it is foreseen to have 130 million, exceeding the population of Russia. Nigeria which was 33 million in 1950 will reach 300 million in 2050, the present population of the United States. Africa, the poorest continent, will go from 8% of the world's population in 2006 to 25% in 2050, hosting a population equaling the world's population in 1950.

With the increase in population in the poorest countries and a desirable increase in their wealth, the quantity of greenhouse gases introduced into the atmosphere will also increase along with that the temperature of the planet. Today the average emission of CO_2 in Asia and Africa is about 1.5 tons per inhabitant per year, while that in the developed countries has gone from 7 to 18 tons per inhabitant (Appendix A.2). It is not difficult to predict that the demand for energy will double in the next 30 years, and if we do not change the kind of energy we use, the emission of greenhouse gases will double as well.

Moreover forests are reduced at an impressive rate to leave space for agriculture and other human activities. It is estimated that every year between 10 and 13 million hectares of forest disappear, and with that the ability of nature to absorb CO_2 and produce oxygen is diminished. The report of the

---

6 Present and past non anthropogenic CO_2 degassing from the solid earth.: *D. M. Kerrick Review of Geophysics, 2001 Vol. 39, No. 4, pp. 565–585.*

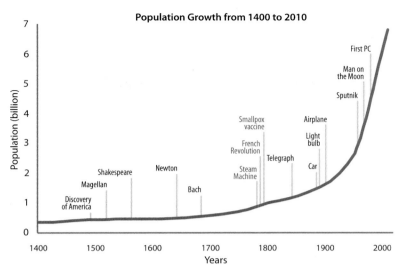

Population Growth from 1400 to 2010

Population (billion) — axis labels: 0, 1, 2, 3, 4, 5, 6, 7

Years — axis labels: 1400, 1500, 1600, 1700, 1800, 1900, 2000

Event labels on the graph: Discovery of America, Magellan, Shakespeare, Newton, Bach, Steam Machine, French Revolution, Smallpox vaccine, Telegraph, Light bulb, Airplane, Car, Sputnik, Man on the Moon, First PC

Figure 7.4: *The rapid growth of the world's population began around 1800, coinciding with three events (in red) that occurred in a period of 10 years. These events are considered the symbols of the change which made this growth possible. In chronological order, they are: the invention of the steam engine, indicating the beginning of the industrial revolution, which enormously increased resources at the disposal of humans; the French Revolution, which gave rise to democratic states, governed in the interest of the populace; and, finally, the smallpox vaccine, which symbolizes the birth of modern medicine.*

IPCC predicts that in less than 100 years half of the today existing forests will disappear. Deforestation and other human activities have increased soil erosion by a factor of 3 in the last century, destroying 2 billion hectares, an area equal to that of the United States and Canada combined.

Data from atmospheric satellites indicate that the sea level is rising by 3 mm per year[7] and that the rate is accelerating: the level rose by 20 cm during the last century and by 3 cm in the last 10 years. According to the 2007 report of the IPCC, the estimated increase by 2100 should be between 30 and 150 cm.[8]

Since the end of the last glaciation 18,000 years ago, the sea level has increased by 150–200 m. That was initially caused by the melting of the ice that covered a large part of America and northern Europe. Today, with ice restricted to the colder Arctic zones, near half of the increase in the ocean level is due to

---

7 *It is not easy to evaluate the increase in sea level, as it is difficult to distinguish between the causes for a variation in the volume of the see waters, because it can happen for the increase in sea temperature and for the melting of ice, or instead of continental drift that changes the volume of the container in which the oceans fill. Only with satellites that accurately follow the movement of land and ocean mass, and the quantities of ice staying on the land, it is possible to evaluate the phenomenon.*

8 *If there is no immediate intervention, the first victim of the rising sea levels will be Venice, whose situation is aggravated by the flooding of the soil, such that in the last century, the relative variation of the sea level was 35 cm. The first effect has been a dizzying increase in water levels: in the decade 1920–1929 there were five episodes in which the water levels rose higher than 1 m (an average of one every two years), while 70 years later, in a period just as long (1990–1999), there were 105 (10.5 per year).*

the rises in temperature which expands the volume of ocean waters. The increase is about 80 cm for every degree in water temperature. The present melting of ice in Greenland and Antarctica (the ice of the Arctic is floating and its melting does not add to the volume of the oceans) contributes to the remaining half.

The average temperature of the Earth has increased by 0.8°C in the last century and the rate of increase seems to be accelerating. According to the IPCC if the temperature of the planet increases by another degree, drought will cause hundreds of millions of people to abandon Africa and South America. With the increase in the temperature, the ice caps on some mountains will disappear, and with that will disappear the principal reserve of water in many regions during warm periods (when agriculture uses the most water), condemning the people to severe drought.[9] The floating icebergs of the North Pole have been reduced by 25% in the last 30 years; if the trend continues they will be reduced by 60% by 2050 and will disappear before the end of this century, taking along the polar bears that live on it. In return, the polar routes, once reserved for airplanes, will be opened to ships. In 2007 the European Space Agency (ESA) stated that satellite photos showed that the Northwest Passage between Europe and Asia, along the northern coast of Canada, was free of ice. It was the first time such a phenomenon has occurred since 1978, when satellite pictures began to be taken.

The meteorology of the planet is changing more rapidly than ever; the number of typhoons registered in the last 15 years have increased. In Mediterranean waters, tropical species are found. Should the increase continue, the Mediterranean panorama will change: the maritime pines could disappear from the coasts, to be replaced by palms. They will be found in northern Europe together with oaks and walnuts, all having emigrated toward fresher climates.

The number of species extinct in the last century, as we saw above, is typical of the rates during mass extinction periods, even if the number of existing species remains the highest in our biological history. There is no evidence of significant extinction in the seas, but many scientists maintain that if the temperature and the level of pollution continue to increase the extinctions could appear all at once because nature "reasons" in an exponential way. As an example the extinction may increase as a power of 10. After a certain period 1 species is extinct, after a similar period, 10 species, then 100, 1000, 10,000.... Initially we would not notice the phenomenon, and then it would increase so rapidly that it would be difficult to intervene.

According to some researchers humans started to influence the climate 10,000 years ago with the beginning of agriculture in the Fertile Crescent, the

---

9 *Rainfalls are mostly in autumn and winter, while water for agriculture is needed in spring and summer. The ice is a natural reservoir preserving the winter rains for summer; its disappearance will condemn large area of the Earth to drought. Large reservoirs can reduce the problem, but they are expensive and will never replace glacier.*

Iraq of today. With agriculture humans declared war on the ecosystem; to remove trees we burned down entire forests, devastating far more land than was necessary, we convert lands to monoculture and we destroy the animal and vegetable species that we do not consider useful. We have deforested vast regions, reducing the capacity of nature to remove greenhouse gases from the atmosphere, which finally results in increased production of these gases.

The rice fields, already well developed 6000 years ago in the East, favor the decomposition of vegetation with production of methane. Also, the practice– existing still today — of burning the scrub to clean and prepare fields contributes to the production of greenhouse gases.

It is estimated that without human activities the average temperature of the planet would be 5–6°C cooler. For this increase, we estimated that only 0.6–1°C are due to the past industrial activities, the rest is due to agriculture. This is confirmed by an examination of the gas trapped in the polar ice (Figure 7.9) showing an anomalous increase in greenhouse gases in the last 10,000 years. The effect would not have been noticed in the past because, on account of the glacial cycle (Section 9.8) 12,000 years ago, a new glaciation should have started. The greenhouse effect caused by humans would have counteracted the cooling, maintaining a relatively stable temperature. This effect has been positive in the past; a reduction of about 5°C in the average temperature would have rendered large areas of the planet uninhabitable and subtract large parts of arable land from North America, Europe and Asia.

7.6 The Anthropocene

In 2002 the authoritative journal *Nature* published an article by the 1995 Nobel Prize winner (for atmospheric chemistry) Paul Crutzen, in which he proposed a new epoch in the Neozoic[10] (the era in which we live; Figure 7.1) — the Anthropocene.[11] A period characterized by the growth of *Homo sapiens,* by his strong influence on environment, and by the decline of other species caused by its presence everywhere on the Earth. In his article Crutzen proposed 1784 as the starting year of the period; it is the year in which James Watt constructed the first steam engine, making the beginning of the industrial era (Figure 7.4).

Crutzen started from the considerations mentioned in the previous section. He observed that the human impact on the environment will increase in the years to come, due to the increase in the world population and, as a conse- quence, of the increase of the area occupied by humans and of pollution. He noted that these changes had never occurred in the history of the Earth and

---

10 *The Neozoic is divided into two periods: the* Pleistocene, *from* −1.8 *million to* −100 000 *years; and the* Holocene *from* −100,000 *years to our times. To these two periods, Crutzen proposed adding a third, the* Anthropocene.

11 *We suggest the book* Welcome to the Anthropocene, *by Paul J. Crutzen (Mondadori, 2005).*

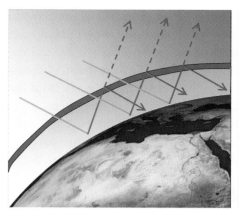

Figure 7.5: The greenhouse effect (left) is caused by gas molecules (mainly CO_2, CH_4, H_2O) located high in the atmosphere. These molecules are transparent to the impinging sunlight, whereas they absorb the infrared radiation emitted by the Earth (as the glass of greenhouses does). Sunlight (yellow) crosses the atmosphere and heats the Earth. The Earth in turn emits the absorbed energy in the infrared (red lines), which is partly scattered in space (dashed lines) and partly sent back to the ground (red solid lines) by the greenhouse gases. Today the equilibrium between the impinging energy and the one dispersed into space is averaged at 15°C. With the rise in greenhouse gases, the energy that is sent back increases and so does the temperature of the Earth.

Aerosols (right) generally have the opposite effect. They are particles emitted by volcanoes, sandstorms, industrial activities, heating, airplanes, diesel engines etc. They wrap the Earth with a layer that absorbs and reflects sunlight, preventing it from reaching the ground. Thus their effect is generally the opposite of that of the greenhouse gases (although there are exceptions). The Earth's temperature decreases with the increase in their concentration. A great concentration of aerosol can lead to a cold darkness that can cause the destruction of the marine and terrestrial flora. The dinosaurs' extinction (Section 6.9) happened 65 million years ago, was caused by the obscurity that wrapped the Earth, because of the dust raised by the meteorite's impact and by fires sparked off everywhere on the Earth.

that they could not be stopped (barring an unwanted catastrophe which reduces the population of the planet). He concluded by saying that humans, being the only responsible for the phenomena (the Anthropocene is caused by us…), have the possibility of studying what is happening, and then attempting to manage the change in order to contain the damage. We are not facing an inevitable natural event, like an earthquake a meteorite impact or a volcanic eruption, but an event caused by humans that humans, for their survival, must seek to control.

To help in understanding how it is possible that a small quantity of gas and dust can modify the thermal balance of the Earth, the caption of Figure 7.5 explains the mechanism by which the greenhouse effect induces warming and the aerosol causes cooling.[12] The two phenomena have opposite effects on the temperature of the Earth, and both can be dramatic in the survival of life on a planet. To understand what happens, one can think of an automobile left under the Sun with its windows closed. The temperature inside can easily

---

12 *An example of how dramatic can the the cooling casued by dust is, the great night that led to the dinosaurs' extinction (Section 6.8).*

arrive at unsupportable values well above those outside, because it is not possible to dissipate the heat that the Sun supplies. To render the vehicle livable, one must open the windows to exchange heat with the cooler air outside, but not too much or the temperature will drop more than desired (on the Earth we may have a glaciation).

This is the role of the atmosphere — to control the exchange of heat with outer space, which is very cold; altering the control could easily bring the temperature of the Earth outside the limits of Figure 6.3. Today the concentration of greenhouse gases is the highest in the last million years at least, and keeps our planet from cooling itself. The consequence is the "global warming" which we read about in the newspapers.

To try to contrast global warming, several techniques are studied to increase the aerosol by introducing into the atmosphere microscopic particles that reflect light. A possibility is sulfur crystals seeded by stratospheric balloons, or salt crystals drawn from the sea by immense machinery capable of spreading them to great heights. Other studies regard the possibility of covering the deserts with reflective materials. All these are operations, because they are too costly. Crutzen noticed that to have a detectable effect on the climate the cost would be more than US$ 50 billion and the results of these interventions would last for only a few years, after which the suspended particles would fall to the Earth (increasing soil pollution) and the reflective materials would have to be replaced, ruined by atmospheric agents. The cost to keep all these thing in place for centuries will overcome the resources of our planet.

7.7 Polar Ices

The problem of global warming was posed by scientists more than 50 years ago, but it was not taken much into consideration until the evidence from measurements on the polar ice and the clear alteration of the climate in these years left little doubt of the change going on. An example of results obtained from the study of the ice is shown in Figure 7.6 which reports measurements of the concentration of greenhouse gases in the atmosphere from 1850 till today. If we compare this figure with Figure 6.5, we see that humans are going back in the work done by the first micro-organisms. We are releasing into the air, by means of agricultural and industrial activities, CO_2 and CH_4 in quantities greater than what the ecosystem can break down. The equilibrium between the produced greenhouse gases and what nature can deal with is broken. The result is the increase in the temperature of the planet (Figure 7.7).

The measurements in Figures 7.6 and 7.7 are very precise and were obtained by chemical examinations of the gas contained in the ice caps of Greenland and Antarctica taken from deep drilling. The ice, when it is formed, entraps

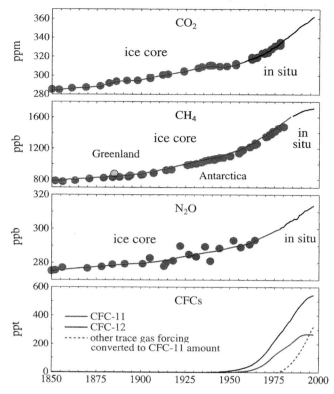

Figure 7.6: *Concentration of greenhouse gases of human origin from the beginning of the industrial revolution up till today. We can see that the increase in CFCs (gases used for refrigerators and air conditioners) stopped in 2000, for worldwide elimination of their use (Section 7.11). The abbreviations "ppm", "ppb" and "ppt" stand respectively for "parts per million", "parts per billion" and "parts per trillion". [© Charles Langmuir, Climate change and greenhouse gases, EOS, 80, 453, 28 September 1999. (Reproduced with permission from American Geophysical Union).]*

microscopic capsules of air that remains unaltered for even millions of years, until the ice melts again. From the chemical analysis of the gas and particles so trapped, is it possible to reconstruct the atmosphere's composition at the moment in which the snow fell; the deeper we go in the glacier the farther they go back to the past.

To investigate the ancient past of the Earth's atmosphere, the best sites are those in which there are very deep layers of ice, above plateaus where the ice does not move rapidly and in places where it does not snow too much so that at a few kilometer's depth it is possible to find very ancient snows. The ideal sites for these investigations are the great plateaus of Greenland and above all of Antarctica where the thickness of the ice can be more than 3 km. The most interesting site for these studies is the Italo-French base at Dome C in Antarctica (Figure 7.8), which is 3233 m above sea level, of which 3000 m are ice.

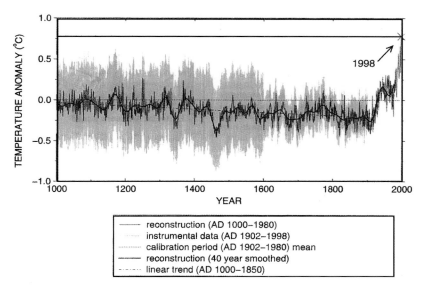

Figure 7.7: Temperature fluctuations in the northern hemisphere from the year 1000 up till today. The intense temperature increase recorded in the last decades is very clear. [© Charles Langmuir, Climate change and greenhouse gases, EOS, 80, 453, 28 September 1999. (Reproduced with permission from American Geophysical Union).]

Precipitations given the altitude are very rare, not more than 2–3 cm per year. Moreover, being the site at the center of a large "plateau," the ice moves slowly, a few centimeters per year. All these allow the deep drilling at Dome C, supply samples of snow fallen as far back as 800,000 years ago.

In Figure 7.8 (upper left), there is a map of Antarctica with contour lines that indicate the thickness of the ice. Shown on the map are the positions of Dome C,[13] of the South Pole where there is an American base, and the highest point in Antarctica, Dome A, at 4000 m (where China has installed a new base for arctic research). At the lower left there is a photo of the Italo-French base: two cylinders constructed on moveable feet to rise up the base, which, because of its weight, is sinking into the ice. The feet rise up alternatively, so that snow can be accumulated under them and rise up again the edifice.

On the right of Figure 7.8 we see a piece of the ice core; the horizontal lines correspond to layers of ice deposited in successive years. They are so highly identifiable that they can be counted like tree rings. The dark traces are due to dust incorporated in the ice, together with samples of the atmosphere; at the center we see a particularly dark streak, probably due to intense volcanic activity. The method of analysis of gas found in the various layers is so precise that

---

13 *The Italian activities in Antarctica are led by the Programma Nazionale per la Ricerca in Antartica (National Program for Research in Antarctica), managed by a consortium of four Italian research institutions: ENEA, CNR, INGV and OGS, as defined by the February 2002 ministerial decree.*

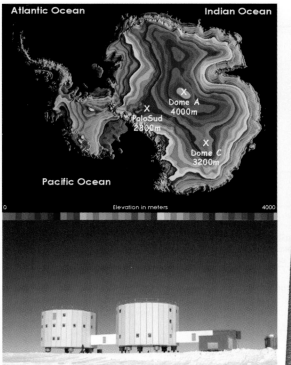

Figure 7.8: Top left: a small map of Antarctica where the position of the Italo-French base at Dome C is indicated, and the base is shown in the lower figure. Right: A sample of ice core taken at Dome C under EPICA (European Project for Ice Coring in Antarctica) project: several layers are clearly identifiable, corresponding to various years. (© Top–SGS; bottom–PNRA S.C.r.l.)

researchers could identify traces of metal produced by the "industrial activities" of ancient Greeks and Romans. They have also found carbon residues 8000 years old that could be attributed to the fires used by our ancestors to remove wood and cultivate the land.[14] It is impressive to discover how tiny traces of metals or carbon generated tens of thousands of kilometers from the pole were transported by winds to the Antarctic plateau, where they have been preserved for hundreds of thousands of years.

In Figure 7.9 we see the results of studies undertaken at the Antarctic Russian base at Vostok; on the horizontal axis, above the figure the depth from which the sample was taken is labeled, below the labels indicate the corresponding ages which go back to — 400,000 years, or 100,000 years before the Neanderthal man appeared on the Earth. The graph shows the concentrations of CO_2 and CH_4 that are reported together with the mean temperature on

---

14 *Different from metals, which are definitely of human origin, carbonaceous residues are more difficult to interpret as they can come from natural fires or from volcanic activities.*

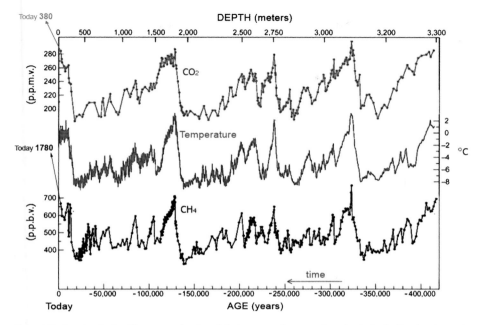

Figure 7.9: Results obtained by an examination of the ice cores taken at various depths in the Russian Antarctic base at Vostok. On the horizontal axis above the graph, the depth where the sample was taken is labeled; on the horizontal axis below the graph is labeled the age of the sample. The three curves show that the concentration of greenhouse gases (CH_4, CO_2) and the temperature (obtained from the concentration of ^{18}O) correlated and they show an oscillation with a period of 100,000 years. (see Section 9.8). The periodicity is interrupted in the present age (left): Petit et al., Nature, 3 June 1999. (© Nature Publishing Group.)

Earth, obtained by measuring the abundance of ^{18}O, an isotope of oxygen whose abundance is very sensitive to the temperature.[15]

The graph shows that the greenhouse gases and the temperature are always correlated, and that warm and cold periods occur with a precise periodicity of 100,000 years (four in the last 400,000 years), suggesting a very regular phenomenon that alternates glaciations and warm periods. A possible explanation of such a precise period is given in Section 9.8.

Figure 7.9 shows that in its glacial cycles, once the Earth's temperature reaches a maximum, it always has a rapid decrease toward a new glacial period, with the exception of the last cycle (leftmost in the figure). In the last cycle, the temperature of the Earth, after the last glaciation culminating 18,000 years ago rose rapidly as in previous cases, reaching a maximum 12,000 years ago.

---

15 The oxygen isotope ^{18}O is 500 times less abundant than ^{16}O, the normal oxygen. Being heavier, ^{18}O forms a water molecule slightly heavier than that formed by ^{16}O. When water evaporates, the lighter molecule evaporates more easily than the heavier ones, so that when there is higher evaporation, in the atmosphere (and hence in the polar ice caps) the concentration of water formed by ^{16}O increases, while in the sea increase the water formed by ^{18}O. Since the oxygen is contained also in marine sediments, from their study it is possible to estimate the temperatures tens of millions of years ago, with a dating that is, however, less accurate than that obtained by counting the ice layers in Figure 7.8.

The temperature should then have dropped again, but that did not happen.[16] As we have seen, a possible cause of this phenomenon could be the rise of agriculture and the consequent emission of greenhouse gases. William F. Ruddiman, a climatologist at the University of Virginia,[17] maintains that humans had a great destructive capacity since the beginning of civilization and that the Neolithic farmers could have destroyed forests with fire far beyond what was necessary in order to have fields to cultivate. To the CO_2 produced by these fires, it was added, 9000 years ago, the CH_4 produced by the first rice fields. Not everyone agrees with this view; there are those who claim that the small world population could not have influenced the climate. There are, however, traces in the polar ice that can be due to these fires, and there is no better explanation to justify the absence of a drop in temperature.

Figure 7.9 also shows that in the last 400,000 years the concentration of CO_2 in the atmosphere has never gone above 280 ppm (parts per million) and that of CH_4 has never gone above 800 ppb (parts per billion). Figure 7.6 reveals that these were the values in 1850, and from that point on the concentration of these gases has done nothing but to increase, because of human activities. The concentrations of CO_2 is today close to 400 ppm and that of CH_4 close to 1800 ppb, the highest ever registered since several tens of millions years. Correlated with these greenhouse gases, the temperature of the Earth is also at its maximum.

The situation would be worse if the Earth were not surrounded by a dense layer of aerosol (Figure 7.5), bad for health at low altitudes but useful for the climate at high altitudes. Aerosols are composed of microscopic grains of powder that reflect the Sun's light, diminishing the insolation of the soil. It can influence the climate and is produced by volcanoes, airplanes flying at high altitudes, and to a lesser extent by diesel motors, by heating and many other industrial activities brought to great heights through winds. Their contribution to the cooling of the planet is so important that some researchers regard with apprehension the policy applied in many countries to reduce the aerosol (which cause lung disease and cancer) without first setting up a serious worldwide policy to reduce greenhouse gases.

7.8 The Climate: An Unstable System

Figure 7.9 shows that variations in the concentration of greenhouse gases and in the temperature occurred also when humans did not interfere with the

---

16 Actually there have been two intense cooling periods at about −12 000 years and then at −8000 years, when the new glaciation should have started. In these periods the Earth's average temperature decreased by 5°C for relatively long times. Some interpret them as unsuccessful attempts at triggering a new glacial period. Also, in recent times substantial climate variations are known. A heating around 1100 permitted Vikings to colonize Greenland; it was followed by a cold period between 1300 and 1800 (Figure 7.7).

17 W. F. Ruddiman, Plows, Plagues, and Petroleum (Princeton University Press, 2007).

environment, with a precise cycle of 100,000 years.[18] This (together with the precision of the cycles) has led to the hypothesis of the existence of various phenomena, other than greenhouse gases, that have the role of clock, in causing such a periodic variations of the temperature. In such a scenario, greenhouse had, at least in the past, the role of amplifying the variations in temperature due to other causes as discussed in Section 9.8.

The mechanism would be of this type: an increase in the temperature, due to a periodic phenomenon, causes an increase in the concentration of greenhouse gases in the atmosphere (because of the degassing from ground and sea, Sections 7.1 and 9.6) which in turn causes additional warming. The temperature continues to increase until another phenomenon reverses the tendency. Such a phenomenon must exist given the cycles of Figure 7.9 (moreover temperature and greenhouse gases cannot increase to infinity). At that point the temperature begins to diminish, reducing the emissions of greenhouse gases from sea and land, which in turn further reduces the temperature. This causes oscillations as seen in Figure 7.9, typical of an unstable system.[19] Human intervention in these circumstances might be very dangerous; an amplification of the natural cycle could bring about dramatic increase in the temperature. In the past it interfered with the cooling of the planet setting off a new glaciation, today it may causes the temperature to increase above the historic maximum, and tomorrow it could put our survival at risk.

There are, therefore, all kinds of reasons for us to get worried about the future of our planet. We have seen that the most reasonable hypothesis for explaining the majority of mass extinctions of the past is a temperature rise, induced by increasing greenhouse gases that are emitted by volcanoes and amplified by the emission of greenhouse gases contained in the soil and in the sea. Today, humankind could replace volcanoes and cause the same catastrophes because, as we have seen the anthropic emission are 100 times those of volcanoes. The extinction that happened 250 million years ago, the worst in the history of our planet, seems to have been caused by such a mechanism (Section 7.2).

We are speaking here of theories; the factors involved are unfortunately many and ill-defined from the quantitative point of view. The predictions are

---

18 *The study of marine sediments shows periodicities, in the temperature and the concentration of greenhouse gases have existed since about 3.5 million years, when there was the closure of the Panama's isthmus that pushed the Gulf Stream toward north Atlantic.*

19 *The stability or instability of a system depends on the time constants of the phenomena in play, instead of leading the system to equilibrium, they make it oscillate around that value. A typical example in biology is the ratio between prey and predators. A large number of prey makes the predators increase, which in a short time reduces greatly the number of prey. At this point the predators remain without food and their number is reduced, allowing the prey to increase again; when the prey are sufficiently increased, the predators start to increase again, and so on. The number of predators and of prey vary in time and oscillate, because the time necessary for preys or predators to increase is longer than those in which the predators succeed in destroying the prey. If these times were the same, the predators and prey would stabilize around an equilibrium value, with no oscillation.*

uncertain, but even in their uncertainty the reasons for worrying are strong. This explains the warning from the scientific community about the necessity to reduce the emission of these gases.

A first step toward limiting greenhouse gases is the Kyoto Protocol, which, while not facing the problem directly, represents the first step toward recognizing it on a global level. The solution will largely depend on the energy sources humankind will choose for the future; to that we devote the next section.

7.9 The Problem of Energy

In the previous section we have seen that there are serious reasons to be concerned about the greenhouse gases that humans emit into the atmosphere. One number suffices to justify the concern: industrial activity produces 30 billion tons of CO_2 per year, 100 times the volcanoes emissions, that compared with 1800 billion tons of atmosphere contents shows that in 70 years we emit an amount of carbon dioxide equal to the atmospheric content.

Moreover, consider the scenario where the presently rich countries becomes the model for the undeveloped ones: if at the end of the century they will develop with a per capita emission like present Europe, the years to emit the same amount of carbon dioxide contained in the atmosphere will become 16; if the model is USA the number of years become 8 (3 for Qatar). The actual situation is even more dramatic because, to the CO_2 emission also the one of other greenhouse gases have to be added; mainly: methane CH_4 (which has 21 times the greenhouse effect of CO_2), nitrous oxide N_2O (produced by combustion processes, by motors, or converted through using nitrated manure, with 270 times the greenhouse effect of CO_2) and sulfur hexafluoride SF_6 (produced by certain industrial activities, with 24 times the greenhouse effect of CO_2). To these gases we must add water vapor, a very effective greenhouse gas whose concentration increases with temperature.

We do not need an accurate evaluation to state that this is not acceptable and that strong changes in our society are necessary. Luckily more than half of the anthropogenic CO_2 emissions are absorbed by the soil and the seas. There are processes which fix the CO_2 in rocks with chemical reactions whose efficiency increases with temperature (Section 9.6); plants in the presence of an increase in the atmospheric CO_2 accelerate their growth and absorb more CO_2 than in the past (that will by large be fixed in the soil); finally nearly 1/3 of the human emission of CO_2 is absorbed into oceans on reaction with water, forming carbonic acid ($H_2O + CO_2 \rightarrow H_2CO_3$); a process positive for reducing the global warming, but not for the marine ecosystems; the increased acidity of water corrodes the shell of mollusks and of the coral barriers, endangering marine wildlife, as happened in the past after great volcanic eruptions.

To these "natural techniques" there are other "artificial" that try to reduce the CO_2 emissions. One, under study, is to augment the growth of algae by fertilizing oceans with iron powder; they will fix the CO_2 to the ocean floor as plants do in the land.

Another technique is to pump the greenhouse gases produced by power stations,[20] where, with the elapsing of time, they can combine chemically with the rocks and will be fixed definitively in the soil. In various places of the world there are experimental plants of this type. Every day thy introduce underground about 10,000 tons of CO_2. Although the quantity is modest, the results are encouraging and projects for mass storage of CO_2 could soon begin. If these projects are successful, it could be possible to bury, by 2020, 10% of the CO_2 produced, and by 2050, 40% — an important contribution[21] that, however, will not be sufficient. The 60% not buried will, within 50 years, be much greater than the today emissions because of the worldwide increase in energy consumption, of which we will speak in the following.

All these processes are not sufficient to eliminate the excess of CO_2 as clearly shown in Figure 7.6. If we want to stop the heating of the Earth, the only solution is by reducing the greenhouse gases emissions.

A reduction that cannot be obtained reducing agriculture which to the contrary is destined to increase in coming years to satisfy the food demand of emerging countries (Section 7.11) even if new technologies may reduce the impact of the agriculture on the environment. It will not even be obtained by reducing the energy consumption, because this will cause the collapse of our *society, which is based on the availability of energy.*

For richer countries, in fact, to reduce energy consumption means to reduce their standard of living. For poorer countries (80% of the world's population), energy is a necessity and cannot be reduced. Instead, it is predicted (and desirable) that they will increase the energy consumption to pursue better conditions of life. These countries have a large population that is growing very fast, and that constitute the largest growth in the production rate of greenhouse gases. Moreover, being poor, they could not invest much in emissions control. The situation is dramatically summarized in Figures 7.10 and 7.11.

On the vertical axis of Figure 7.10, the rate of electrical energy[22] consumption per inhabitant, is reported on an exponential scale (every interval

---

20 *Another possibility might be to put CO_2 into the bottom of oceans, where it is estimated that there could be 37 trillion tons — a solution with some risk, as the processes by which CO_2 may come back to the surface are not well known (Section 9.6).*

21 *The solution is not simple. It requires the separation of the greenhouse gases from the products of combustion and their transfer via dedicated "pipelines" into sites where there can be geologically stored (generally, exhausted natural gas or oil), sites where there are porous rocks surrounded by impermeable rocks that prevent the gas from returning to the surface.*

22 *The same plot could have been done using the total energy instead electrical, but the spread among countries would be lower because the consumption of electricity is not only proportional to the wealth of a country, but also to the development they have reached.*

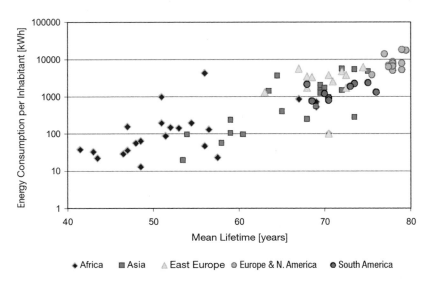

Figure 7.10: *Electrical energy consumption and life expectancy. The curve shows how the life expectancy of a population is correlated with the energy consumption. Note that the ordinate scale (vertical) is exponential: every interval corresponds to an increase by a factor of 10 in the consumption.*

corresponds to a factor of 10), for various countries in the world; while on the horizontal axis the life expectancies of those countries is labeled. The different colors represent different continents.

It is immediately obvious that the mean lifespan and the consumption of electrical energy are strongly correlated. The countries that consume the least energy are in the poorest regions of Africa and Asia, where the populations live to an average of 40 years, while those of rich countries in Europe, America and Japan consume 1000 times more electrical energy and live to an average of 80 years. The explanation is simple: consuming more energy means to have better hospitals, better schools, better food, and a more comfortable and longer life, while consuming less energy means the deprivation of all these things and therefore a poorer and shorter life. The figure shows that *it is not possible to reduce the energy consumption of a population without reducing the quality and expectation of life.*

The graph on the left of Figure 7.11 is taken from the website of the UN and shows in blue the estimates of the growth in population of the richer counties (North America, Europe, Japan, Australia and New Zealand) and in yellow all the other countries of the Earth that are considered poorer. The greatest increase in population will take place in the nations that are today poor, those that consume little energy and have a low life expectancy. Since, as we have said, it is desirable that the inhabitants of these countries reach the same level of welfare as the inhabitants of richer countries, consuming and polluting as much as they do, we must conclude that the consumption of energy is destined to increase dramatically in the coming years (unless there is an event, certainly undesirable, that reduces the world's population).

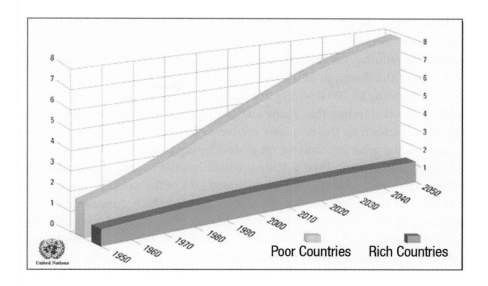

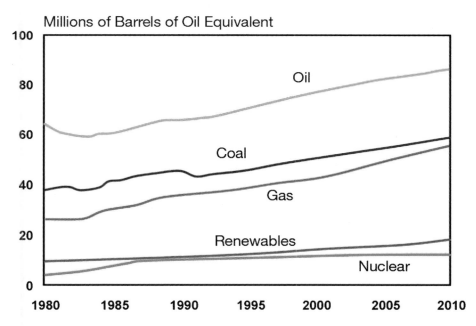

Figure 7.11: Top: World population growth predictions. In blue are the richer countries; in yellow the poorer ones, those that have the lowest energy consumption in Figure 6.10 (Source: ONU). Bottom: The main sources of energy used in the world. Only renewable and nuclear energy do not produce greenhouse gases.

The difficulty that the Kyoto Protocol encountered in its application confirms this analysis, even if it foresees a reduction in emissions of only 5% for developed countries, and no reduction for the poor countries, those destined to increase the most (besides, it was not signed by the United States, India and China, which represent 50% of worldwide emissions).

Moreover, if the Kyoto Protocol were applied, it would only make a ridiculously small contribution. To have a significant effect on the environment a worldwide reduction of more than 80% is required, which is unthinkable today.

A sign of the direction in which the world is going is given by the rise in oil prices beginning in 2004, which stopped in 2008 only because of a worldwide crisis. Notwithstanding this, many analysts predict the prices to begin increasing again as soon as the economy recovers. The principal cause of the price increase is the great demand for oil in developing countries, which grew enormously in these years. First is China, which has the largest increase, and it is followed by India. In 2005, China consumed 10% of the world's energy, putting it in second place, just behind the United States (which consumed 24%). In spite of this, the consumption per capita of energy was only 25 kWh/day, compared to 110 for Italy and 275 for the United States. With these numbers it is easy to predict that the energy demand will increase notably in the future[23] and with that the use of other forms of energy, most prominently coal, the cheapest, but the most pollutive among the fossil fuels.[24]

To understand how the increase in greenhouse gases from these countries, just consider (Appendix A.2) that in 2009 the CO_2 emissions in the United States was 17 tons per inhabitant, while in China it was 5.1, in Brazil 1.7, in Indonesia 1.7 and in India 1.4. It is not hard to predict that these countries, all with rapid development, will increase their emissions by a factor between 3 and 8. Keeping in mind that they represent a little less than half the world's population (China, India, Indonesia and Brazil have a total of 3 billion inhabitants) we realize what kind of future we are constructing.

If we consider only emissions from vehicles (among the main sources of greenhouse gases), in China 14, 000 vehicles were sold per day in 2004, 15,600 in 2005, and 19,800 in 2006, while in 2009 the sales figure was about 35,000 per day (of which one-third were produced locally). In 2009, with about 13 million cars sold in 12 months, China reached the first place in the world's automobile market, surpassing the United States. The American agency Keystone predicts that by 2020 China will have 50 million cars in circulation, followed by India, the second-largest world market, with 30 million cars in circulation (even so, the cars in circulation in India and China will be well below the European and American standards). Keystone foresees that in the

---

23 To have an idea of the fast development that those countries will undergo, it is sufficient to think that in 2004 the GDP (gross domestic product) of China surpassed that of Italy, in 2007 that of Germany's, Japan's by 2010 and is foreseen to surpass that of USA's by 2020, thus becoming the top economic power. For India, the overtaking, of Germany is foreseen in 2020, of Japan in 2030, while it is uncertain if the USA will be surpassed.

24 China, burned in 2006, 2.4 billion tons of coal, with which it produced 70% of its electricity. Every ten days there is in China the inauguration of a new power plant. The aerosols produced by China and India have (it is also a great coal user), changed the region's climate by increasing the Pacific ocean cloudiness by 40% in 10 years.

first quarter of this century there will be more cars sold than in the rest of human history. The market will be driven by the increasing contribution from the emerging countries.[25] The cars have a great effect on greenhouse gases: each car emits 0.18 kg of CO_2 per liter of fuel, an average of 1.8 tons of carbon dioxide for every 10,000 km traveled.[26]

In conclusion, Figures 7.10 and 7.11 show that, if there is no drastic reduction in the world's population, there will be no alternative to the increased consumption of energy in the coming years. Therefore the only possibility of reducing greenhouse gases is to turn in large proportions to "alternative sources," those that do not produce the gases. In other words, the problems of the environment cannot be solved without adopting solutions to the production of energy different from the ones used today.

7.10   A Difficult Choice

The considerations of the previous section show that the solution to the problems of the environment must come mainly from the sources of energy we will adopt in the future. Today the main source of energy are fossil fuels, with which 81% of the energy consumed in the world is produced. The decision to reduce their use have to be implemented by choice, in order to reduce their impact on the environment; in fact we cannot much rely on exhausting the reserves, because the stocks of fossil fuel are still enormous, possibly enough to satisfy the demand in the coming centuries even if the years of low-priced oil are probably at an end.[27]

For oil, there are few regions left for exploration, the largest being the virgin territories of the Arctic and Antarctica, which are protected from exploitation by an international treaty. For the Arctic, Russian geologists estimate the reserves to be a little less than what exists today in the Middle East, while for the Antarctic there is no estimate but a similar figure could be possible. To these reserves of oil that are sufficient for a few decades, we must add "costly" supplies; the growing oil prices of these years are making them competitive. They are the bitumen deposits, also known as "tar sands" or "oil sands," found in various fields around the world, whose reserves are estimated to be twice those of liquid oil. The largest deposits are found in Canada and Venezuela. For Canada the deposits of tar sands are estimated at 175 billion barrels of oil (equal to two-thirds of the reserves of Saudi Arabia). This alone could allow

---

25 *These predictions are taken very seriously by automobile companies which are constructing factories in developing countries to avoid import taxes. Volkswagen, Fiat and Ford, have factories in these countries since many years, and BMW is building a factory in Madras, India, for luxury autos.*
26 *Another large contribution to greenhouse gases is made by maritime traffic, which is estimated to produce 4% of the worldwide emission of CO_2, twice the contribution of airplanes. The amount is destined to increase because of a predicted doubling of maritime transport in the next 30 years.*
27 *David Goodstein,* Out of Gas *(Norton, 2004).*

the country to become the second-largest producer of oil in the world. (The extraction of tar sands will allow Canada to become the largest supplier of oil and refined products to the United States.) The cost of extracting oil from tar sands is about US$40 per barrel, about 10 times that of pumping oil out of wells. No one in the past has exploited those fields, but when the prices of oil rises above US$70 per barrel, extraction becomes reasonable. In the last years new extractive technologies those of shale oil and shale gas are also greatly increasing (probably doubling) the existing reserves.

In any case if the oil will finish, as some predict, or it will become too expensive, coal will be a source of energy, terribly pollutive but cheap and plentiful. Its reserves are immense, enough to satisfy the planet's hunger for energy for centuries, and unlike petroleum the mines are distributed quite uniformly around the world. Moreover carbon can be transformed into oil through the *Fischer Tropic*[28] process used by Germany during the World War II. The cost is again of about US$50 per barrel competitive with the today's cost of oil.

For these reasons, even though coal damages the environment, its use continues to increase: from 1970 to 2009 the use of coal went from 25% to 27% in the primary energy supply (while oil decreased from 46% to 33%) and from 38% to 41% in the production of electricity (while oil descended from 25% to 5%) — an increase that will accelerate in the future, if we do not change our model of society. An example is China, which has in recent years set up hundreds of coal-fired power plants; or Indonesia, which in 2006, having to increase the production of electricity by 40%, decided to build 35 new carbon power plants.

The only solution for reducing greenhouse gases is therefore to decide not to use fossil fuels and instead use sources of energy that do not produce them, i.e. renewable (hydroelectric, solar, wind, geothermal, wave and tides) and nuclear energy. Among these sources of energy that are not producing greenhouse gases, *only nuclear is today capable of substituting fossil fuels*. In the future, with the progress of technology, *solar energy may also have this capacity*. Unfortunately, both of them produce only electricity.

The other sources of energy can make a contribution, but the amount of power they can provide is largely insufficient to substitute for fossil fuel, taking into account the demand for energy which will double in the next 30 years. In fact:

Hydroelectric energy is the only one among the renewable energies that today makes a substantial contribution to worldwide consumption of electrical energy (about 16.2%), accounting for about 2.3% of the total world consumption (recall that electricity is about 1/7 of the total energy consumed in the world). It is,

---

28 *The process consists in combining carbon with hydrogen to form hydrocarbons, the basic molecules of oil. Hydrogen is taken from water though electrolysis, an expensive process that makes this oil expensive.*

however, close to saturation, because most of the exploitable falls have been used. In the period 1970–2009 the contribution of hydroelectric energy went down from 21% to 16% of the total electric energy produced (although in the same period the production of hydroelectric energy has increased by a factor of 2.4), and this percentage will continue to decrease in the future.

Geothermal energy consists in the use of hot water (sometimes introduced artificially underground) to produce steam for electricity, heating systems, greenhouses and certain industrial and thermal activities. Among the renewable energies, geothermal energy is, along with hydroelectric, the most promising. In theory, it could be available at every point on the planet; in the volcanic regions the eat is close to the surface, in the other regions the temperature underground increases by 30°C for every kilometer of depth. Therefore it would be sufficient to have a depth of 5–10 km to reach the temperature necessary to produce energy. In fact, there are big technical problems to be solved,[29] and today's technology permits us to exploit only volcanic zones where heat is near the surface concerning only the 2% of the Earth (Italy and Japan are among the most favored nations). In 2005, geothermal energy enabled the saving of 18 million tons of oil and about a million tons of CO_2. This is a significant amount but negligible compared to the 30 billion tons produced in the world. For the coming decades there are no predictable developments that could permit geothermal energy to substitute for fossil fuels.

But even if all the technical problems will be solved, geothermal energy will not have the power to satisfy the world's energy demand. The energy consumed in the world in 2009 was $1.4 \ 10^5$ TWh (81% of which produced by fossil fuels), while that produced inside the Earth by radioactive materials (Section 9.9 and Appendix 2) is $3.8 \ 10^5$ TWh; this means that the energy used to power volcanoes, earthquakes and continental drift is only three times what is used by humans and, before the end of this century, the two numbers will probably be equal. Since it is unthinkable to use a large fraction of this energy (condemning life to disappear from the Earth, as we will see in Chapter 9), it is easy to see why this contribution will remain marginal.

Wind energy is another promising renewable source of energy. In 2010, it satisfied the 2.5% of the worldwide electrical demand (it was 0.54% in 2004) that is 0.08% of the total energy demand). The most optimistic estimates predict that in 2050 it can arrive at 15% of the electrical demand, a contribution similar to that of hydroelectric, amounting to 2% of the world demand for energy. It does not seem possible to far exceed this figure: if Italy wants to use wind to satisfy its demand for electricity, it would need 80,000 generators 25–50 m high and fed with a constant supply of wind. With the levels of wind present in the territory the number of generators should increase by

---

29 *Water often contains corrosive materials which, with the aid of high temperature, damage the apparatus.*

about a factor of 10, and that is just for electricity. To the difficulty of placing such a large number of generators must be added the problems posed by the environmental groups, who are opposing their use because of the noise they produce and of the difficulty of inserting them in the panorama.[30] In conclusion, it is estimated that renewable energy (aside from solar energy) could in the future contribute to 30%–40% of the electrical energy production. Today they supply the 19.5% of the world electrical production, mainly through the hydroelectric which contributes about 16.2%, while all the other renewables contribute to the 3.3% (mainly of the geothermal and wind power. Even if we adopt the most optimistic figures, of the 40%, of electricity produced with the renewables the question remains: Where does the other 60% come from? If we continue to burn fossil fuels, the emission of greenhouse gases will be greater than they are today, because the global demand for energy will double in the next 30 years and will continue to increase in the future.

If we want to eliminate fossil fuels, there are no other solution than solar and nuclear (fission today, fusion in the future), which will become the principal sources of energy in the future when, within a few centuries, the reserves of fossil fuels will be exhausted. Today, however, the use of solar energy to produce electricity is very costly.

It is hoped that the production of electrical energy from solar energy will become competitive soon, together with it nuclear fusion, (the process of Figure 4.1 by which the Sun produces its energy). Unfortunately the most optimistic estimate 10–20 years to have a production of electricity from the sun competitive with that of coal and 50 years for the first experimental fusion power plant.

Nuclear energy from fission the only possibility we have today of substituting fossil fuels in the production of electricity. The most serious problem with nuclear energy is that they can produce materials that can be used to construct nuclear weapons — an example is the concern we have today with the nuclear plants in Iran.

To this problem, we have to add the public's reactions to the presence of nuclear reactors in their territories, and the problem of storage of waste products. There are many people, especially scientists, who maintain that the storage of waste is less dramatic than it seems. The waste products of fossil fuels are a problem because they are diffused into the environment while the waste products of nuclear reactors are concentrated as solid waste, which is certainly dangerous but, contrary to those of fossil fuels, they can be stored and controlled.

The problem of nuclear waste would exist even without the nuclear plants; they are produced in medicine and industry, and they are more dangerous than the waste from nuclear reactors because they are spread out in the

---

30 A solution could be to make fields of generators on the sea far from the coast, or up in the sky anchored to balloons — promising technologies in part to be developed.

territory and, therefore, less controllable.[31] The real risk of employing nuclear energy, as we have said, is in the production of materials that can be used to construct atomic bombs, and in its spread to "untrustworthy" nations like Iran or North Korea. This is the risk that must be balanced against the increase in the temperature of the Earth.

It is, however, difficult (if not impossible) to think of a worldwide reduction in fossil fuels without thinking of using nuclear energy on large scales. This will be the necessary choice in the years to come. We must therefore decide if it will be less dangerous to continue to use fossil fuels, or to reduce their use by employing nuclear energy wherever possible.

The case of Italy shows that the choice is between these two options. For this country, the exit from nuclear power has meant a continuous increase in the use of fossil fuels, with the exception of the electricity imported from abroad, which is largely produced through nuclear means — all this in spite of the billions of Euros invested in renewable energy.

For Italy, the exit from nuclear energy has also meant the failure to keep the promise of Kyoto; signing this treaty, in December 1997 Italy was engaged to reduce the emission of CO_2 by 6.5% before 2012. Instead of reducing the emissions, Italy has continued to increase it, to the point that in 2005 to keep its promise, Italy would have had to reduce the emission by 13% (and 17% in 2009). For Italy, as for most other industrialized nations, honoring the promise of Kyoto seems difficult.

Large-scale adoptions of nuclear energy can help in maintaining the Kyoto promise, but it would not solve the environmental problem, unless other technologies are adopted in parallel. Assuming, for instance, that all of the electrical energy in the world will be produced using techniques that do not produce greenhouse gases, the environmental problem will remain unsolved because electricity is only 1/7 of worldwide energy consumption. As an example, let us compare France and Italy, two countries that have adopted opposite environmental policies for the production of electricity. France has decided to produce all electricity that cannot be obtained from renewables, (mainly hydroelectric) from nuclear energy; it has 50 nuclear plants producing electricity (which is also used by Italy), with no greenhouse gases emissions. Italy, after Chernobyl, adopted the opposite solution — all the electricity that cannot be obtained hydroelectrically is obtained from thermal plants, with emission of greenhouse gases. In 2009 the production of CO_2 in France was 6.2 tons per inhabitant; in Italy, in the same period it was 7.6 tons per inhabitant, which becomes 8.2 if

---

31 The ideal sites for the storage of dangerous materials (nuclear and chemical) are places where geologists can guarantee that for millions years water will not seep through. Water might disperse this material in the territory polluting it for a long time. An example of safe sites are the salt mines at great depths underground. The presence of salt (water-soluble), guarantee that in those mines there has been no water flowing through for millions of years, and geologists can calculate when it may happen for the future.

we take into account the electricity imported from abroad (the number would be even higher if Italy had not made ample use of methane, the fossil fuel that emits the lowest amount of CO_2).

Differing energy choices mean that France emits about 2 tons of CO_2 per inhabitant less than Italy. It is an important contribution, but only ¼ of the total emission of greenhouse gases. The reason is that many industrial activities, heating systems, cars and airplanes can only use fossil fuels. Substituting with other sources of energy is not be easy and will require technologies like the use of hydrogen, which are not sufficiently developed and will not be available in another decade or two.[32] Moreover we should not commit the error by thinking that hydrogen is a source of energy, because this gas does not exist in nature and must be produced through the consumption of other energies. Hydrogen will therefore be an advantage to the environment only if it is produced without greenhouse gases.

There is not much choice: the worldwide average production of CO_2 is, in 2009, 4.3 tons per inhabitant per year, that of OECD countries is 9.8 tons while the United States produces 17 tons per inhabitant per year. To live in a sustainable world, in agreement with our environment, requires reducing the worldwide average to less than 1 ton per inhabitant per year — unthinkable with the present model of society.

7.11 What Future Will We Have?

In early 2006, the noted English scientist James Lovelock published, at the age of 87, a book with the aggressive title *The Revenge of Gaia*, in which he argues that by now it is too late to remedy the damage done by man. Global warming is by now unstoppable and he predicts that by 2100 it will cause billions of deaths. Lovelock proposed in the 1970s the theory that the Earth is a system (Gaia) that behaves like a living thing, and it regulates itself to produce the best conditions for the species that inhabit the planet. "The revenge of Gaia" is caused by the impossibility of repairing the damages created by humans, hence the necessity of eliminating the species responsible for them.

Not everyone agrees with this extreme vision. Many scientists think that there is still time to intervene, like the Nobel laureate Paul Crutzen (Section 7.6), holding that in any case it is worth trying. In years to come the effects of global warming will be so evident that public opinion will push governments to do the right choice. We will also understand better the cause of thermal cycles and will be able to study well-targeted interventions.

---

32 *For example, hydrogen could be a good substitute of oil for transportation. Today we store hydrogen in heavy and dangerous high-pressure tanks that cannot be used in airplanes. If we use liquid hydrogen, it requires a temperature of −250°C and to store the energy content of 1 liter of gasoline, a refrigerated tank of 4 liters is needed.*

Our recent past shows a positive example — that of ozone, the first and up to now only example in which human intervention on a global scale has solved an environmental problem. The ozone is depleted by certain gases (the CFCs in Figure 7.6), of which the best known is Freon, used in various applications such as refrigerators and air conditioners. Their emission into the atmosphere caused the famous hole in the ozone layer, through which radiation dangerous to health penetrated. The hole grew from year to year, causing increasing worries. The problem was posed by scientists in the 1970s, and in 1987 the Montreal Protocol was signed, requiring the signatories (practically all of the industrialized countries) to reduce the use of CFCs gradually. After 13 years, the total elimination of this emission was realized, and in 2000, the hole stopped growing. This result shows that, if there is the political will, intervention is possible. Ozone was an easy case, however; there were other gases that were not damaging the environment and not costly to be used in its place. For the greenhouse gases the scenario is different: alternative energies are insufficient and nuclear energy entails risks that many do not want to take; moreover it produces only electricity and is insufficient to stop the increasing in the use of fossil fuels. The ability of Gaia to recover from the damage humans have caused, could leave enough time to intervene. It is in any case necessary that good decisions be adopted and in a hurry.

This chapter concludes with a strong signal of alarm for the future of our species — an opportunistic species (like all the other living things) capable of dominating and modifying the environment in which it lives, more than any other living species has ever done in the past. Humans have their future in their hands, a possibility that was not given to other species that have become extinct in the past. Will this be the reason for our survival or the cause of our extinction? The answer to this question should be obtained in this century and, in the search for it, we should not forget what has been said before in this book: *Homo sapiens* has existed on the Earth for "only" 0.15 million years, a very short time compared to the 3500 million years for bacteria and the 135 million years for dinosaurs. If humans disappear, this will only be a small footnote in the biological history of the planet, which might not be noticeable in a graph like that in Figure 7.3, which might one day be made by other "intelligent" species.

In human history there are many examples of cultures that have become extinct. They reasoned as we did in the past; they thought that they were eternal, at the center of the world; they exploited the environment foolishly, and in a short time, they disappeared. A book by Jared Diamond (author of *Guns, Germs and Steel*, which won the Pulitzer Prize in 1997) titled *Collapse* tells a few stories that happened in ancient and modern times. In these stories exploitation of territories has brought great civilizations to the destruction of resources, leading to their decline and disappearance. Emblematic is the story of Easter Island, where the inhabitants foolishly cut down all the trees, eliminating the primary material from which they constructed boats, condemning themselves, a Polynesian people who took a great part of their resources from the sea, to isolation and decline (arriving, for the shortage of food, to cannibalism). Other stories tell of peoples, with fewer resources than those that collapsed, living for millennia in hostile places, like the Arctic, applying environmental policies and very careful controls of the population grow.

In the places where the environment was destroyed, life was reborn through the great capacity of nature to recore. Recovery that occurred after the elimination of those that had caused the destructions, which in the case of the whole planet could be the entire human species, just as Lovelock predicted....

There are, however, reasons for optimism. Humans, as we have seen, are the result of an extraordinary cultural evolution (Section 5.7) that has given them a unique rationality in the animal kingdom. We have the capacity to analyze problems, predict the future and suggest for solutions. It has yet to be seen how much we will use this capacity to save Gaia and avoid our own extinction, or whether we will instead be dominated by our animal nature, being opportunistic, egotistical and dominated by the greed to plunder, which will condemn us to suicide. The problem, as we have already said, is the exponential increase in population which nature has always resolved with extinctions.

Unfortunately, the task of predicting and of preventing the future are carried out by different people. To predict what will happen depends on scientists, following the rules of logic. To prevent catastrophes with the right actions is in the hands of politicians and depends on the consensus of the electors, acting on a brief timescale (no politician will defend an action that in 30 years will save the world and in one month will let him lose elections). It is therefore important that the largest number of people become conscious of these problems in order to influence politicians, as it should happens in democracy.

If that will happen, in a century from now we will say that we had a happy time in which the rationality of humans were allied with Gaia and Gaia has "renounced" its "revenge." To arrive at this point we must be convinced that humans have the future in their hands, and if there will be another extinction it will not be caused by a volcano or an asteroid or a particularly intense glaciation, but by the choice of a species that was thinking of being intelligent: it will be the failure of our cultural evolution.

An Inhabitable Planet

8.1 The Habitable Zone in the Galaxy

Imagine yourself being on a spaceship coming from a civilization far away that has to explore our galaxy in search of habitable planets — a story we could place anywhere during the last 2 billion years, when the atmosphere of the Earth had gained enough oxygen to support species like ours,[1] or in the next 5 billion years, after which the Earth will be swallowed up by the Sun, which will become a red giant. Our story then could occur at any time of the 7 billion years during which our planet can be considered habitable. It is a period that cannot be any longer, because the Sun's nuclear fuel will, by then, have exhausted, but it could be shortened by a cosmic or environmental catastrophy among those we have spoken of.

Just as in every story we can use a little fantasy, so we permit our spaceship to overcome the limit of the speed of light and jump from one point in the universe to another in a few seconds. It is a condition necessary for such a voyage, without it hundreds of billions of years would be necessary to complete it (Section 11.5).

Before the last jump which will bring the spaceship close to the planetary system to be explored, the captain studies our galaxy with the powerful telescope on board, to decide where to begin the exploration. The image he sees is not very different from Figure 8.1: a galaxy nearly 90,000 light-years in diameter, at the

---

1 Oxygen, as we have seen in Chapter 6, did not exist when the Earth was formed. The first organisms that inhabited the Earth produced it.

Figure 8.1: NGC 2997, a spiral galaxy similar to ours. Our galaxy has a diameter of approximately 90,000 light years; the Sun is approximately 28,000 light years from the center, where the arrow points, rotating around the center of the galaxy with a speed of about 220 km/s (Section 2.9). Viewed from the position of the Sun, our galaxy appears as a luminous stripe crossing the sky — called the Milky Way — since ancient times (Figure 8.2). © ESO PR Photo, March 1999.

center of which there is a luminous nucleus from which great arms are departing, forming a spiral. The measurements of the spectrometers on board will permit the captain to evaluate (Doppler effect, Appendix A.3) the velocities at which the stars are moving and show that the whole galaxy is rotating around its center like planets do around a star. Examining the image more closely, he sees that the luminosity has its maximum at the center, and then it diminishes toward the outside.

Because the luminosity is given by stars, the image shows that stars are numerous in the center and that their density decreases outward irregularly, with higher density along the spiral arms.

It is a fascinating image that our captain, once he enters the galaxy, will no longer see. When the spaceship has entered into the galaxy, all stars of the disks will appear on a single luminous stripe, the one the that crosses our sky, which we call the Milky Way (Figure 8.2).

Using the instruments on board, working in the infrared and the radio wavelengths, sensitive to cold objects (Appendix A.2), the captain measures the distribution of dust and gases in the galaxy, tracing the great condensations where the stars will be born (the clouds in Figures 3.1 and 3.4). He thus has at his disposal two pieces of information that complement each other: the luminosity of the galaxy in the visible, which shows where the stars are, and the luminosity in the infrared and the radio, which shows where future stars will be born.

Figure 8.2: *The Milky Way. The image, composed from a number of photographs, shows the disk of our galaxy as it appears from the Earth. The visible black regions on the disk correspond to zones of powder and dust near the Earth. They appear to be dark because they block out the stars of the disk behind them and indicate the giant molecular clouds where the stars are born (Chapter 3). Courtesy of Lund Observatory, Sweden.*

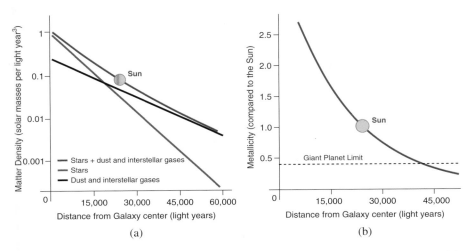

Figure 8.3: *(a) Variation of the density of stars, dust and gas as a function of the distance from the center of the galaxy. Moving away from the center there are fewer stars and less material to form them. (b) Metallicity (the abundance of elements heavier than helium relative to that of the sun) as a function of the distance from the center of the galaxy. The graph shows that the metallicity decreases 10 times going from the center toward the edge of the galaxy. The limit below which no giant planets are found is also shown in the figure.*

The results of the measurement, elaborated by the computers on board, are summarized in Figure 8.3; the captain expects this result because he knows that, qualitatively, they are true for all the galaxies of the universe. In the figure, with two curves of different colors, are shown: the mean density of stars (taken from Figure 8.1) and that of cold matter (taken from measurements in the infrared and radio on the dark zones in Figure 8.2). Along the vertical axis, the values of the mean density are shown, and on the horizontal axis, the distance from the center of the galaxy. For example, Figure 8.3(a) tells us that at 28,000 light-years from the center of the galaxy, where the Sun is, in a volume of one cubed light year there is on the average 0.03 solar masses in the form of stars and 0.06 solar masses in the form of dust and gas. The graph shows that, moving away from the center of the galaxy, the number of stars and the quantity of dust and gas are both diminishing. This is an expected result: the graph says in fact that in the central luminous zone more stars are born because there is more dust and gas.

At the center of the galaxy there is thus a larger density of matter, and where the matter is denser, massive stars are born more easily — those which explode as supernovae producing heavy atoms (Section 4.5). We therefore expect a larger density of heavy elements at the center of the galaxy than in the external zones. The result is confirmed by the measurement of the chemical composition of the matter (obtained by spectrometers on board, Section 4.9) summarized in the graph Figure 8.3(b). On the horizontal axis, once again, the distance from the center of the galaxy is shown and, on the vertical axis, the "*metallicity*," that is the *abundance* of elements heavier than helium relative to their *abundances* in the Sun: metallicity 1 means to have the same abundance of the Sun; 2 the double; 0.5 the half.

In conclusion, the two graphs show to our intrepid explorer that the abundances of heavy elements diminishes as we go from the center of the galaxy outward, as do the stars, the dust and cold gas. With these data in hand the captain plans where to begin his search for a habitable planet, because he knows very well that evolved forms of life cannot be developed anywhere in the galaxy.

The central zone of the galaxy

The density of stars at the center of the galaxy renders the zone extremely dangerous for the ship, and the captain knows that the probability for life to survive in that zone is practically zero, for two reasons. The first is that the high density makes probable the coming of stars into close contact with another, with catastrophic effects on planetary systems that could exist. The reciprocal attraction of gravity can displace the planets from their orbits, making them to fall into a star, or sending them into the external regions of the planetary system where life cannot survive. Moreover, if two stars pass by each other, they would displace from their orbits asteroids and comets contained in the planetary system (in the Solar System they are in the Kuiper Belt and the Oort Cloud, Section 8.2). Therefore during the lifetime of a planetary system there will be a high probability for a large number of comets or asteroids of large diameter to strike planets, with disastrous effect for any form of life that could exist on them (Section 10.3). In Chapter 6, we have seen the damage caused by the impact of an asteroid 10 km in diameter, which caused the extinction of the dinosaurs. The dimensions of asteroids in a planetary system can reach hundreds of kilometers, (in the Solar System they are on stable and relatively safe orbits) enough to sterilize a planet like the Earth (Figure 10.5).

The second reason for the difficulty of biological organisms to survive in the central region of a galaxy is that, as we have already said, very massive stars are formed there, much more frequently than in the external zones. These stars end their lives as supernovae (Section 4.5) which explode, emitting streams of high energy particles, X-rays and gamma rays that destroy every form of life within several light-years. Some of these stars end their lives as black holes (Section 4.5). The largest of them is found right at the center of our galaxy, with a mass of about 3 million suns. In its surroundings tens of black holes were found, although with more modest dimensions, and presumably others will be discovered in the future. These objects are most dangerous for the poor planets that may be found in their vicinity. Every time they digest matter they emit beams of X-rays and gamma rays that can be much more intense than those produced by the explosion of supernovae. In other galaxies, fortunately not in ours, these phenomena have an intensity that is unmatched in the universe. Some of these beams are highly collimated (like the beam of a lighthouse), becoming true "death rays" that reach great distances, sterilizing everything they encounter. These are phenomena yet to be fully understood, but their existence is sufficient to render "uninhabitable" a large region around the center of the galaxy.

The intermediate zone

Even the planetary systems that could exist in the intermediate zone of the galaxy, i.e. the zone containing the Sun, are not completely free from the catastrophic phenomena we have described. Being far from the center of the galaxy the probability of destructive phenomena is reduced, but not eliminated entirely. For example, in the vicinity of the Earth there are giant stars that will explode as supernovae. How dangerous are they? It is a point that the scientists will further clarify in the coming years. Also, the risk of an impact by a comet or asteroid is small for the Earth, but not completely zero (Section 10.3). We expect that once every 400 million years a star will arrive near the Oort cloud, increasing the risk for new comets to be displaced from their orbits and diverted to the interior of the solar system.

The external zone

Problems with finding habitable regions could also exist in the external zone of the galaxy. As the instruments on board have shown (Figure 8.3a), in that region there is a low density of dust and gas, so that few stars are born there. Thus, there are fewer heavy elements than in the interior regions of the galaxy. With fewer heavy elements, our captain thinks, it could be difficult to form planets heavy enough to be habitable (although on board there is no unanimity on this point). The terrestrial astronomers would agree with the captain, at least for the giant planets that have been found around stars; with the techniques described in Section 3.9, they are only found around stars rich in heavy elements. In Figure 8.3(b), the horizontal dotted line represents the limit under which no giant planets have been found. This result is also confirmed by the modern theory that, to form a giant planet, a rocky core larger than the Earth is required, which is difficult to construct with few heavy elements.

Does this mean that there cannot be smaller planets? Are there stars without planets? Probably not: to form a star a disk is needed; these disks in the end form planets that can exist even without giant planets.

So it is reasonable to assume that there exist planetary systems even in the external parts of the galaxy, around stars with few heavy elements. Containing only relatively small planets, they would be different from the Solar system, in which there could be life, even if, as we will see in Section 10.2, the absence of a large planet like Jupiter could make difficult the survival of life on a planet like the Earth.

There is another element our explorer must take into account to decide where to search for habitable planets. As we will see in the next chapter, life needs a planet that is geologically active with continental drift, volcanoes and earthquakes. The energy that drives this activity comes from radioactive materials that heat the interior of the Earth. These materials are produced in supernovae. If, moving outward in the galaxy, there are less supernovae, there will also be less radioactive material and, therefore, less geologically alive planets capable of supporting life.

To conclude, the measurements by the instruments on board confirm that, in our galaxy, like in all other galaxies, there is a central part where the distances among stars and the frequent explosion of supernovae render impossible the survival of any form of life and an external part where the existence of life is improbable. The search for life should be concentrated on the intermediate part of the galaxy, that wherein the Sun lies.

8.2   The Reservoir of the Comets

After all these measurements, the captain chooses the region to begin his exploration and, with a final jump, arrives close to a yellow star (which by chance is our Sun…) and begins to enter into the planetary system to be explored.

The first objects our explorer encounters are those of the Oort cloud, i.e. the frontier of the Solar System. The cloud is composed of billions of objects that

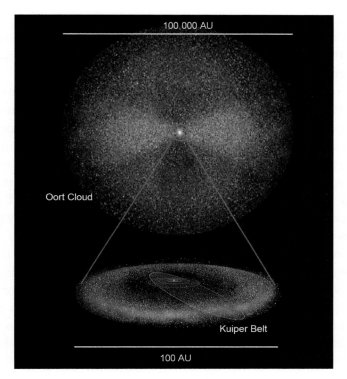

Figure 8.4: Top: The Oort cloud, whose diameter is approximately 100,000 AU, similar to that of the parental cloud (Figure 3.1). From this cloud comes the long periods of comets (greater than 200 years). Inside it is the Kuiper belt, which extends from the orbit of Neptune to about 100 AU, the remaining part of the disk from which the Solar System was built. From the Kuiper belt comes comets with shorter periods (less than 200 years), like Halley's comet, whose orbit is shown in the figure. At the center of the Kuiper belt there is the Solar System. (Adapted from NASA images, STScI-PR02.)

rotate around the Sun. The diameter of the orbits goes from 30,000 to 100,000 AU,[2] almost half the distance between the Sun and the nearest star (Proxima Centauri), filling up a great spherical cloud (Figure 8.4) that surrounds the solar system. The Sun, viewed from one of these objects, looks just like another star, distinguishable only with difficulty.

The diameters of the orbits of these objects are so large that the time needed to complete one orbit is very long: tens of thousands of years, and in extreme cases a million years.[3] The majority of objects in the cloud do not get into the inner part of the Solar System where the planets orbit. A few, however, because of perturbations due to stars that passed by close to the Sun, may move to elliptical orbits that take them close to the Sun. They are comets that have very long periods, so long that our civilization did not exist the last time they

2 Distances in the Solar System are measured in units of AU. 1 AU corresponds to 149.6 million kilometers and is the distance between the Sun and the Earth (Appendix B).
3 By Kepler's third law, the square of the period of one orbit is proportional to the cube of the orbit's diameter.

crossed the Solar System. We know about their existence and the orbit they will follow only when they are close enough to the Sun (and the Earth) to be seen. For this reason, they are potentially the most dangerous objects in the Solar System (Section 10.3).

No one on the Earth has seen the Oort cloud or the objects that constitute it; they are too far and too cold to be observed with the existing instrumentation. The existence of the cloud was, however, postulated by the Dutch astronomer Jean Oort, in 1950, to explain the existence of the long periods comets, and is today accepted by everyone. Oort noted that these comets, with periods from 200 years to a million years have elliptical orbits like those of planets. He concluded that they were objects orbiting around the Sun, and not stray objects that entered the Solar System. He also noted that when they presented themselves suddenly in the Solar System, they did not come from a particular direction in the sky and were not in the ecliptic plane in which all the planets are rotating. Comparing their orbits, he saw that they were coming from an immense spherical cloud surrounding the solar system (Figure 8.4). By the frequency of their appearances astronomers calculate that in the cloud there must be at least 10,000 comets of large dimension. It is hypothesized that these comets, at least the larger ones (Section 3.8), were formed in the interior of the Solar System and were then thrown into the Oort cloud by the giant planets. In those distant regions at the edge of the parental cloud (Section 3.1) the density of matter was in fact too low to permit the formation of objects this large (Section 3.8).

In Figure 8.5, we see the comet Hale–Bopp (named after the two astronomers who first saw it), which appeared in our sky in 1996. Its orbit, represented in Figure 8.5, is practically perpendicular to the ecliptic planes where the planets' orbits are located. Hale–Bopp has a diameter of about 40 km and a mass estimated at 40 billion tons, enough to cause the extinction of the majority of

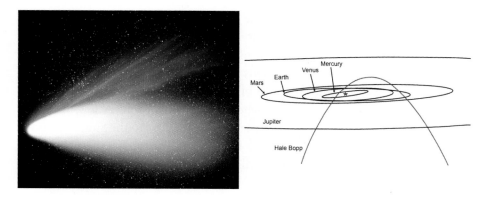

Figure 8.5: *On the left is the comet Hale–Bopp, and on the right is the orbit it follows, crossing the Solar System, almost perpendicular to the ecliptic plane (where the orbits of the planets are). (© H. Mikuz & B. Kambic, Crni Vrh Observatory.)*

the existing species if it were to impact the Earth. This comet is four times larger (which means at least 50 times more massive) than the asteroid (or comet) that caused the extinction of the dinosaurs (Section 6.9).

Another comet from the Oort cloud that crossed our sky in the same year was the Hyakutake. With a diameter of only 3 km, it was big enough to destroy our civilization. Fortunately the risk of impact by a comet of this type is very low (but it is not zero, as we will see in Chapter 10). Those are objects which we have spoken of in the previous section that could have been thrown into the Solar System by a star passing in its vicinity, into an orbit that may collide with the Earth (but no passage of a star near the Oort cloud is expected for millions of years, so there is no problem for our future…).

After passing through the Oort cloud, the captain descends into the ecliptic plane where the orbits of the planets are. Before beginning the exploration, he must cross another reservoir of comets, the Kuiper belt (the lower part of Figure 8.4), which contains the rest of the protoplanetary disk, from which our Solar System was born (Section 3.6). The Kuiper belt is much more populated than the Oort cloud. It is estimated that there are 35,000 objects with diameter over 100 km. The belt extends from the orbit of Pluto out to about 100 AU (100 times the distance from the Earth to the Sun). The objects in it are therefore much closer to us than those in the Oort cloud. There are HST images of a few of these. One of the best known is Chiton, which has a diameter of 170 km. From the Kuiper belt come all of the comets with a short period (less than 200 years), which, unlike those from the Oort cloud, are all in the ecliptic plane.

Given the nearness of these objects, their orbits can be perturbed by giant planets, most of all Jupiter and Saturn, which attract them into the Solar System. These comets are potentially less dangerous for the Earth than those from the Oort cloud, because traveling through their orbits in relatively short times, they can be followed for many years before a possible collision with the Earth. With the technology at our disposal today (Section 10.4) it is thought possible to displace them into less dangerous orbits. The most famous of such objects is Halley's comet (whose periodicity was discovered in the 17th century by Edmond Halley, the Newton's friend). It has a nucleus of 15×8 km and appears every 75 years. Its orbit is represented in Figure 8.4. Its last appearance was in 1986; the next will be in 2061.

8.3 The External Planets

Having departed from the icy world of the comets, our captain continues his voyage toward the interior of the Solar System, searching for habitable planets.

At the threshold of the Solar System our explorer encounters Pluto, which for years was considered a planet, and was the last one discovered by terrestrial

Figure 8.6: Pluto and its satellite Charon which was discovered in 1978 — it was so close to Pluto that the telescopes available then could not distinguish the two objects. (Photo by HST © NASA & R. Albrecht, ESA/ESO.)

astronomers.[4] In effect it is an object difficult to observe because of its great distance and small size. Pluto measures only 2320 km in diameter, the distance between Naples and London. It is also a double "planet", because its satellite Charon is only slightly smaller, 1270 km in diameter (Figure 8.6).

Because of its dimensions and of its rocky constitution, which are different from those of the successive planets, Pluto is considered by many planetologists to be the first object in the Kuiper belt, rather than the last planet of the Solar System. The hypothesis was reinforced after 2000, when six similar objects have been discovered. One of these, Eris, is even bigger than Pluto. These discoveries furnished the proof that the Kuiper belt contained numerous objects larger than 1000 km and, if these objects are called "planets", there is a risk of finding many more "planets" tens or even hundreds.

In order not to let the number of planets increase, at the IAU meeting in Prague in 2006 it was decided to downgrade Pluto to simply a trans-Neptune object, fixing permanently at eight the number of planets of the Solar System. For the nostalgic of Pluto as a planet, a new category of objects called nano-planets was created. In 2009 this category was populated by nine objects: Orcus, Pluto, Varuna, El 61, Quaoar, FY9, Eris and Sedna, to which the asteroid Ceres was added, being similar in dimensions and composition. To discover and study this new type of objects, NASA decided in 2006 to launch the satellite "New Horizons", which, after nine years of travel (in 2015), will pass at a distance of 9000 km from the Pluto surface; after having studied it, the probe

---

4 *Pluto was discovered in 1930, by the American astronomer C. Tombaugh. He was looking for a large planet that could explain the perturbations observed in the orbits of Uranus and Neptune. Its mass is one-hundredth of that of the giant planet (Table 8.1); it turned out to be too small and the distance too big to justify these perturbations. (See Figure 8.7 for its actual dimensions.) The following observations have shown that these measured perturbations were not real. Pluto was therefore discovered by chance.*

Table 8.1 The principal parameters of the planets (including Pluto) of the Solar System. The number of satellites of the giant planets (penultimate column) could increase in the future with new discoveries. The parentheses show the number of satellites discovered prior to the Pioneer 10 (1973) and *Voyager 1* missions (1977). For Mercury, having a face always pointed toward the Sun and the other toward extrasolar space, the temperatures of both are given.

Planet	Distance from Sun [million km]	Diameter [km]	Period [years]	Mass/ M_{Earth}	Number of satellites	Mean temperature [°C]
Pluto	5900	2306	247.70	0.002	4 (1)	−233
Neptune	4498	49 528	164.79	17.23	13 (2)	−214
Uranus	2869	49 000	84.01	14.54	27 (5)	−215
Saturn	1427	120 200	29.46	95.14	62 (10)	−178
Jupiter	778	142 600	11.86	317.83	66 (12)	−150
Mars	228	6790	1.88	0.11	2	−60
Earth	150	12 756	1.00	1.00	1	15
Venus	108	12 140	0.61	0.89	0	450
Mercury	58	4850	0.24	0.055	0	−173 + 470

will continue its voyage toward the Kuiper belt to make a census of the objects that populate the belt and discover new nanoplanets.

The distances between the Sun and Pluto or the other nanoplanets, make these objects uninhabitable. The mean temperature on Pluto is −233°C (Table 8.1), and the others further away have even lower temperatures. No form of life is possible under these conditions; therefore the captain decides to continue his voyage toward the inner Solar System.

The next four planets, Uranus, Neptune, Saturn and Jupiter (Table 8.1 and Figure 8.7), are the giants of the Solar System. They are all very cold and have a liquid surface; their atmospheres, near the surface, because of their high pressures, have the characteristics of liquids so that it is hard to tell where the surface of the planet ends and the atmosphere begins. The atmospheres of these planets are composed mostly of hydrogen (between 85% and 95%) and helium (between 5% and 15%), which are the most abundant elements in the universe (Figure 4.7). The low temperature and great mass of these planets have kept these light elements from evaporating, as happened on the rocky planets (Mars, Earth, Venus and Mercury), which are closer to the Sun. There are also many molecules, among them CO, CO_2 and finally methane (CH_4), to which Uranus and Neptune (with a concentration of about 2% methane in their atmospheres) owe their characteristic azure appearance (Figure 8.7).

It seems very difficult for any form of life to exist in these environments. In spite of this, on some moons, like Titan (of Saturn), or Europa (of Jupiter),

Figure 8.7: The giant planets of the Solar System (from the left): Uranus, Neptune, Saturn and Jupiter. Below them are the rocky planets (from the left): Mars, Earth, Venus and Mercury. Their relative dimensions are real. (© NASA.)

there could be relatively warm underground cavities with large lakes of liquid water that may host some primordial form of life. This hypothesis appears today probable, after what is said on the origins of life in Chapter 5, and after the discovery of psychrophilic organisms mentioned in Section 6.4. In any case, it seems impossible that in such environment forms of evolved animal life like those we have on the Earth can develop.

Leaving Jupiter, the spaceship proceeds with caution through the asteroid belt (Figure 8.8), which separates the external zone of the giant planets from the inner zone where the rocky planets are. This region is composed of planetesimals which can reach dimensions of hundreds of kilometers (the largest, Ceres, has a diameter of 800 km, while others, like Vesta and Pallas are around 500 km in diameter). The asteroid belt consists of planetesimals formed in the initial phase of the Solar System, survived until today, because in that region of the Solar System no planets were born. It is not clear why the process did not take place. It is hypothesized that proximity to Jupiter, with its large mass, had in some way blocked the aggregation of planetesimals. The same hypothesis is used to explain why Mars is so small, about a tenth of the mass of the Earth and Venus.

8.4 Mars

Past the asteroid belt, the captain finally arrives at the first of the rocky planets: Mars. This planet is close enough to the Sun (Table 8.1) to have liquid water and, therefore, it could host some evolved forms of life and therefore it could be classified as habitable. The planet however appears cold, without oceans, with a surface marked with numerous craters, resembling our Moon (Figure 8.9).

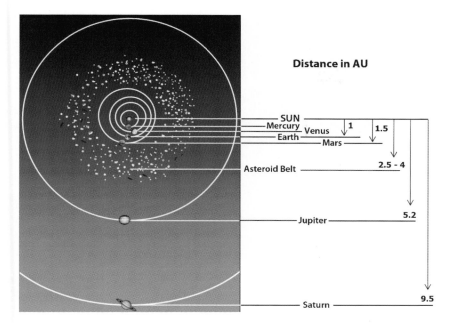

Distance in AU

SUN
Mercury — Venus — 1
Earth — Mars — 1.5
Asteroid Belt — 2.5 - 4

Jupiter — 5.2

Saturn — 9.5

Figure 8.8: The planets of the Solar System, inside the orbit of Saturn. The distances are given in units of UA (Appendix B). The region of asteroids between Jupiter and Mars, like the Kuiper belt, is a residue of the disk from which the Solar System was formed.

Figure 8.9: A composite image of the Martian surface obtained in 1978 by the photos of the Viking mission of NASA. Its surface is similar to the Moon's, with the total absence of mountain ranges like those on the Earth. (© US Geological Survey & NASA.)

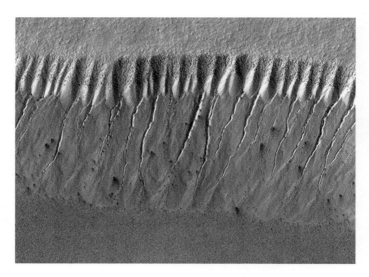

Figure 8.10: Evident traces of erosion on the surface of Mars. The image was obtained by the camera of the Mars Global Surveyor in July 1999; the area is 2.8 × 2.1 km² at 70.7 S and 355.7 W. (© NASA Green DataBase JPL.I

This aspect influences the captain to be pessimistic about the possibility that the planet has ever hosted evolved forms of life. Mountain ranges like the Alps, Apennines, Andes and Himalayas do not exist on the planet. The only reliefs on Mars are craters due to dead volcanoes or meteorite impacts.

The lack of mountains on Mars means that on this planet there has been no movement of continental plates that, for billions of years, have modeled the surface of the Earth and, at points of contact, have created those corrugations on the surface of the Earth that are the mountains, a phenomenon to which the Earth owes the volcanism that maintained its dense atmosphere (Section 9.5) and the return of the land eroded by atmospheric phenomena and transported by rivers to the seas (Section 9.1). On Mars there are traces of erosion but they are very modest (Figure 8.10). This allows the captain to understand that meteorological phenomena are practically nonexistent; the atmosphere is so tenuous that any evolved form of life is impossible.

Although on Mars there are no mountain ranges like those on the Earth, there are many (dead) volcanoes with lakes of lava and ash. In some cases they are enormous compared with those on the Earth. For example, Mount Olympus the largest Martian volcano, is 26 km high (three times the height of Mount Everest) and has a base of 600 km (the distance between San Francisco and Los Angeles). It is much higher than the volcanoes of the Earth because on Mars erosion has been moderate. On the Earth such a structure could exist for only a few million years, the time needed to be leveled down by erosion (Section 9.1).

The European probe *Mars Express*, launched in 2004, photographed much of the Martian surface. The study of these photos shows that a number of these volcanoes have been active in recent times. In some places volcanoes seem to

Figure 8.11: Martian ice pack. Image of an equatorial zone (5°N, 150°E) obtained in February 2005 by the Mars Express. The opinion of scientists is that it represents an icy sea covered by volcanic ash and dust. Five million years ago (the age is given by the number of meteorite impacts) in this piece of land, there existed liquid water. Then the water got frozen, incorporating the preexisting blocks of ice that were floating on the surface (the darker pieces on the surface). On the Earth the ice pack is formed in the same way and has a similar aspect except for the color that is white; here it is brown because the ice is covered by the volcanic dust. (© ESA/DLR/FU Berlin, G. Neukum.)

have been active 2 million years ago, a very brief time with respect to the life of the planet (the Earth was already inhabited by *Homo habilis*). In other zones, especially in the southern hemisphere, volcanic activity seems to have finished long ago — some estimates say more than a billion years.

Mars seems to be not very uniform in its volcanic activity. The age of volcanoes is deduced from the impact craters on the land (like those in Figure 8.11). In fact it is possible to estimate with good precision the number and dimensions of asteroids hitting the planet's surface in a given period (Figure 10.5). Therefore, counting the craters and measuring their diameters, it is possible to estimate how long a particular area has been bombarded by meteorites, i.e. when it has solidified. In comparison, with Mars the lunar surface has a much more tortured aspect, having been bombarded for thousands of years.

If in some zones of Mars there was a recent volcanic activity, we can suppose that life has existed on the planet until a few million years ago because, as we will see in Section 10.5, the gas erupted from volcanoes can form on a planet a much denser atmosphere than that of Mars today. A denser atmosphere can produce a greenhouse effect sufficient to bring the temperature between the values shown in Figure 6.3 and make possible the birth of life. We can therefore understand why the search for life is one of the chief objectives of the current missions on Mars. To look for traces of life, even fossil, would be an important test of the theories in Chapter 5.

Among the results recently obtained, there is the certainty *that water, the element essential for life, exists on the red planet*. On Mars, the water found is less abundant than that existing on the Earth, but it would be enough to fill (if the temperature were higher than it is now) lakes, rivers and seas. Even in the photos, in 1999, of the American probe Mars Global Surveyor, one finds structures that would be difficult to explain without admitting the erosion of the soil due to the flow of a liquid. Figure 8.10 shows an area of 1.3×2 km with furrows on the border of a slope due to erosion. Its analogy with terrestrial photos of mountainous deserts is impressive. The high latitude of those zones (on the Earth they would be in Antarctica) and the low temperature of Mars make it impossible to explain these furrows as due to water without assuming an underground heat source. The analysis of furrows in the terrain suggested to scientists in NASA that they are not older than 10 years, so we are speaking of phenomena that are still active today. What caused them? A small crack? Landslides? Liquids flowed out of the soil and immediately turned into ice or absorbed back into the land? Is it possible that the furrows were caused by water? If so, what was heating it? Were they the same radioactive materials found today on the Earth? And if there is liquid water does this mean that some form of life exists today? These are mysteries that Mars may, one day, reveal to the probes that are studying the planet.

There are by now numerous photos taken by the Mars Express showing the existence of water on the planet. To view them, visit the website of the mission at ESA. Among these photos are the frozen seas of Figure 8.11, which are estimated to have been liquid as recently as 5 million years ago. The age is deduced from the impact craters, of which we have spoken before, measured in a region larger than what is shown in the figure.

In Figure 8.12 we see other traces of water. On the left is a furrow on the Martian surface that has the morphology of a river bed, while on the right there are two nearby craters, produced by the impact of two asteroids which form an SOA of "hourglass." In the photo we see a glacier (consisting presumably of solidified water) that descends from the upper crater to the lower, 500 m below, after breaking through by erosion the edges of the second crater in the plane below. It is an image that reminds us of terrestrial mountain forms apart from the reddish Martian sand.

An unassailable proof of the existence of water on Mars is given by the spectrometers (Section 4.9) of the Mars Express (Figure 8.13), which are sensitive to the bands of iced water (they measure a radiation that only solid water can produce). The photo shows the Martian North Pole; the white we see is due to ice not covered by volcanic ash. The zones that appear darker in the photo could also be due to ice, covered by ashes.

An estimate of how much water is contained in the polar icecap of Mars has the result that, if it were redistributed across the entire surface of the planet, it

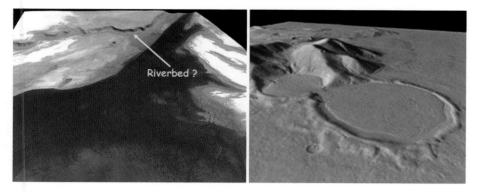

Figure 8.12: *Left: Traces that could be interpreted as a riverbed. Right: Photo of a Martian glacier streaming from an upper crater to another 500 m below. The two craters are due to the impact of two asteroids; the upper one has a diameter of 9 km, the lower one 17 km. (© ESA/DLR/FU Berlin, G. Neukum.)*

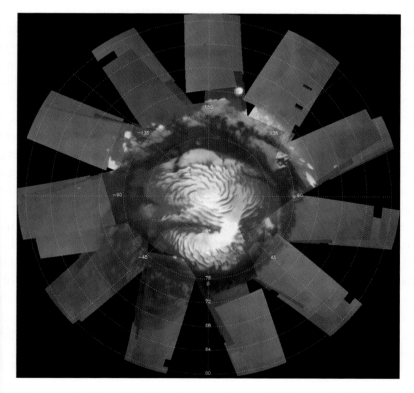

Figure 8.13: *Image of the North Pole of Mars obtained at the end of 2004 from the Italian-French spectrometer OMEGA on board the Mars Express, capable of determining the composition of the surface of Mars from the properties of light reflected. It shows the presence of solid water (in white). The darker zones are thought also to be ice, covered by volcanic ash carried by the wind. (© ESA, ASI & G. Bellucci, INAF/IFSI, Roma.)*

would cover Mars with a 25 m deep layer, while on the Earth the polar icecaps water together with the oceans would cover our planet to a depth of about 4 km (Figure 9.1). If water will not be found underground in the form of permafrost,[5] Mars will seem to have less water than the Earth. In the Earth water occurs in one part per thousand of the total mass while in Mars the proportion would be less than one part in a hundred thousand.

There are, however, indications that in the distant past there was more water on Mars than today. The study of the southern hemisphere valleys, in the oldest part of the planet, shows that long ago there were great rivers. Moreover in both hemispheres linear structures thousands of kilometers long are observed. They are all at the same height as should be the banks of ancient oceans. If that will be proven, the quantity of water that can be estimated from the dimensions of these basins could cover the planet up to a depth of 100 m — four times the water we find today, but still much less than what is on the Earth. If the existence of these rivers and oceans is confirmed, we must conclude that Mars, in a distant past, had a denser, warmer atmosphere, with liquid water on the surface — the conditions necessary for the birth of life.

But where has the missing water gone? A part is certainly trapped underground at a depth that could be of the magnitude of kilometers. Other could be trapped in the rocks, chemically bound to other composites. The difference with the Earth remains in any case large. There are scientists who hypothesize that on the red planet there was less water at the beginning and that the process that formed the Solar System for some unknown reason favored the Earth over Mars. Another hypothesis is that in the first phase of life on the planet the impact of a large meteorite had caused much of the water to evaporate (Section 6.2.) Other hypotheses say that the water on Mars is in the same proportion as that on the Earth and that the amount bound in rocks or trapped underground is much greater than one would guess.

The Martian soil therefore holds many mysteries, possibly among them traces of ancient forms of life. If the furrows in Figure 8.10 are recent and are due to liquid water, then some form of life may have survived up to the present. Hints on the existence of life on Mars, even in the past, would constitute a proof of the thesis in Chapter 5, that life germinates at every point of the universe where liquid water exists, and disappears as soon as liquid water disappears. Until today, all experiments conducted to search for past or present forms of life on Mars have failed, but the exploration of the red planet has just begun.

8.5 Venus

Continuing his voyage, our explorer finally arrives at a habitable planet — the Earth! With seas, rivers, forests, meadows, snow-topped mountains, and with an extraordinary blue sky....

---

5 *Permafrost is a mixture of ice, soil and rocks that remains all year below 0°C, even during the summer. It is found in many regions of the Earth — in the Alpine valleys remaining after glaciations, and in many subpolar regions, like Greenland, Iceland and Siberia. (In these places, inside the permafrost, perfectly preserved, remains of ancient animals like mammoths have been found.)*

With the discovery of the Earth, the voyage of our brave explorer is finished. The next planet, Venus, could be habitable because it is far enough from the Sun to have a temperate climate, but the spectrometers on board indicate a very strong greenhouse effect that raises the temperature to 480°C, creating conditions impossible for any form of biological life. The following planet, Mercury, also has a temperature of over 400°C, but this is not a surprise for the captain — the planet is close to the Sun and that is the temperature he would expect to find. After registering the data gathered in his logbook, our captain leaves the Solar System for a new exploration. In that system he has found only one habitable planet, the Earth, even if two others could have had the potential to be that: Mars with a more efficient greenhouse effect, and Venus with a less dense atmosphere and a much weaker greenhouse effect.

The mean temperature of Venus, given its distance from the Sun, should be around 27°C. It would be considered habitable (Figure 6.3). The planet is surrounded, however, by a dense blanket of CO_2 that causes a greenhouse effect intense enough to bring the temperature up to 480°C. This makes Venus the hottest planet in the solar system, even hotter than Mercury, which is closer to the Sun but does not have an atmosphere and therefore does not have the greenhouse effect. Venus is an infernal world with an atmospheric pressure 100 times higher than that of the Earth and physical properties similar to those of a liquid — an ocean of hot carbon dioxide. Other than CO_2 in the atmosphere, there is sulfur, and a little amount of water (1/10,000) that, reacting with the sulfur, produces sulfuric acid.

In its first years of life the Earth should have been similar to Venus, with a temperature estimated at around 1000°C and an atmosphere containing, in the form of vapor, the water that today fills the oceans (Section 6.2). The density was probably higher than what we measure today on Venus. Then the destinies of the two planets went apart. Our planet cooled; there were long periods of rain that reduced the greenhouse gases in the atmosphere and formed the oceans. On Venus, the atmosphere remained dense and hot, and water evaporated, rendering the planet forever uninhabitable.

The dense clouds which envelope Venus prevented us for many years from studying its surface. No one knew what might exist below — volcanoes or, if different from Mars, tectonic activity like that of the Earth. In 1990 the planet was reached by the American probe Magellan, which studied its surface with a radar (Appendix A.4) capable of penetrating the atmosphere. In such a way we obtained the first images of its surface, which have been put together in a unique panoramic picture (Figure 8.14) where we see Venus as if our view were not blocked by its atmosphere.

The images show that there are no mountain ranges on this planet and thus there is no continental drift like that of the Earth. *The Earth remains the only planet in the Solar System with great mountain ranges* like the Alps, the Himalayas

Figure 8.14: The surface of Venus as it appears in the radar exploration of the probe Magellan of NASA. It is an infernal world with a temperature of 480°C and a dense atmosphere of carbon dioxide, sulfur and sulfuric acid. (© NASA, JPL.)

or the Andes. The mountains and the highlands that are observed on Venus are all of volcanic origin (like Mount Maxwell), the highest of which measures 11 km. This mountain indicates that on Venus there is little erosion; the reason is again the lack of liquid water. There are instead the remains of giant volcanic eruptions that in a distant past remodeled the surface of the planet. With respect to Mars, there are fewer craters due to asteroid impacts, the reason being that the smaller asteroids were destroyed during their passage through the dense atmosphere. The ongoing erosion has smeared craters smaller than a kilometer, while the older ones, among them the larger craters, have been covered by the volcanic eruptions.

Water probably existed on Venus at its beginning with quantities comparable to those on the Earth. Then the molecule, more fragile than CO_2, was dissociated by sunlight (a process that also happens on the Earth) into hydrogen and oxygen, and differently from what happens on the Earth, for the high temperature they reached the escape velocity and drifted away.

It will take some time before we understand why Venus, similar to the Earth and composed of similar materials, had such a different destiny. The explanation may be found in its different distance from the Sun or in the different angle of its rotation axis (Section 10.1). We must also understand why the atmosphere of Venus contains so much carbon dioxide, and whether it is possible that something similar can happen on the Earth. The greenhouse effect on Venus has raised its temperature to 480°C, but a much lesser temperature is enough to make a planet uninhabitable. We have seen that some of the most terrible extinctions of the past were due to CO_2 from volcanoes (Section 7.2). Can such eruptions be possible again on the Earth? Or can CO_2 produced by human activity contribute to a similar effect?

8.6   The Earth: A Habitable Planet

What make the Earth a habitable planet? The answer is its distance from the Sun, combined favorably with the greenhouse effect. They have guaranteed stable temperatures that has permitted the development of life, although with the oscillations that led to glaciations (Section 9.8).

Without the greenhouse effect, remember, the mean temperature of our planet would be −18°C while, because of it, the temperature is now 15°C. A contribution of 33°C that has guaranteed the development of life till now. It is a value that must not increase too much, in order not to trigger runaway processes that will increase the CO_2 in the atmosphere, bringing the Earth to a condition similar to that of Venus, or diminish too much and turn the Earth into an icy planet like Mars. The thermal cycles of the past have alternated glaciations with periods of heating; in extreme cases they have led to mass extinctions. We have seen that the mass extinctions of our past had all been caused by large variations in the temperature and were probably correlated with (if not caused by) large variations of greenhouse gases contained in the atmosphere as shown in Figure 7.9.

It is therefore the chemical composition and the density of the atmosphere of a planet (the thin layer in Figure 8.15) that will characterize the habitability of a planet, that makes possible the existence of life on the Earth and not on Mars or Venus. *A chemical composition that, if measured from a distance, would tell us whether a planet is habitable and whether there are biological forms of life on it.*

Figure 8.16 shows the infrared[6] spectra of Venus, Earth and Mars, three planets with an atmosphere; only one of them is habitable, with an evolved form of animal life. The figure tells us that in the atmospheres of all three planets there is carbon dioxide. This means that this molecule survives to extreme conditions like −100°C in the atmosphere of Mars and 480°C in the atmosphere of Venus, an interval exceeding by far what we consider possible for the

---

6 *We are speaking of spectra different from those we saw in Section 4.9. Corresponding to the various chemical compositions, instead of a peak we observe a dip, which is the opposite of a peak. This is called an absorption spectrum, unlike those in Figure 4.12, which are emission spectra. How does a spectrum of this type come about? The surface of the planet, because of its temperature (Appendix A.2), emits infrared radiation which, in order to leave the planet, has to pass through the atmosphere, where it encounters molecules that absorb light according to its characteristic colors (Section 4.9). The molecules, absorbing the radiation, become excited, and re-emit in all directions. This causes a reduction in the luminous intensity of the characteristic colors of the molecules of the atmosphere, producing the dip we see in the figure. The deeper the dip is, the more molecules of that type there are in the atmosphere of the planet.*

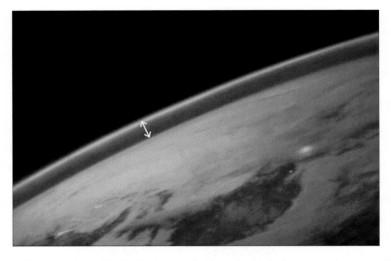

Figure 8.15: The atmosphere of the Earth, photographed from the shuttle. Its mass is about one millionth of that of the Earth (5.13 x 10¹⁸ kg). Much of it (two-thirds) is contained in its first 8000 m, and it is therefore a very thin layer. The greenhouse gases it contains raise the average temperature by 33 degree from –18°C to 15°C, rendering possible life on our planet. A reduction in the greenhouse gases would take the Earth to a glaciation, an increase could cause a heating that, as we have seen (Section 7.1), could be deadly for most of the biological species of our planet. (© Clementine, NRL & NASA.)

existence of any form of life. The peak of CO_2 is much deeper than those of other molecules, making it easier to be measured; *therefore CO_2 is ideal for understanding whether a planet has an atmosphere.*

In the atmospheric spectrum of the Earth, however, there are two other types of molecules that are not found in the spectra of Mars and Venus: water (H_2O) and oxygen (in the spectrum we see the molecule of ozone, O_3). They indicate that there is life on the Earth but not on the other two planets.

As we have seen (Section 4.9), *water* is the essential element for life — without it, life cannot exist; there are organisms that can live without oxygen and without light, at extreme temperatures (Section 6.4), but without water they cannot survive. Water is important because it is fundamental in the chemical reactions with which biological organisms are working. It is an excellent solvent and an abundant source of hydrogen for many important biochemical reactions like photosynthesis (Section 6.5). For these reactions water has to be in the liquid state, when it is in the gaseous (vapor) or solid (ice) state, these properties are greatly reduced, if they do not disappear completely. Large quantities of water in the atmosphere, enough to be measured from long distances, can exist only on a planet with liquid water, as in the oceans of the Earth. The Martian ice, for example, does not produce enough water vapor to show its presence in the spectra of Figure 8.16. A hot planet like Venus, instead, has little water in the atmosphere because, as we have seen (Section 8.5), it was lost. Therefore, the *measurement of water in the atmosphere of a planet indicates that life can exist there* (it is not,

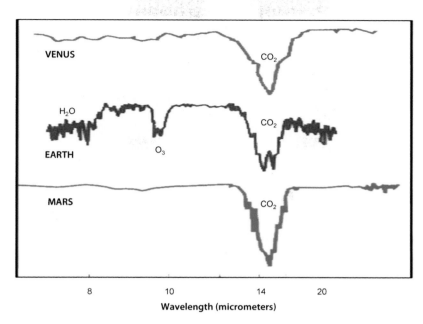

Figure 8.16: The absorption spectra of three planets of the Solar System; all have an atmosphere and are at a distance from the Sun such that it could support life. Mars is the farthest from the Sun, with a distance 1.5 times that between the Earth and our star, and it has a temperature that can drop to −100°C; it is not habitable because it is too cold. Venus is closer, at a distance 0.7 times that between the Earth and the Sun, and its temperature reaches 480°C, it is not habitable because it is too hot. All three planets show the presence of carbon dioxide (CO_2), indicating the existence of an atmosphere, but only the Earth, where life exists, reveals the presence of water and oxygen. Water on Mars, unlike on the Earth, is not detectable in the atmosphere because it would be iced for the low Martian temperature.

however, a proof that life does exist). It shows that the planet has plenty of liquid water and the temperature is in the interval that permits life to exist.

Oxygen is the second element measured in the spectrum of the Earth, while it is absent from the spectra of the other two planets. The measurement of oxygen in the atmosphere of a planet indicates that on the planet life exists. As we have seen in Figures 4.9 and 4.10, oxygen is one of the most abundant elements in the universe. Nevertheless, it is very difficult to find it free, not bound to other atoms because it is one of the most reactive elements that exist in nature. It has a great affinity to bind chemically to other elements, oxidizing them. Oxygen combines with everything, including metals and rocks, and is consumed by biological organisms (which would die without it). If there had not been a process that continually creates oxygen in atomic form, life would have disappeared rapidly from the atmosphere of the Earth. On our planet the process that continually creates oxygen is photosynthesis; it is the process on which all evolved forms of life are based. Therefore, only on a planet in which photosynthesis is active, can so great an abundance of oxygen exist in the atmosphere to be observed from long distances. Evidence of oxygen in the atmosphere of a planet is thus the proof that on this planet forms of life similar to ours exist, possibly not developed, and possibly not intelligent, but life capable of reproducing, evolving, and adapting to its environment. Oxygen exists in the atmosphere of the Earth as O_2 and O in quantities greater than that of O_3 represented in Figure 8.16, but they are difficult to observe even if they are abundant.

To summarize, Figure 8.16 shows that there are three fundamental indicators in the atmosphere of a planet: CO_2, H_2O and O_3. They indicate:

for CO_2 the existence of an atmosphere;
for H_2O the existence of the conditions necessary for the birth of life;
for O_3 (or O_2 and O) the existence of life.

As we will see in Chapter 12, what we will seek to measure in the atmospheres of extra-solar planets in the coming years are these three molecules. A difficult measurement indeed, but possible today.

8.7 The Habitable Zone of the Planetary System

We have seen that water is essential for life, and fortunately it is abundant in the universe (it is the third molecule in abundance, after H_2 and CO). Water is formed in the interstellar space through the chemical reactions that take place on the grains of dust and on the hot gases surrounding the stars under formation (Section 4.9). Once formed, water condenses into particles of ice and joins the planets through the parental cloud (Figure 3.1) and the disks. The comets, composed of the material from which the Solar System was formed, are the proof: they contain great quantities of water. Water should therefore exist everywhere in the Solar System, even on the Moon, in all the places which are shielded from direct solar radiation, where it has not evaporated. Only in the hottest two planets, Mercury and Venus, must water be largely lost and dispersed into space because of their high temperatures.

Even if water has fallen on all of the planets (where it exists in different quantities according to the planet's history) it is not yet possible for life to be born on all of them. Life needs (Section 8.6) water in the liquid form, and we know that on the Earth (at the pressure of 1 atmosphere) water is liquid between 0°C and 100°C. This small interval of temperature imposes a very strict limit on the distance a planet can have from the star about which it rotates. It defines for every planetary system a habitable zone, wherein the planet is close enough to the star, to prevent water from freezing, and far enough for it not to evaporate. The greenhouse effect, as we have seen, can change the temperature of a planet making, for the habitable zone, a relatively large band.

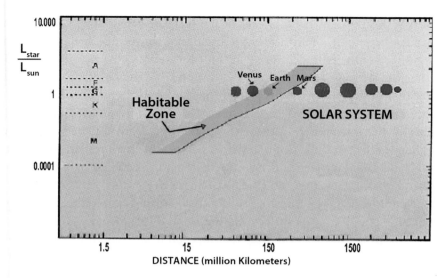

Figure 8.17: *The habitable zone. Life requires water in the liquid state. On the planets that are too close to the star, water evaporates, and on those too far it freezes. In the figure the distance interval of a planet from its star that permits liquid water is illustrated as a function of the luminosity of the star (the width of this zone is due to the greenhouse effect, which elevates the temperature of a planet). The Earth is inside the zone; Venus and Mars are at its borders. (From the homepage of the Darwin Mission.)*

In Figure 8.17, the habitable zone is shown for planets orbiting around stars of various luminosities. On the vertical axis the luminosity of the star relative to the Sun is labeled, and on the horizontal one the distance between the planet and the star. The yellow band shows the distance from the star a planet must have in order for water to be liquid. This band is called the habitable zone, because only inside it life does exist. The habitable band is limited, on the highest luminosities, by stars that have four times the mass of the Sun, because, as we have seen (Section 4.3), stars more massive and luminous do not live long enough to allow life to develop. It is also limited, on the lowest luminosities, by stars of a few tenths of the solar mass, because stars with such low mass evolve so slowly that from the moment at which the galaxy was born they had not enough time to become stars. The Solar System is shown in the figure with its planets; the Earth is at the center of the band, while Mars and Venus are at its borders. This explains in part the good fortune the Earth for hosting biological forms; in the case of climatic perturbations, the Earth has a greater margin to recover the situation and revert to equilibrium. On planets like Venus and Mars it is easier to move them outside the strip causing irreversible phenomena.[7]

---

7 *We have already mentioned possible irreversible phenomena that according to some scientists could also happen to the Earth (Section 7.1). A glaciation that covers the Earth with ice, reflecting sunlight, could keep it icy forever, just as a strong heating that causes much of the oceans to evaporate could leave it forever hot, as what happened on Venus.*

Mars is on the right hand border of the band of habitability, it would not take much to enter it. In its history there could be a greenhouse effect efficient enough to raise the temperature by the bit necessary to create conditions suitable for life. We have seen that forms of primordial life could have existed on that planet and could exist even today underground where there is liquid water, heated by residual volcanic activity. One day it might even be possible to modify the atmosphere of Mars by making it denser, so increasing the greenhouse effect and making the planet colonizable by humans. Projects of this sort are found in science fiction; to do it, man would first have to land on Mars and possibly return again (Section 11.5).

For Venus, which is on the other side of the band, its temperature is too high to make life possible on its surface. Colonization of the planet seems impossible, even with the most fervid imagination; even if we could initiate a process that reduces the greenhouse gases, thus permitting the planet to cool, the problem is that the planet has lost its water for forever.

With this chapter, we have concluded the description of the conditions that make a planet habitable, and above all what has to be measured to evaluate the possibility or even the existence of life on a planet. Before discussing, in the last two chapters, the attempts to put us in contact with "intelligent" life on other planets and what we will do to find inhabitable planets even if they are not inhabited by "intelligent" life, we will come across, in the next two chapters, a few aspects that contribute to the evaluation of the number of habitable planets: the geological aspects (continental drift), which are fundamental for keeping a planet like the Earth habitable for such a long time, and the planetary aspects (the presence of Jupiter and our Moon), which have protected our planet in the course of time. We will mention in the end the risk for life to be destroyed by cosmic events.

The Importance
of Continental Drift

<u>Chapter 9</u>

9.1   A World of Water

We have seen that water is in abundance in the universe (Section 4.9); on the Earth it constitutes one part per thousand by mass and is contained mainly in the oceans. It arrived on our planet together with the materials that have built the Solar System, mixed with all the elements of the Earth. It was existing predominantly beneath the soil; light and mostly gaseous (the interior of the Earth is hot even today), it reemerged on the surface during the first phases of the life of our planet (Sections 3.8 and 6.2) while the heavy materials, especially metals, descended toward the center of the Earth to form the ferrous nucleus from which comes terrestrial magnetism.

Part of the primitive water evaporated and was dispersed into space, and part of it — mainly what was brought to the Earth by comets at the end of the accretion phase, when the Earth, almost formed, was cooling — remained as vapor in the fiery atmosphere of the planet. It is estimated that 10% of the objects falling on the proto-Earth between −4500 and −3800 million years were comets, and that the water they contained was sufficient to fill all the oceans. Water has therefore been present on the surface of the Earth from the beginning, as it is witnessed by the discovery of marine sediments 3.6 billion years old (Section 6.3).

We do not know much about the first lands that emerged out of water and about the first continents because they have all been destroyed with the passage of time. Traces of them remain in the sedimentary deposits spread out on the continents, which indicate that the land had to exist since the birth of the oceans, at least in the surroundings of the volcanoes (Figure 6.1) But from these materials, fragmented by time, it is impossible to reconstruct how the lands were, how they have been formed and how they were extended. Today, we can reconstruct, with a reasonable approximation, the evolution of the continents after Pangaea, the megacontinent from which the today emerged lands rose. We can recount, with good precision, only the history of the last 250 million years, a very brief period compared to the history of the Earth. On what happened before Pangaea, we can only formulate hypotheses; the continents certainly merged and separated many times in the course of the 3.8 billion years of the history of the Earth. Because the inside of our planet was hotter than what is today, we can suppose that the tectonic movements were also greater than they are today, but the data for reconstructing that past are very uncertain.

We have seen that the oldest rock found on the Earth contained also fossils (Section 6.1) that are very useful for reconstructing the variations in climate, to study the life forms and to follow their evolution, but they are ill-suited as a source of information on the geological structure of the past.

In spite of the lack of experimental data, we can say that in the beginning land consisted only of volcanoes. There have also been many craters resulting from meteorite impacts; the surface of the Earth was, in many aspects, like that of the Moon or Mars, but covered by water. The mountains that we have today did not exist, because continental drift, the mechanism that formed them had not enough time to build them.

Today the aspect of the emerged land is very different from the first years when life formed on the Earth. Figure 9.1 illustrates the profile of the of their elevation today. It is discontinuous: it goes from the 8848 m (above sea level) of Mount Everest down to the 11,521 m (below sea level) of the Mariana Trench. A difference of 20 km that cannot be very old. On the Earth there is a dense atmosphere that causes a strong erosion which reduces the differences; it levels the mountains and fills the trenches of the oceans with sediment, so that such large differences in the Earth's profile cannot last very long.

The atmosphere is the main cause for erosion. Even the hardest rocks are excavated by water, shattered by ice and attacked chemically by acid rain produced by volcanoes and by human activity. Gases combined with water produce acids (sulfuric, carbonic, nitric) that attack rocks, turning them into materials soluble in water. Even gravity plays a role in the destruction of terrestrial soil; the gravitational attraction of the Earth makes rocks roll down reducing the height of the mountains. Everywhere we find traces of erosion (Figure 9.2). Nothing survives very long on the surface of our planet, unlike what happens on the Moon or on Mars, where there is no atmosphere, or only very tenuous ones, in which ancient craters are conserved.

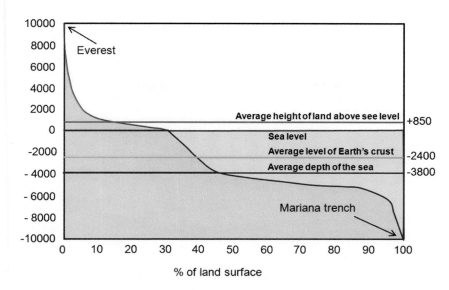

Figure 9.1: *The average values of the seas depth and lands height as they are today. The oceans cover 71% of the globe with a mean depth of 3800 m, while the emerged land covers the remaining 29% with a mean height of 850 m. The volume of the emerged land is a small fraction (a tenth) of that occupied by the oceans. The large unevenness we see in the figure has been formed by continental drift, without which land would not exist.*

In the lower part of Figure 9.2 the quantities of debris that the principal rivers in the world carry to the sea are shown. A large part of this material (a fifth of the total) comes from three rivers — the Ganges, the Yellow river and the Brahmaputra. All three coming from the Himalayas (the peak in Figure 9.1): fifty million years old, they are the most recent and the highest mountain chains on the Earth, as well as the most attacked by erosion.

Every year, rivers carry to the sea 60 km³ of debris, enough to destroy dry land in less than 20 million years; this is a fraction of the life of the planet, well under the 60 million years that separate us from the extinction of the dinosaurs! Hypothesizing that the erosion we see today was active since the beginning, we estimate that a layer of sediments 100 km high (3 times the thickness of the Earth's crust) should be at the bottom of the seas. More than enough to fill the trench of Figure 9.1. Of this enormous layer there is barely any trace at the seabed because, as we will see, (Section 9.4) differently from the continents, the seabed is continuously recycled into the Earth's mantle.

If we examine the mountains of the Earth, we find that they are all recent. The peak in Figure 9.1, Mount Everest, is less than 50 million years old, a time to be compared to the 3800 million years that separates us from the time when the Earth cooled. Mount Everest was formed together with the Himalayan range when India, after a drift of 5000 km, collided with Asia (Figure 9.4(b)), and that happened when the dinosaurs were already extinct. The skeletons of dinosaurs we see in museums come from a world where India was an island

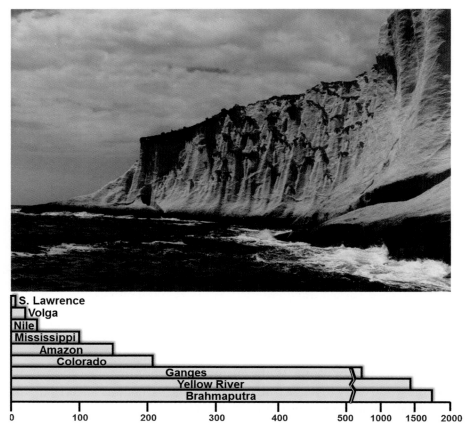

Figure 9.2: Erosion. Top: The Pontine archipelago is an example of the destructive power of erosion. The volcanic eruptions, that formed the islands, ended about one million years ago, and today less than one-tenth of the original islands remains. Bottom: Masses of debris carried to the sea by the most important rivers on the Earth (in millions of tons per year). A fifth of this material comes from three rivers (the Ganges, the Yellow River and the Brahmaputra); all three descend from the Himalayas (the peak in Figure 9.1).

and the Himalayas did not exist. All the mountains of the Earth are young. The alpine arc and the Italian Apennines were born from the northward movement of the African plate, which began 150 million years ago when the Mediterranean Sea was a gulf in the ocean (Tethys, Figure 9.4(a)) and Africa was separated from Asia across the Arabian Peninsula. In the same way, the Andes are the result of the Americas moving westward — a movement that started 250 million years ago, when they were detached from Europe and Africa.

All this has been caused by *continental drift*,[1] which has redesigned the surface of the Earth many times. It is a phenomenon unique in the Solar System; no other planet does anything like it, nor does any have mountains like the

---

1 As we will see in Section 9.4, the motions do not regard single continents, but much larger zones, the plates into which the surface of the Earth is divided.

Earth. For this reason the geological description of our planet does not go back to a time very far away. The oldest continents have largely been destroyed.

The American geologist D. Howel estimates that, because of continental drift, the volume of the emerged Earth *increases today at the rate of about 600 km³ per year*, more than enough to compensate for the 60 km³ that the rivers carry to the sea each year by erosion. The numbers seem enormous; however, they represent an infinitesimal of the volume of the Earth,[2] but were sufficient, over billions of years, to construct the continents on which we live.

The Earth is the only planet in the Solar System with such strong erosion, because no other planet has an atmosphere like ours. As we will see, this too is due to the continental drift because, at the points where the land masses (the plates) collide or move apart, the Earth fractures, volcanoes are born, and these volcanoes erupt the gases that maintain the atmosphere as it is permitting the survival of life on the Earth. The erupted gases put in circulation molecules like CO_2, H_2O, CH_4 and CO, which we have seen (in Section 4.8) are rare on our planet and are fundamental for biological life. These molecules also contribute to maintain an efficient greenhouse effect, without which the Earth would have the temperature of the Moon, $-18°C$, and would be largely frozen.

There are thus three contributions of continental drift to the existence of life on the Earth: the restitution of emerged land that the erosion destroys, the continuous reconstruction of the atmosphere by volcanism which restores rare molecules essential for life and the maintenance of a temperature suitable for life on the surface of the planet (Figure 6.3).

In the beginning, when the inner parts of the Earth were hotter, the tectonics must have been more active than it is today, and the emerged Earth must have grown faster than it does today. It is estimated that it took 1.5 billion years to construct half of the emerged Earth.

As we will see in the following, the Earth's crust, or the external part of the planet, is divided into plates, of which the continents are the parts that have emerged from the oceans. The plates cover the entire surface of the Earth; they move independently colliding or wedging one under the other. The energy needed to move the plates (Section 9.9) comes from radioactive material contained inside the Earth, produced by an ancient supernova that exploded before the Earth was formed. The radioactive material decays with time, diminishing in quantity and reducing its capacity to heat the interior of the planet. It is estimated that the heat produced by this material has already diminished by a factor of 3 from the time Earth was born, and will continue to diminish in the future. With the decrease of heat produced in the mantle, the movements of the continental plates will decrease and, with that, volcanism

---

2 *A small number also with respect to the 2000 km³ of icebergs that break away every year from the Antarctic.*

will decrease, the earthquakes' strength will be reduced and less mountains will be born.

In the end, if the Earth will live long enough, the heat generated in its interior will no longer be sufficient to move the continents, and with them the volcanism and the earthquakes will stop. Our planet will become geologically inactive, like the other planets of the Solar System. If that will happen, erosion will reduce the height of the mountains, the land will be invaded by water and, in the end the Earth will return to a planet submerged under its oceans. This will not be the only effect: with the reduction of the volcanism there will be less greenhouse gases and the temperature of the planet will fall. On a planet covered with ice, life as we know will not survive. Fortunately, it will take a very long time for all of this to happen.

9.2   The Structure of the Earth

The inner part of the Earth is a world inaccessible to humans for direct exploration and will probably continue to remain so. We have taken samples of the lunar soil, and, probably in the future, we will have those of comets, of the martian soil and, one day, we may leave the Solar System, but, despite any great technological progress, the direct exploration into the center of the Earth will remain, very likely, impossible. The reason is the high pressure (millions of atmospheres) and the high temperature (thousands of degrees, enough to melt metals) that exist inside our planet. The deepest wells that humans have dug are about 10 km deep, negligible if they are compared with the 6400 km to the center of the Earth.

In spite of the impossibility of reaching the center of the Earth, we have a precise idea of its structure thanks to a means of exploration provided by the Earth itself: the earthquakes.

The Yugoslavian geologist Andrea Mohorovicic (1857–1936) was the first to realize their extraordinary potential. At the beginning of the 20th century, by studying the record of an earthquake in the Balkans taken at various distances from the seismic epicenter, he discovered that the velocity at which the seismic wave propagates in the Earth had a discontinuity at a depth of 10–30 km, depending on the zone of the Earth being considered. He thus discovered the discontinuity between the crust and the mantle, which in his honor was called Moho. With these measurements Mohorovicic paved the way to study the interior of the Earth.

Earthquakes are caused by the sudden breaking or slippage of rocks deep below the surface which provokes pressure waves similar to sound waves, except that they are at lower frequencies. Like all the waves, the seismic ones move at a velocity that depends on the physical characteristics of the medium through which they move (Appendix A.1): the density (the velocity is highest in solids, lower in liquids, and still lower in gases), the rigidity (the velocity is

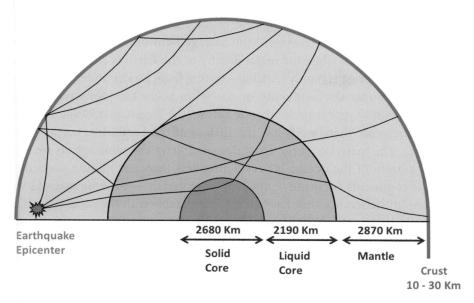

Figure 9.3: Cross-section of the Earth, with the paths of seismic waves that permitted the identification of four different zones of discontinuity in their transmission. The four zones are the crust, the mantle, the liquid core and the solid core.

highest in metals, lower in rocks and lower still in conglomerates and sandy materials) and the temperature (for each material the velocity decreases as the temperature decreases).

Seismic waves, like all other waves, change direction every time the physical characteristics (density and/or temperature and/or rigidity) of the medium through which they propagate change, as happens to a beam of light that changes direction when it passes from air to glass. Figure 9.3 shows schematically how a seismic wave propagates through the Earth. Measurements of velocity (the time a wave takes to reach different points on the Earth's surface) show that all the paths have a strong upward curvature,[3] as one expects since the density and temperature increase going downward. They also show sudden changes in the direction of propagation at four different depths. That happens because at these points, the mechanical characteristics of the Earth change suddenly. The four zones are called: the crust, the mantle, the external core and the internal core.

The *crust* is the external layer of the Earth and has a thickness of 10–20 km. It is the surface in contact with the atmosphere and is a rare place, an oasis in a hostile universe, the only place we know, where life is possible. A few kilometers above and below the crust the environment makes impossible any form of

---

3 If in a material the density and temperature change continuously (as happens in the Earth where the temperature and pressure increase with depth) the wave that crosses it follows a curve, i.e. a slight change of direction for every small change of density and temperature.

biological life. The crust is the lightest of the four layers that constitute our planet: its mean density is less than 3 (on average 1 cm³ of crust weighs a bit less than 3 cm³ of water), while the mean density of the Earth is close to 5.5. The thickness of the crust is highly variable — going from 10 km under the oceans to 20–30 km under the continents, up to 70 km below the Himalayas whose weight causes it to sink into the mantle below (the largest part of the mass of the mountains are therefore below the surface of the Earth, like the icebergs in the sea). The materials of the crust do not all have the same age; as we will see, the bottom of the oceans is very young, less than 100 million years old, while the continents are made of much older materials — in some cases, as we have seen in Section 6.2, they have an age comparable with that of the Earth.

Beneath the crust there is the *mantle*, 2870 km thick, constituting five-sixths of the volume of the Earth and containing most of its mass. The mantle is relatively hot; its temperature varies from a few hundred degrees near the crust to a few thousand degrees near the core, where the pressure is more than a million atmospheres. Seismic waves indicate that the mantle is solid, with a few parts, those that furnish lava to volcanoes, being almost fluid. Continental drift arises from the interaction between the crust and the mantle and presupposes that the mantle is rather flexible, like highly viscous fluid. The mantle can therefore be described as a solid that, for a few phenomena that occur on a timescale of millions of years, behaves like a fluid.

Inside the mantle there is the *core*. Measurements show that it has an external part that is liquid, and an internal part that is solid. The core is the densest part of the Earth; its mass is so high that it must be made of metals, predominantly iron (one of the most abundant elements in the universe; Figure 4.9) and nickel. The Earth's magnetism is due to the ferrous core and the Earth's rotation. The internal part of the core is solid because of the enormous pressure is subjected to — nearly four million times higher than the atmospheric one. This enormous pressure compresses the material so much that, in spite of the high temperature (almost 3000°), the material rearranges itself in a structure similar to that of the crystal lattice of a solid.

The structure we have described is what is expected from the history of our planet. When it was born, the Earth was a ball of molten rock, because of the energy released by the material falling on it. During that phase the matter differentiated according to its mass; the lightest elements went to the surface, forming the atmosphere, the oceans and the crust, while the heavier elements like iron and nickel descended toward the center, forming the ferrous core of the Earth.

9.3 Continental Drift

The theory of continental drift has ancient origins. The first formulation is due to a great exponent of scientific rationalism (Section 3.1), the English philosopher Francis Bacon (1561–1626). The Minister of Justice in the reign of Elizabeth I,[4] studying the first maps of English and Dutch sailors representing the coasts of Africa and South America, noted that the forms of the two continents fitted together like the pieces of a puzzle and wrote in *Novum Organum* (1620) that it was not possible that this happened by chance; the only explanation for their shape was that the two continents had been united in the past. In the following years, other similarities were observed between the coasts of different continents and remains of mountain ranges going from one continent to another were found. Also, the discovery of identical fossils on distant continents where the species are today different (as one would expect from evolution happening in separate environments) pushed many naturalists and travelers of the 17th and 18th centuries to take into consideration the possibility that the continents were moving around.

Credit for the first formulation of the theory goes, however, to the American geologist Frank Taylor (1860–1938) who published it in 1910 without much success. A few years later in 1912, the German scientist Alfred Wegener (1880–1930) published his famous book *The Origin of the Continents and the Oceans* (paraphrasing Darwin's *The Origin of Species*), which is considered the basis of modern geology. Wegener was not a geologist, but a meteorologist who arrived at the theory that revolutionized geology from the studies of the climates of the past. He studied glaciations and, looking for traces of the very old ones, he found that the same glaciation was occurring at different latitudes on different continents. On one continent it stopped before the tropical regions, while on another it descended to the Equator. He asked himself if the explanation would not be found in the fact that, when those glaciations occurred the continents were in positions different from those of today. Searching for a proof of this intuition, he collected and classified all the writings of the naturalists of the past centuries, and finally published his theory.

According to this theory, 250 million years ago (a short time compared to the 3800 million years of the Earth), on our planet there was a single great continent: Pangaea (which owes its name to Taylor and Wegener).

This great continent began to break (Figure 9.4), and from the fractures water gushed out and the oceans were born. The continents began to move in different directions: the Americas toward the west, Antarctica toward the south, and Australia and India toward the north-northeast. The theory was for years considered a curious deviation, a fascinating idea but too strange to be possible. The measurement of seismic waves, of which we spoke in the previous section, showed that the mantle was solid and this was considered the proof that the continents could not move. To move they needed some

---

4 *Among the legend of the time it has been told that he was the natural son of Queen Elizabeth I, and the author of the works of Shakespeare.*

Pangea (250 millions years ago)

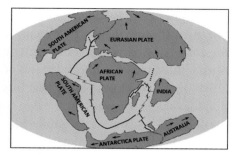

End Creataceus (70 millions years ago)

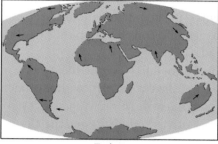

Today

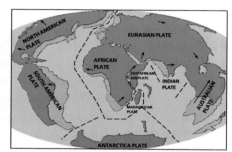

Possible situation in 50 millions years

Figure 9.4: The surface of the Earth has changed radically in the course of time. (a) 250 million years ago, there existed Pangaea, a single great continent in which we recognize the Gulf of Tethys, which will become the Mediterranean. (b) 70 million years ago (when the dinosaurs had already inhabited the Earth for 150 million years), the continents were well differentiated, India was still an island, and Australia and Antarctica were united. The black dotted lines represent the ocean's ridge, the fracture from which the matter of the mantle reaches the surface; the arrows indicate the directions of the continents' movement. (c) The present Earth (see also Figure 9.6). The arrows indicate directions of the continental movement we measure today. (d) The continents as they could be in 50 million years: the Mediterranean Sea will largely disappear, squeezed by Africa and Australia will be connected to the Asian continent. The parts in a lighter color are the increase in the emerged land.

fluid beneath them on which to float, something the measurements were excluding.

The supporters of the theory, starting with Wegener, were not discouraged (Wegener died in 1930 in Greenland, where he was searching proof of his theory); they continued to gather evidence of the existence of Pangaea, and in the end the theory was proven to be correct. Starting from 1960s the experimental data collected showed uncontestedly the existence of Pangaea. Wegener's theory was a fact — although still not explained by science, it was irrefutable. A fact without which life on the Earth would not have been possible.

In favor of the theory there are numerous indirect proofs. As already noted, traces of ancient mountain belts have been found in Africa and South America. They seem to have belonged to the same range that crossed the two continents when they were united. Other traces of mountain ranges have been found between Africa and India, and Antarctica and Australia.

Another proof is furnished by evolution, traced from fossils that, as noted by the naturalists in the 19th century, were the same on all continents up to 200 million years ago, and then became differentiated. It was as if they had lived in the same environment until a certain moment, and then continued their lives in different places, no longer interacting with one another. Another indirect proof comes from fossils found in the arctic regions belonging to species needing tropical temperatures to live. These facts are difficult to explain unless we admit that the continents have been, in the past, in different positions than those of today.

The decisive element that dispelled all doubts about continental drift comes from *paleomagnetism*, a science that allows to reconstruct the direction of the magnetic field in various periods of the geological history of the Earth. The Earth behaves like a giant magnet that orients compass needles toward the north everywhere on it. Paleomagnetism looks for rocks containing small fragments of iron. When they were formed (erupted by volcanoes) they melted, allowing iron fragments to orient themselves along the Earth's magnetic field, like many compass needles. When the rocks solidified, these iron fragments were blocked in the direction of the magnetic field at that time, transmitting us its direction. The study of these natural compasses shows that rocks of the same age were indicating different directions in different continents; as if at the same time the north was different in Africa, America or Australia, like if the continents had rotated independently of each other. Replacing the continents in the positions given by Wegener's theory, the magnets were all oriented toward a common north. This was considered by everyone the indisputable proof of the theory, a certainty of our past that we can explain by considering a solid, rigid mantle that behaves as a liquid on the timescale of millions of years.

Today we can also add a direct proof of continental drift. The satellite techniques of measuring distances are so accurate that we can directly measure the displacement of continents (the values range from a few millimeters to a few centimeters per year), eliminating every bit of remaining skepticism. For example, the velocity at which India moves toward the Euro-Asiatic plate is 4 cm per year.

In the next section we will briefly speak of the plate tectonic, the model that best explains the continental drift.

9.4 Plate Tectonics

According to the model of plate tectonics, the entire surface of the planet is divided into plates, which move relative to one another, either approaching or moving apart. In the points of collision, the great corrugations that form the mountain ranges are created (they do not exist on other planets of the Solar System). Above the plates we find the continents which are composed of relatively light material (Section 9.2), and therefore they float on the mantle, which is made of heavier, denser and hotter material.

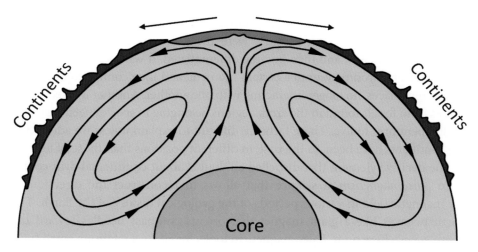

Figure 9.5: *According to the model of plate tectonic, the movement of the continents is due to the convective motion of the mantle on which the continents, being lighter, float.*

According to this model, the mantle, in spite of being solid, behaves like a fluid heated from below (Figure 9.5). Having to transfer the heat produced inside the Earth to the surface, it triggers a process of convection no different from what happens in a pot of water on a fire. The material of the mantle heated near the core, increases in volume (like water at the bottom of the pot), becomes lighter and rises to the surface, where is pushed aside by new material that rises, it moves horizontally, and drags the continents with it. In this horizontal motion the mantle cools and increases in density. Its weight, after about 100 million years, is greater than that of the hot material beneath it, and then the mantle and crust sink again toward the interior of the Earth. Geologists call this layer in horizontal movement the *lithosphere*; it has a thickness of about 100 km. The lithosphere consists therefore of the crust and the upper part of the mantle. Pangaea was fractured in the places where matter from the mantle rose; the horizontal motion of the lithosphere is what causes the continents to move apart or to approach each other. Sea water fills the cavity left behind by the movement of the continents, and so new oceans are born. For this reason the ocean floors are young, because they are made of material newly arrived from the mantle that will descend again inside the mantle in less than 100 million years. Exploration of the bottom of the oceans in recent times has confirmed this description: at the center of the oceans there is a long range of mountains and volcanoes, called the ocean ridge, easily identifiable, going from north to south in the Atlantic and Pacific (Figure 9.6). Along the ocean ridge matter from the mantle comes to the surface. It is where most of the underwater volcanoes are found.

When the lithosphere gets deep, it carries with it the ocean floor, and this is the reason why at the bottom of the ocean we do not find traces of the

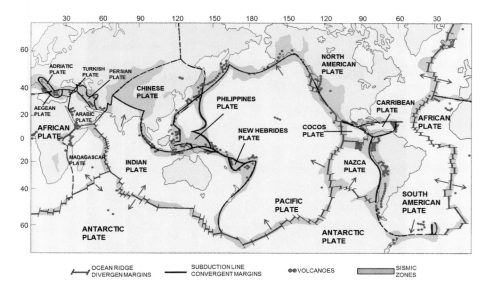

Figure 9.6: *The various plates into which the crust of the Earth is divided. The blue lines are the ocean ridges, the diverging margins of these plates from which the ocean floors are born. The continuous green lines are the zones in which the plates impact one another, they are the converging margins of the plates (the faults), the zones with the most active volcanoes and the strongest earthquakes. Note how the volcanoes follow these fault lines.*

sediments 100 km thick that would have been formed in 3.8 billion years: they have been carried into the mantle. This movement of the lithosphere does not disturb the continents, which are floating on the mantle.

The interaction of the lithosphere with mantle is felt more on the edge of the continental plates, where, as we will see, most of the volcanoes are located and most earthquakes happen. The centers of the continents, instead, are less influenced by these phenomena and so they conserve traces of the oldest materials of the Earth, through which it is possible to reconstruct its history back to its origins (Figure 6.2).

Figure 9.6 shows the plates into which the surface of the Earth is divided. Note on the upper right, where the Atlantic crosses Iceland, the only point at which the ocean ridge emerges from the sea (the island emerged from the sea only 20 million years ago). To this, we owe the volcanic activity that makes Iceland a true geological rarity. Figure 9.7 depicts the "ocean ridge" crossing the island with two borders that are moving apart of about 1 cm per year, leaving space for a future ocean. Two-hundred-and-fifty million years ago, when Pangaea fractured, the Atlantic must have looked similar. In other parts of the island, very long strips of volcanoes (Figure 9.8) are found similar to those existing along the ocean ridge.

In many places on the Earth we observe fractures of the continents. The Red Sea is an example of a sea just formed; 70 million years ago (Figure 9.4(b)) it

Figure 9.7: The birth of a new ocean: the ocean ridge as it surfaces in Iceland. In such a way the Atlantic Ocean was born 250 million years ago.

did not exist. In the same zone, the movement of the Madagascar plate with respect to the African one is giving rise to a new sea. Whoever has been in Tanzania, Kenya, Somalia or Eritrea has seen the Rift Valley, a great crack on the surface of the Earth, oriented from north to south, that crosses a large part of East Africa and is associated with numerous volcanoes. The crack is much larger than that in Iceland (Figure 9.7); at some points it is so deep that in its interior, lakes, that one can recognize in maps of southern Africa, have formed. They are all rather long, narrow and oriented from north to south. Within 50 million years (Figure 9.4(d)) in that piece of Africa, in place of the Rift Valley there will be a new born sea, which will not differ much from the Red Sea. The Red Sea, instead, will become a closed sea, like the Caspian Sea, as a result of the Madagascar plate motion. The Indian plate, which today has

Figure 9.8: The volcano Lakagígar (with craters of Laki) in Iceland is a long cleft oriented along the ocean ridge. On 8 June 1783, it became active, erupting lava to from a slit 25 km long. The eruption lasted eight months and poured on the island 15 km³ of basaltic lava that covered a zone of 600 km². The toxic cloud emitted by the volcano caused the death of 10,000 Icelanders of 80% of sheep and 50% of the remaining cattle and resulted in the poisoning of the island's meadows for many years. The fine dust carried by the erupted gas in the high atmosphere reduced the average temperature of the Earth by 1° (Figure 7.5). (© Fred Kamphues/ Mill House.)

advanced 2500 km inside Asia, will disappear almost completely. Himalayas will notably reduce its peaks by the effect of erosion, and will be reduced to small hills.

Australia should move north together with Borneo until it reaches Asia to form a new mountain range, a future Himalayas. The Mediterranean, what today remains of the ancient gulf of Tetide will disappear together with Italy, crushed between the African plate and the European one which, today are approaching at a speed of 7 cm per year. In its place there will be a great mountain range similar to the Alps due to the encounter between the two continents: some lakes will witness its past existence. There will also be new emerged land, mainly on the coasts opposite the direction in which continents move, like the Atlantic coast of America. The new land is represented in clear colors in Figure 9.4(d). Only Antarctica will not move from the position it is in today.

On a longer timescale, between 100 and 200 million years, America should rejoin the Euro-Asiatic plate, which should already be united with Africa and Australia. At that point, all (or almost all) the emerged land should be united, in a new supercontinent, a new "Pangea." The climate of this new supercontinent will probably be desertic in its interior with terrible hurricanes on the coast (for the great extension of the oceans, they should be much more powerful than they are today).

For Italy, great changes will be noted in 2 million years. In place of the Alps there will be hills, while at the place of the Adriatic Sea young mountains have begun to grow, created by the collision between Africa and Europe. Corsica and Sardinia will be united with Liguria in a long peninsula, and the Po will probably have changed its course, going from East to West finishing in a sea north of Corsica in what will remain of the Mediterranean. In 5 million years, the Mediterranean will exist only west of Sardinia.

9.5 The Importance of Earthquakes and Volcanoes for Life

Figure 9.9 represents the cycle of the lithosphere on the surface of the Earth. To the left is the line of underwater mountains of the oceanic ridge, from which the matter from the mantle arrives at the surface. From that line, the marine seabed diverges and pulls away the continental plates. On the right of the figure there is the zone where the lithosphere, having become cold and dense, wedges again into the mantle. The process by which the lithosphere plunges into the mantle is called *subduction* by geologists, and it is dramatic but vital for the Earth: dramatic because this phenomenon causes the worst earthquakes and the most destructive volcanic eruptions; vital because, as we will see, without this process life would not exist on the Earth.

The fracture zone between two plates in motion is called the fault, and along the fault rocky masses of various natures and ages come into contact with each other resulting in transverse and vertical displacements. Very often the displacement of rocky blocks does not happen at a well defined point and so geologists speak of "fault zones", where one finds numerous secondary faults attached to a principal fault.

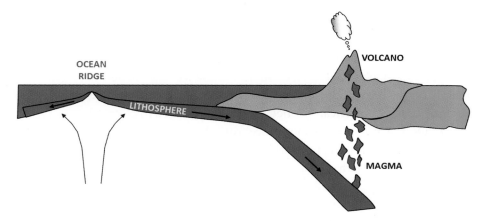

Figure 9.9: Subduction: the descent of the lithosphere under the mantle causes earthquakes and volcanic eruptions through which materials precious to life, like carbon and oxygen, are restored to the atmosphere and to the terrestrial crust.

Along the faults earthquakes occur because the motion of the surface does not happen with continuity. Between the plates in motion there is in fact strong friction that blocks the movement until the stored energy manages to move it again. The movement thus occurs in a shot and when it happens in a few seconds the energy stored over decades or centuries along the fault is liberated, causing earthquakes that can be terrible. Of the four strongest earthquakes registered by modern seismographs (the only ones above the 9 level on the Richter scale)[5]: one was in Chile in 1960, the second in Alaska in 1964 both connected with the subduction of the Pacific lithosphere under the American continent, the third, which provoked the disastrous tsunami of the Pacific on 26 December 2004, is connected with the subduction of the Indian plate under the China plate (Figure 9.6) and the last which struck Japan on 11 March 2011 with a terrible tsunami was again the subduction of the pacific lithosphere under Asia.[6]

The intensity of earthquakes depends, therefore, on how much energy is accumulated along the fault (or how big the friction is between the plates) and on an unpredictable factor — whether the energy will be released in one big earthquake or in a series of smaller ones. Because the energy accumulated in a fault is known to geologist's, the maximum intensity of an earthquake is known, even if one cannot say when it will happen (like the famous Big One, long awaited on the San Andreas' fault, which runs through California and the city of San Francisco) and how much energy will be released. There also exist "silent" earthquakes, ignored for years by geologists and only recently discovered. They can discharge their energy gradually, without causing damage. Water trapped beneath the soil seems to be an important factor in determining the seismic intensity: a large amount of water causes the lithosphere to be flexible, reducing friction and making large earthquakes less likely.

Also associated with the movements of continental plates is the phenomenon of volcanism, that happens at points of fracture of the terrestrial crust, along the ocean ridge where the lithosphere rises to the surface and in subduction zones where the lithosphere sinks into the mantle. Nearly all the underwater volcanism of the Earth is tied to the oceanic ridge which develops for about 70,000 km at various depths, from Iceland, where the ridge reaches the surface (Figures 9.7 and 9.8) down to 5000 m beneath the surface of the sea. It is a volcanism difficult to study, because until today it has not been possible

---

5 Nine degrees on the Richter scale corresponds to the release in a few seconds of energy equivalent to 32 billion tons of TNT, equal to the simultaneous explosion of two million atomic bombs like the one that exploded over Hiroshima.
6 In the course of the 20th century, earthquakes caused about 3.2 million deaths, versus 2.4 million due to all other natural disasters. Earthquakes are dangerous because they strike without warning, and for that reason they kill more people than any other natural cause. Science is not capable of finding a way to predict the place and the hour of earthquakes, but only what the maximum intensity can be. The only possible prevention against earthquakes is to build houses that can survive to its strength.

to build instruments able to approach such volcanoes. It is estimated that on the average every year they erupt more than 12 km³ of lava.

Subduction is responsible for a different kind of volcanism that occurs mainly on the surface. The famous ring of fire of the Pacific (Figure 9.6) is due, for example, to the wedging of the Pacific lithosphere to the west and north under America and the Aleutian Islands, to the east and south under Asia and Oceania. The same phenomenon has produced the Italian volcanoes which are aligned along the fault that separates the European plate from the African one.

It is not clear why these volcanoes are associated with subduction.

In any case, it is our good fortune that these volcanoes exist. They are localized near the point where the lithosphere descends into the mantle and are generally 120–180 km above the descending lithosphere (Figure 9.9).

It is precisely because these volcanoes are located at the beginning of the descent of the lithosphere into the mantle that they play an important role in the existence of life on the Earth. Many substances of the crust which are necessary for life, like oxygen, carbon and water (which is fixated in the rock), are transported by rivers to the oceans with the sediments. These substances would be buried at the bottom of the sea for hundreds of millions of years, if not forever. On the contrary, with subduction they enter the mantle, which through the heat separates gas and water from the rocks and sends them back into the atmosphere through the volcanoes, thus returning them to the biological cycle. Without this process the surface of the Earth would be dramatically deprived of these elements necessary for life. The absence of subduction on Mars has probably caused the disappearance of these elements and of water on the surface. The volcanism on Mars (Section 8.4) has probably been different from that on the Earth; in the absence of subduction it has discharged the heat contained in the interior of the planet, recycling only a small part of the water and the organic material that could have existed on the planet. Without the movement of the continental plates the atmosphere of the Earth would have become like the one of Mars, much more tenuous.

9.6   The Importance of the CO_2

We have seen that without subduction and volcanoes the atmosphere would be depleted of the greenhouse gases (especially CO, CO_2 and CH_4) and the Earth would have a much lower temperature, similar to the Moon, which, being at the same distance from the Sun, receives a similar amount of heat. The Moon has a mean surface temperature of $-18°C$, against a mean surface temperature of the Earth of $15°C$ — a difference of $33°C$ given by the greenhouse effect, which allows to understand what would happen if the atmosphere would not hold the heat of the Sun.

The atmosphere regulates the temperature of the Earth very closely, and guarantees that, for much of the planet, the temperature is always between $5°C$

and 40°C, well inside the 0–100°C interval, for which water is liquid. It is an interval essential for the survival of most of the animal species (Figure 6.3).

According to some researchers, this extraordinary result is due to the CO_2, which, aside from being a greenhouse gas, acts as a natural thermostat put in place to protect the climate of the Earth. The CO_2 in fact, reacts with the silicates of the rocks ($CaSiO_3$), transforming them into calcium carbonate ($CaCO_3$)[7]; it is also absorbed by the plants and the phytoplankton of the sea to grow their structures (Section 6.5) — two mechanisms that oppose changes in the temperature.

The reaction between CO_2 and silicate is in fact very sensitive to the temperature, so that when the Earth warms up the reaction becomes more efficient and more CO_2 is subtracted from the atmosphere to react and form calcium carbonate, causing a depletion of the atmospheric CO_2 and a reduction in the greenhouse effect opposing the increase in temperature. Vice versa, when the Earth cools the reaction becomes less efficient, and the CO_2 in the atmosphere increases, causing a greater greenhouse effect, and then increasing the temperature. In other words, this reaction modifies the concentration of CO_2 in the atmosphere in a way that opposes changes in the temperature of the Earth.

Plants and phytoplankton behave in a similar way; during their life they absorb CO_2, which, at the end of their life cycle, will be deposited in the soil or in the seabed. In the end part of it will go into the deposits of petroleum and coal.

It has been observed that the speed at which vegetables grow depends sensitively on the quantity of CO_2 in the atmosphere; it increases with an increasing quantity of CO_2 in the air and it decreases with a decreasing quantity. Plants in greenhouses in which the CO_2 is kept artificially high, develop more than those kept in a normal greenhouse. The CO_2 behaves as fertilizer; an increase in its atmospheric content makes the forest more luxuriant.

In 2009, these laboratory results were confirmed on the field. An article published in the journal Nature, based on 80 samples taken from African forests, allows to estimate that the carbon captured by the trees has increased in the last 40 years by 340 million tons per year. Something similar should be happening in the forests of the Amazon and Siberia, and therefore the net effect is estimated to be an increase in CO_2 depletion by more than a billion tons per year.

This effect obviously do not compensate for the deforestation caused by agriculture due to the increase in the world population. Its efficiency is much

---

7 *The reaction that transforms a silicate ($CaSiO_3$) into a limestone ($CaCO_3$) is $CaSiO_3 + CO_2 = CaCO_3 + SiO_2$. For anyone who wants to know more on this point, we suggest: J. Walker, P. Hays and J. Kasting, 1981, Journal of Geophysical Research 86, 9776.*

reduced today because of the decrease in forests, and is believed by many researchers to be one of the causes of the ongoing extinctions (Section 7.4).

These two natural thermostats would be in crisis if the volcanoes did not release back rapidly into circulation the greenhouse gases that marine sediment had subtracted from the biosphere. The control of climate by means of CO_2 is therefore the third gift from our planet to us, after the continental drift which restores the land destroyed by erosion, and the volcanoes which restore elements essential for life that would otherwise remain buried underground. Geology does not fully explain these phenomena, but we owe them our existence.[8]

These thermostats act slowly and are not able to control relatively rapid variations in the climate which in the past led to glaciations, of which we will speak in the following, even though they have contributed to reducing their effects and to restabilizing the temperature. They are not able to compensate for the climatic variations caused by volcanic eruptions that, as we have seen (Section 7.1), have caused mass extinctions in the past. They certainly cannot compensate for the climatic variations that humans are causing with the great quantity of greenhouse gases emitted into the atmosphere (Section 7.4) and the reduction in the forests.

The greenhouse gases would have a terribly negative effect if they brought the temperature of the Earth outside the interval of survival in Figure 6.3. A risk could come from the large quantity of gas (most notably CO_2 and CH_4) contained in depths of the oceans where gases coming from submarine volcanoes and from sediment have been accumulating over time. It is a concentration that has always existed and that exceptional events can liberate. In Section 7.2 we have seen that the best explanation for the worst known extinction, which happened 250 million years ago in the Permian, is attributed to the liberation of these gases.

A similar phenomenon, although of rather modest dimensions, happened from time to time in the volcanic lakes of Cameroon. Lying deeply in them, great quantities of CO_2 arising from underground is accumulated. Some time it occurs that the gas rises rapidly to the surface, with dramatic consequences for the local population. The worst of these events happened in 1986 at Lake Nyos in Cameroon and killed 1800 people. The process that liberated the gas from the bottom of the lake is similar to what could have happened in the Permian (Section 7.2). The gases, and mainly CO_2, have the property that they can be absorbed by water (an example is CO_2 in sparkling water); in quantities that increase with increasing pressure and decreasing temperature.

---

8 *Regarding the excitements of the Martian exploration there are people who fantasize about the possibility of setting in place mechanisms for emitting greenhouse gases, above all CO_2, into the atmosphere of the planet to raise its temperature. This idea is not absurd — humans have acquired a great capacity to produce greenhouse gases. A good idea would be to export those in excess on the Earth to Mars.*

Lake Nyos is very deep, like many other volcanic lakes. At a depth of 200 m, with a pressure of 20 atmospheres and a temperature of 5–10°C, a concentration of more than 15 liters of gas per liter of water was measured. On 12 August 1986, the concentration must have been similar if not greater. Then in the depths of the lake there was some instability; some of the water from the depth rose a few meters toward surface where the water was warmer and at a lower pressure. In those conditions the gas was liberated from water, increased its volume and began to bubble toward the surface, creating an upward current which claimed more of the deep water that liberated more CO_2. A pump was then put in motion that was able to empty in a few minutes the lake of the gas contained in its depth. On the surface a great cloud of CO_2 was formed that, being heavier than air, descended into the valley, reaching a distance of 23 km from the crater before being dispersed by the wind. On the path, this cloud killed by asphyxiation (CO_2 is not toxic) every person or animal it encountered. One thousand and eight hundred people and tens of thousands of animals died. Then the wind dispersed the CO_2, and convection carried it into the upper atmosphere, where it contributed to the greenhouse effect.[9] Today, thanks to the intervention of the UN, a mechanism that controls the degassing of the lake has been put in place (Figure 9.10). It can be activated or deactivated with a tap at the bottom of the lake. The magnificence of the jet in the figure shows how efficient the process can be once it starts.

In the deep cold water of the oceans, especially in polar regions, there are vast quantities of greenhouse gases (mainly CO_2 and CH_4) produced by submarine volcanoes and the degassing of sediment. We are speaking of an enormous quantity, trillions of tons of gas, greater than the emissions that come from the Earth over thousands of years. Every year a part of this gas rises to the surface and contributes to the greenhouse effect.[10] Can a rapid increase in the temperature of the Earth put large quantities of this gas in circulation? Can it create fountains similar to that in Figure 9.10 but with gigantic dimensions involving the gas contained in all the depths of oceans? Unfortunately, we are speaking of physical and chemical phenomena difficult to evaluate, but if today it should happen on a worldwide scale the effect on the climate would be devastating.

---

9 *You should not be amazed that the CO_2, although heavier than air, could climb to great heights. The lower atmosphere is mixed by convective currents. As an example, in the upper atmosphere, fine dust and sand from the Sahara are found (transported by winds to the north of Europe); their weight is much greater than that of a molecule of CO_2.*
10 *Among the explanations given for the disappearance of boats in the Bermuda Triangle, there is that of the bubbles of gas rising suddenly from the bottom of the sea. The mixture of gas and water would not have the density to keep a ship afloat and the ship would therefore sink. For a test, a ship was sunk by making air bubbling under it.*

Figure 9.10: The jet of CO_2 on Lake Nyos: a fountain of sparkling water. Thanks to a UN project, the CO_2 is extracted from the bottom of the lake in a controlled manner, to avoid tragedies. The majestic sight of the jet shows how efficient the degassing mechanism can be once it is primed. See the website on Lake Nyos. (© AGU, M. Halbwachs et al., 2004, EOS 85, 30, 281—285.)

Many of the mass extinctions of the past (Section 7.1), if not all, are explained by climatic variations which the reaction mechanism of the Earth has not moderated in times necessary for the survival of biological species. This should be a warning against reckless emission of greenhouse gases that humans are creating, because, once triggered, these mechanisms often are self-feeding and cause changes in the climate that nobody can stop.

9.7 Glaciations

We have seen that the climate of the Earth is unstable and can easily be modified. These changes are based on the theories explaining glaciations, which, as we have seen in Figure 7.7, have shown a periodicity of 100,000 years in the last millions years. The study of sediments shows that in the Southern Hemisphere the glaciations were less intense than in the north. During the last glaciation, as an example, the temperature of the Southern Ocean was only 2°C lower than it is today.

The last glaciation had its maximum 18,000 years ago, in Europe the ice arrived to the south of Berlin, covering England and a good part of France. In America they came down to New York and covered all of Canada. The sea level was then 120–150 m below what it is now; a good part of the Adriatic sea was dry; New Guinea was attached to Australia; Asia and America were joined together through the Bering strait[11]; many islands (like Malta) were attached to the nearby continents.

---

11 *The Bering Strait divides Asia from America by a channel 95 km wide and less than 50 m in depth. During this glaciation it was almost 100 m above sea level, creating a bridge through which our species crossed to reach America.*

According to the cycles in Figure 7.9, we should already be in a new glacial period. The warm interglacial period (the relatively narrow peaks in Figure 7.9) should have ended about 12,000 years ago, and then the temperature of the Earth should have descended, rendering uninhabitable some of the most populated zones of the Earth. It is therefore important to understand the cause of glaciations, the cause of this extraordinary periodicity and why it occurs differently in the two hemispheres.

Figure 7.9 shows that the temperature and greenhouse gases are strongly correlated: in the preceding section we saw that the greenhouse gasses can be a cause (the greenhouse effect) or a consequence (degassing of the soil in marine depths; Figure 9.10) of the increase in temperature. In any case they play a central role in the climate cycle of our planet, and every climate model must take them into account. The mechanism by which the greenhouse gases could provoke thermal cycles is simple: a phase of warming (like what we are experiencing now) is caused by an increase in volcanic activity and/or excessive use of fossil fuels. A phase of cooling comes instead with their diminishing through the processes described in the previous section that subtract greenhouse gases from the atmosphere. We have seen that the temperature of the Earth, given its distance from the Sun, should be $-18°C$; a reduction in greenhouse gases can thus produce a glaciation.

Marine currents and winds redistribute heat among various latitudes, rendering the tropical zones less warm and the northern zones more temperate. The quantity of water transported by the marine currents is enormous; it is estimated to be about a million cubic meters per second (ten times that of all of the Earth's rivers combined), and because the water has a thermal capacity among the highest in nature,[12] it is easy to realize that the quantity of heat transported is enough to modify the climate of the Earth. To understand how much heat is carried by the currents, we look at the distribution of arctic ice, which, in Canada, descend to the latitude of about 73° north (the latitude of northern Scandinavia), while in Europe, because of the Gulf Stream, they reach the latitude of 82°N, 900 km further north, permitting navigation up to the island of Swalbard (from which the unfortunate expedition of Nobile started).

According to the most-accepted models, the mechanism that moves the marine currents could be obstructed by the melting of ice in the northern zones, which by blocking the currents can lead to a new glaciation.

To understand how this can happen, we have to understand the mechanism called the "ocean conveyor belt" (Figure 9.11), a circulation that remixes the

---

12 For this reason water is used for most heating or cooling circuits. The thermal capacity of water, by volume, is 3000 times greater than that of air, so it is not difficult to understand how it can modify the temperature.

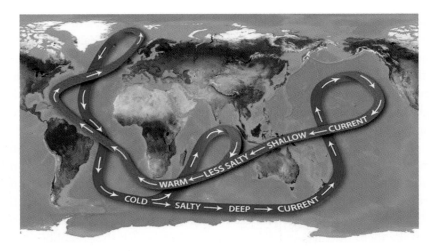

Figure 9.11: The thermohaline circulation that continually mixes ocean waters and transfers heat to the northern zones of the Earth. (© F. Perroni & NASA visible Earth.)

waters of the oceans and transfers, with great efficiency, heat from the Equator to the northern regions.

This mechanism is not very different from the thermal siphon used in many houses: there is a heater (the Sun) that heats the water in the equatorial zones, and there are the users (the cold northern zones) where the warm water arrives and deposits its heat, and then returns the coldness to the Equator to be heated again and to begin another cycle.

The variations in density are the driver of the currents (like in a thermal siphon) which, in the case of marine currents, are caused not only by the variations in temperature but also by the variations in salinity. For this reason the global circulation of the oceans is called *thermohaline* because it is governed by the temperature (*thermo*) and by the salinity (*haline*). The density of the water increases if the temperature decreases and if the salinity increases, i.e. cold salty water is denser than warm sweet water.

The principal motor of this circulation is the Gulf Stream (Figure 9.11), fed by the hot and not very salty water of the equatorial zone that moves toward higher latitudes, where the winds (which in the North Atlantic blow toward Europe) cool it. The same winds cause also a strong evaporation that increases the salinity (only sweet water evaporates) and cools it further. Proceeding north, approaching to the arctic, the water freezes, and because only sweet water freezes, the salinity of the remaining liquid part increases further[13] and in the Arctic the surface water is so cold and salty that it becomes denser than the water below it (less salty), and it sinks attracting new warm water from the Equator. The cold water then returns south through the Atlantic Ocean

---

13 *Sometimes, inside the ice of the polar ice pack, bubbles of liquid water are observed. They are not frozen because of their high salinity.*

(because the passage from the arctic to the Pacific Ocean is closed by the shallow Bering Straits); it moves slowly south toward the Indian and the Pacific Oceans.[14] Along this path waters will be divided into thousands of trickles, following the jagged structures at the bottom of the sea, it mixes with the warmer and less salty water and returns to the surface to start a new cycle.

Once the mechanism that feeds the currents is understood, it is also understood that if the Gulf Stream slows down or stops, it will slow down or stop the entire oceanic circulation. This could happen if on their way toward the North Atlantic the equatorial waters, which were not very salty, are further diluted by the sweet water coming from the melting polar ice. They become less dense and will remain floating on the deeper and saltier waters. If they do not sink they will not call up warm water from the tropics and the motor of the currents will stop. With the currents stopped, or at least reduced, the winters become more extreme and the summers less hot. The snow that falls during the winters does not melt entirely during the summer, and the ice descends every year to lower latitudes and a new glaciation begins.

With the passing of the millennia, the phenomenon is inverted: part of the ocean waters goes into the glaciers that cover the continents, causing the sea level to drop. With less water the oceans become saltier and the motor of the currents slowly increases; the climate is not as cold and we move toward a new warm period. There is more land in the Northern Hemisphere, and so the circulation of currents in the Southern Hemisphere encounters less obstacles. This difference in distribution and morphology of emerged land in the two hemispheres could be the reason why the glaciation is more intense in the Northern Hemisphere than in the south. This may also be the reason why the most terrible glaciations never turned the Earth into a ball of ice (from which our planet could not heat again, Section 7.1); in the Southern Hemisphere, the currents have always been able to find the heat to melt the northern ice.

According to some models, the start of a glaciation in some cases can be very fast; it can happen by a phenomenon called *the collapse of the Arctic ice*. According to this mechanism, the rising sea level caused by the heating of the Earth raises the icy platform of the coast of Antarctica and of Greenland. These platforms are a buttress for the ice of the Arctic continents, which, being no longer held up, slide into the sea. This would cause a further rise of the seas that would increase the sliding of continental ice into the oceans and in a short time the ocean level may increase in tens of meters.[15]

---

14 *This part of the path is less easy to measure than that of the warm surface current, because it is very diluted, but it obviously exists because the great mass of water that comes from the north Atlantic must somehow return.*
15 *Also, ice floating on the North Pole would melt rapidly in this phase, if they still exist. The warm currents of the Pacific Ocean, moving northward, are today stopped by the shallow seabed of the Bering Strait; a rise in the sea level would augment the entry of warm water to the Arctic, accelerating its melting. Staying afloat they will not contribute to the ocean level rise.*

This would be a collapse from which humanity could not defend itself. The construction of dams, as is done in Holland (and would be done in Venice), could save a few cities, but the coastal zones where the largest part of the population lives would be submerged. The collapse of our society would happen long before the collapse of the polar ice.

Mixing greenhouse gas and the ocean currents is possible to build mechanisms that alternate hot and cold periods. But we will not succeed in explaining the periodicity of 100,000 years as seen in Figure 7.9. It is too precise to be explained by imprecise mechanisms like greenhouse gases and marine currents. We should, for example, hypothesize that every 100,000 years there are violent volcanic eruptions that increase the greenhouse gas and initiate the cycle. This idea is difficult to sustain, because today we are at a maximum temperature and there is no trace of violent eruptions.

9.8 Glacial Cycles

The existence of thermal cycles in the climate of the Earth has been known since the 19th century. They were known by observing samples of geological layers and marine sediment, although there was not the impressive evidence in Figure 7.9. To explain these cycles, the Serb mathematician Milutin Milankovitch (1879–1958) proposed in 1920 a theory that did not have much success then, but is taken seriously today because it is the only one that explains so precisely the observed periodicity. The theory starts from realizing that a precise periodicity cannot be tied to casual phenomena (such as an increase in CO_2 due to volcanic eruptions). Instead, it requires a mechanical movement like that of the pendulum of a clock, or like the periodic motions of the heavenly bodies. There are many examples of periodic motions connected to heavenly bodies: the alternation of the seasons is due to the motion of the Earth around the Sun, that of day and night to the rotation of the Earth on its axis, the tides to the motion of the Moon around the Earth, and the emission of pulsars to their rapid rotation (Section 4.5).

Therefore, Milankovitch searched for a motion with a period long enough to justify the cycles and suggested that the climate cycles might be due to a change in insolation, induced by periodic variations in the orbit of the Earth. He identified three motions: the periodic variation of the eccentricity of the Earth's orbit (the Earth moves around the Sun in an ellipse very close to being a circle, whose eccentricity varies periodically by 3%), the variation of the inclination of the Earth's axis, and the precession of the equinoxes. These three motions have periodicities of approximately 100,000, 40,000 and 23,000 years; they are found in the geological data and also in Figure 7.9. The periodicity of 100 000 years is the most intense, as one can see from the figure.

When the theory of Milankovitch was postulated it did not have much success because it did not seem possible to cause important climatic changes like glaciations without changing the heat that the Earth receives from the Sun on the average (the orbit changes so little that the heat that arrives on the Earth, averaged over a year, is practically constant). For this reason the theory has been forgotten for 50 years.

Other hypotheses were made to explain the periodicity of glaciations, and they are always based on astronomical phenomena. One of them hypothesizes a large asteroid/planet called Nemesis (after the Greek goddess of troubles[16]), which passed near the Earth periodically, causing phenomena (volcanic eruptions, earthquakes, *etc.*) that may lead to climatic changes. It is a thesis that, with modern knowledge of the Solar System and of the geological past, is no longer sustainable. A similar hypothesis was proposed more recently by Fred Hoyle (whom we have already met in Sections 2.1 and 5.10); he suggested that the glaciations were provoked by a comet of long period (from the Oort cloud; Section 8.2) that enters in the inner part of the Solar System with a period of 100,000 years, leaving fine dust particles that increased the Earth's reflectivity to the sunlight (Figure 7.5), and causing a cooling that brought on glaciation.

In addition, this theory has no experimental evidence (which would be difficult to find) nor does it explain the existence of periods less than 100,000 years (unless we suppose that there are three comets…) and it is today lightly regarded. Moreover recent studies of sediments on seabed have shown that between 3.5 millions years and a million years ago the period of glaciations was 40,000 years, a change that cannot be explained by a change in the comet orbit even if it is not understood why the dominant effect is changed one million years ago.

Today the Milankovitch theory is accepted even if with some difficulty. It is accepted because of its excellent agreement with the data in Figure 7.9, because there is no better explanation for such precise thermal cycles, and because no other model gives the observed periodicity of 23,000 and 40,000 years.

Although the orbital motion proposed by Milankovitch does not change the average heat the Earth receives from the Sun during the year, it changes the way the heat is distributed among the various seasons and between various latitudes. When the Earth reaches its maximum eccentricity at the perihelion (the point nearest to the Sun), the heat that arrives at the Earth is 20% greater than what it receives six months later at aphelion (the point farthest from the Sun). Every 100,000 years there is a match between the Earth being at perihelion and the winter of the Northern Hemisphere. Under these conditions in the north (where much of the land is) there will be warmer winters and cooler summers, while in the south the reverse happens. According to some models this can cause a glaciation (without changing the energy received from the Sun

---

16 *Following the principle that a happy period should always be followed by an unhappy one, the goddess Nemesis was in charge of the retribution of troubles to compensate for any lucky period that might have occurred.*

through the year). In fact, if the summers are cooler, the snow that falls during the winter at high latitudes will not melt completely in the summer, increasing the reflectivity of the Earth, which thus absorbs less heat. In this manner, from year to year, the portion of land covered by snow increases, the Earth becomes colder and colder, and a glaciation begins (more intense in the Northern Hemisphere). The mechanism also works the other way around, when we have in the Northern Hemisphere warmer summers and colder winters. The ice melts more during the summer, the Earth absorbs more heat, and a hot period approaches.

The Milankovitch cycles do not explain completely the observed glacial cycles[17] and do not seem sufficient by themselves to cause the glaciations. It is therefore hypothesized that the mechanism responsible for the glaciations involves other phenomena like greenhouse gases and marine currents which are much more efficient at heating and cooling the Earth. One seeks to explain glaciations with models that show how the variation in the distribution of solar energy of the Milankovitch cycles can trigger other phenomena that produce glaciations in the end.

A possible scheme (not demonstrated) for explaining the climate cycles, including the greenhouse effect, marine currents, and Milankovitch's cycles, could be the following. An increase in temperature due to astronomical phenomena causes an increase in the degassing of CO_2 from marine sediments and an increase in water vapor. The increase in CO_2 and H_2O in the atmosphere causes the temperature to rise further. The polar icecaps begin to melt, causing a decrease in the salinity of the sea, thereby stopping the motor of the warm marine currents. Without warm currents the zone at high latitudes (especially in the Northern Hemisphere) becomes covered with ice. The increase of ice on the Earth contributes to glaciation because it reflects solar light (Section 7.1) and favors a cooling of the Earth.

The glaciations remove water from the oceans, which are lowered by 150–200 m. The reduction in the ocean levels increases the salinity of the sea and puts back into circulation the marine currents, bringing warm water to the high latitudes; the reduced ocean levels could also enable the release of the gas trapped under the bottom of the oceans, increasing the greenhouse effect. The ice, because of the warm currents, begin to melt; the Earth, having less ice, reflects less light from the Sun and absorbs more heat, and thus returns to a warm period. The glaciations would thus be an event caused periodically by

---

17 *The theory predicts different effects for various latitudes that are not observed. Moreover, the strongest effect would be the periodic variation of the Earth's axis (between 22.1° and 24.5°), with a period of 40,000 years, which we had till a million years ago and not the one at 100,000 years we had during the last million years (Figure 7.9).*

greenhouse gases and marine currents. Therefore, astronomical phenomena would be used to explain the regularity of the cycles, a kind of "clock".

Following these cycles, ten thousand years ago the Earth should have started a new phase of cooling. We have seen (Section 7.4) that this did not happen, probably because of the greenhouse gases produced in the past by fires, agriculture and other human activities.

9.9   The Source of Heat

The Earth is therefore a planet geologically alive, with earthquakes, volcanoes and tectonic plates in motion. Like all living organisms it needs energy to continue to modify its surface. We have already said that this energy is supplied by radioactive materials contained in the mantle and the core, in particular the nuclei of ^{238}U, ^{235}U, ^{232}Th and ^{40}K (recall that the number on the upper left indicates their atomic mass, respectively 238, 235, 232 and 40 times the mass of a hydrogen atom), of which we have already spoken in Section 6.1. They are unstable atoms that decay, forming a family of atoms that are ever lighter; in each decay they emit energy. This energy is the heat that warms the inner parts of the Earth and permits our planet to be geologically alive.

Geologists estimate, with good precision, the heat produced under the soil by radioactive elements, measuring the temperature at various depths. Knowing the thermal conductivity of the Earth, one can easily calculate the flux of heat that rises from the Earth. The highest values are measured in the depths of the oceans where the crust is thinnest, and go from 60 mW (thousandths W) per square meter in the southern Atlantic to 95 mW per square meter in the northern Pacific. For the continents the values are lower (the crust is thicker): they go from 50 mW per square meter in Africa to the hottest point in Australia where it is calculated to be 60 mW per square meter. They are all very low compared with the 1336 W per square meter that comes from the Sun (about half of it is reflected back to space). The flux of heat that comes from underground radioactivity is approximately 3000 times less than what comes from the Sun! It is, however, not negligible. Multiplied by the surface of the Earth it is more than 40 trillion watts, about three times power consumed in the world, considering all sources of energy (Section 7.7 and Appendix B).

In Figure 9.12 the contributions from the principal radioactive materials heating the Earth are shown, with their total contribution. On the horizontal axis we have the age, from 5 billion years ago, when this material was still in the parental cloud, until today; under the horizontal axis, with arrows, the dates of the principal phenomena that have characterized the evolution of life on the Earth are given.

On the vertical axis, instead, the heat produced on the average by each kilogram of mass of the Earth (i.e. the total heat produced by radioactive materials

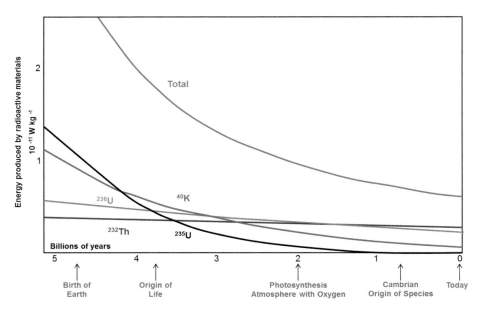

Figure 9.12: The energy produced by the principal radioactive elements present beneath the soil of the Earth, from its beginning until today. The energy is given over every kilogram of mass of the Earth. The total heat is also given. A few elements, like ^{235}U and ^{40}K, have a brief lifetime compared to that of the Earth; their contribution was high at the beginning but is now negligible. Under the abscissa the principal steps in the history of the Earth are shown as told in Chapter 6. (Adapted from Geodynamics by Turcotte and Shubert, 1982, John Wiley & Sons.)

divided by the mass of the Earth) is shown. The values have been calculated, starting with the abundance of the radioactive elements found today on the Earth, and taking account of their mean lifespan (Section 6.1). The graph shows that, at the moment of the birth of life on the Earth, about 3.5 billion years ago, the heating was three times what it is today and mostly due to ^{235}U and ^{40}K. These elements have relatively short lifespans, so that they are now practically exhausted. Today the heating is principally due to ^{238}U and ^{232}Th, which have very long lives: the former halves its heating capability in 4.47 billion years, the latter in 14 billion years. We are speaking of times long with respect to the life of the Earth, so we can stay secured: the heating in the inner parts of the Earth is destined to diminish very slowly, and our planet will remain alive for billion years.

From the ratios among these abundances we can estimate how long before the birth of the Solar System the supernova that produced the radioactive material exploded. The value is very imprecise, because the radioactive material contained in the Earth could have originated from more supernovae occurring at differing times, although the relatively short lifespans of ^{235}U and ^{40}K reduce the uncertainty. The estimates give values between 300 and 800 million years before the birth of the Earth.

9.10 Why Only the Earth?

Why is the Earth the only planet in the Solar System that is geologically active? It is not easy to answer this question. The Earth is the only planet we can study, we have nothing with which to compare it, and the nearby planets Venus and Mars are geologically inactive. On the other hand, we have not even understood why the lithosphere behaves like a fluid (the model of convection; Section 9.4) and it is even more difficult to explain why these phenomena do not occur on the other planets.

We know what the source of energy on the Earth is: it comes from the radioactive material produced by a supernova. But this material arrived on the Earth together with the material from which the Solar System was formed, and it is therefore reasonable to believe that the concentration of radioactive material will be the same on all rocky planets of the Solar System,[18] that therefore the heat produced by radioactive material should be the same for all the rocky bodies of the Solar System. The temperature measurement beneath the lunar soil (the only rocky body we have been able to study up to now) has confirmed the same phenomenon for the Moon. Soon the same measurements will be made on Mars, and it is expected that the result will be similar. The recent results that give a relatively recent volcanism for Mars (Section 8.4) confirm this hypothesis. When these measurements are made, we will probably state that the heat produced inside each planet is roughly proportional to its mass. (The larger the mass is, the more radioactive material will be there.)[19] This explains why the Earth is active and Mars is not.

The mass of Mars is a little more than one-tenth that of the Earth, meaning that the heat produced by Mars is between one-tenth and one-thirtieth of that produced inside the Earth. Is it too little to move the Martian continents? Is the friction too strong and the energy too small to cause convective motions like the Earth's mantle? It is difficult to respond to these questions but the measurements that will be made on Mars could help to understand. The heat produced by radioactive materials on Mars, even if it could not move the continents, did maintain a volcanism on Mars that was active until a few million years ago (Section 8.4). Was that volcanism sufficient to produce an atmosphere like the Earth's? Probably, the Martian atmosphere, when the volcanoes were active, was much denser than it is at present. It must, however, have been different from what is on the Earth. The lack of continental drift and the consequent

---

18 *The giant planets are mainly made up of hydrogen and helium, and radioactive elements are proportionately much lesser.*

19 *In effect one would also have to take into account that there is a difference between the mean densities of these planets. For example, the Earth has a mean density 5.52 times that of water while that of Mars is 3.95. A difference of 30% in density means that there is 30% less heavy materials, and therefore perhaps 30% less radioactivity. It is an effect that has to be multiplied by a factor of 10 due to the difference in mass, so that the heat produced inside Mars could be 10–30 times less than that in the Earth.*

subduction has kept the planet from recycling organic matter (if it has existed) and water with the same efficiency as our planet.

With the passage of time, these materials were more and more depleted, blocked in sediments and carbonaceous rocks underground that without subduction and volcanoes were not recycled. Perhaps, for this reason, Mars is today a dead planet, even if once it could have host to embryonic forms of life.

Some studies[20] have suggested that the absence of geological activity on other planets of the Solar System can be attributed to the absence of water beneath the soil. On the Earth water provides the lithosphere with the flexibility necessary for bending and insinuating itself into the mantle (Figure 9.9). Water would also reduce the friction between the lithosphere and the mantle, facilitating its slippage. On planets without liquid water, friction could be strong enough to block the lithosphere from sliding.

Water, as we know, was present in the parental cloud (Figure 3.1), and therefore like radioactive material, should be present on all the planets of the Solar System. The explanation for the absence of continental drift on Venus and Mercury could have been that they are too close to the Sun and have lost most or all of their primordial water. That, however, is not true of Mars, where water exists (Figures 8.10–8.13). Measurements seem to show that on Mars the quantity of water is not comparable to that on the Earth. If, in the end, we will find that there is on Mars less water than on the Earth, we must understand why. Has it been less from the beginning because the planetesimals that formed Mars had less water than those that formed the Earth? Or perhaps there was the same amount of water as on the Earth, but a large meteorite (tens of kilometers in diameter) evaporated part of the water before volcanic eruptions covered the traces of the impact? Or perhaps, in the first phase of the life of the planet, when the atmosphere was very hot (Section 6.2), the small mass of Mars could not keep the water with the same efficiency as the Earth?

One final hypothesis could be that the absence of water on Mars is due to the lack of continental drift. Accordingly, the water on Mars exists in quantities comparable to those on the Earth, but much of it is beneath in the soil in carbonaceous form, and there being no subduction, it was only put into circulation by volcanoes. The lack of water could therefore be a consequence of the geological inactivity of the planet. An instrument on the *Mars Express* should manage to measure with its radar the depths of the planet and deduce how much of the water is buried in the soil. This will in part provide an answer to the question.

To conclude, if not much water is found on Mars, this could explain the lack of continental drift on that planet. If instead much water is found beneath in the soil, the explanation could be that the mass is too small and the heat

---

20 *V. Solamatov and L. Moresi, 1997, Geo. Res. Lett. 24, 1907–1910.*

produced in its nucleus too little to move continents. For Venus, instead, the reason for the lack of water is the temperature: the water has evaporated, and with little water the lithosphere became rigid and did not move anymore. For Mercury, both these reasons apply: the tiny mass and the lack of water.

One could quickly arrive at the conclusion that, for a planet to be habitable, aside from being in the zone of habitability (Figure 8.14) and having enough water on its surface, it must have a mass similar to that of the Earth to be able to produce enough heat in its interior to move continents and sustain an atmosphere like that of the Earth.

9.11   A Living Planet

The discussion we have made can be summarized in a single sentence: for life to exist and develop it needs to have a geologically alive planet. Life needs water and an atmosphere that acts as a thermostat. But an atmosphere means erosion, so there must be something that replenishes the eroded land. The surface of our planet is the result of two opposing forces: erosion that tends to destroy emerged land and continental drift that tends to restore it. We have seen that at the beginning our planet was a world covered by water, with a few volcanoes that emerged from the waves; then, as a result of the motion of the plates, the first mountains were formed; the first continents grew with time and will continue to grow as long as the inner parts of the Earth remain hot. The source of this heat is the radioactive materials, a source destined to remain active, fortunately, for a very long time.

When the inner parts of the Earth become cold, volcanism will disappear along with earthquakes and continental drift. At that point our planet will be geologically dead; in few tens of million years, erosion will destroy the emerged land and the Earth will return to its origin: a great planet covered with water. A planet that, unlike what it was originally, will be without volcanism, and therefore after a while greenhouse gases will disappear.

Then the temperature will decrease and the planet will become a great snowball where the conditions necessary for life will cease to exist. That will happen only if the Earth cools before the 5 billion years that separate us from the moment when the Sun will become a red giant, encompassing the Earth in its fiery atmosphere. If life is fortunate enough to survive so long, it will end up either in ice or in fire.

To see how likely it is to find a habitable planet in the universe, we must therefore know how probable it is that a planet is geologically active. The exploration of Mars may contribute important answers.

The Earth: A Rare Planet?

Chapter 10

10.1 The Importance of the Moon

Some scientists claim that among the ingredients that made possible the existence and the evolution of life on our planet, the presence of a Moon which has stabilized Earth's axis of rotation and a large planet like Jupiter that has protected our planet from disastrous asteroid impacts have to be considered.

The Moon has a relatively large mass with respect to the Earth. In the Solar System there exists only one similar case — that of Pluto and its satellite Charon (Section 8.3). The angle of 23° (Figure 10.1) between the axis of rotation of the Earth and the plane of the orbit was probably caused by the impact of a planetesimal as large as Mars, possibly the one that had flung into space the material that formed the Moon (Section 3.8). According to current models, the angle of the rotation axis of the Earth remains stable because the Moon, a relatively large satellite, has damped every perturbation that could have changed it.[1] It is argued that if the axis of the Earth were not stabilized by the Moon, it might rotate randomly, arriving even at 90° (with the poles in the place of the Equator) in periods of the order of tens of thousands of years. Such a displacement in such short times would cause climatic upheavals that would render difficult the existence of any evolved life on the Earth.

---

1 A satellite is tied to its planet by gravity. For a first approximation the two bodies can be seen as a unique system, like a single spinning top which has a rotation axis much stable than that of a simple sphere (the planet without a massive satellite).

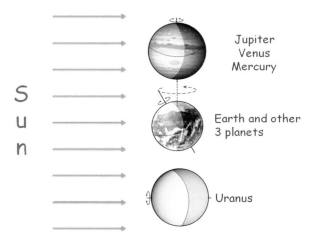

Figure 10.1: Of the eight planets of the Solar System, four have an axis of rotation similar to the Earth, which through the changing of the seasons permits the distribution of the Sun's heat on the entire surface. Three of them, Jupiter, Venus and Mercury, have axes practically perpendicular to the plane of the orbit and, finally, Uranus has the axis parallel to the plane of its orbit. The precession of the Earth's rotation axis is indicated with the dashed curve.

These theories are considered reasonable even though, observing the inclinations of the axes of rotation of the planets of the Solar System (in which only the Earth has a large satellite that stabilizes it), one sees that the Earth, Mars, Saturn and Neptune have their angles between 23° and 28°, while those for Mercury, Venus and Jupiter are below 3°, and for Uranus it is 98°. Of the eight planets, seven have an angle below 28° and four have an angle similar to the Earth. This indicates a certain stability of the axis of rotation; otherwise the planets that are not stabilized by a large satellite would have axes more randomly oriented than the ones observed.

Uranus, with its 98° angle is the only exception (its five satellites move in a plane perpendicular to its orbit[2]) — an exception attributed to the impact of a planetesimal of the dimensions of the Earth that happened in the early stages of the Solar System. If this explanation is correct, it would also confirm the relative stability of the axes about which the planets rotate. This case too suggests a certain stability in the direction of the rotation axis: from the formation of the Solar System until today, there would not have been much change in the direction of the Uranus rotation axis.

The 23° inclination of the axis of rotation of the Earth is very important for the climate of the planet. It permits the alternation of the seasons. Without this angle or with a much smaller angle like for Venus and Jupiter, the Earth would be less livable. The equatorial regions would be much hotter, because they

---

2 *The pole of Uranus has a day lasting 21 years, with a night of equal length at the other pole (1/4 of its 84 years' orbital period).*

would have the Sun directly above them all year round and would probably be beyond the limits for survival of evolved animals, as traced in Figure 6.3. The zones at high latitude, like Europe and North America would be much colder because they would have the Sun low on their horizon throughout the year, and the ice of the Arctic and of Antarctica would be much more extended than it actually is.

The situation of the Earth would be much worse if its axis of rotation were in the plane of the orbit, as in the case of Uranus. In that case because of the conservation of angular momentum (Section 3.6) the axis of rotation of the planet would always point in the same direction in the sky, so that for three months of the year one pole would be torrid pointing toward the Sun, leaving the other pole frozen in the shade for the next three months the Sun would be directly above the Equator, then for the following three more months it would be above the other pole, and so on. In this manner all parts of the planet would be subjected to periods of high temperatures alternated with freezing periods. A planet in the habitable zone of a planetary system (Section 8.7) would nevertheless be uninhabitable, because it would, for most of the time, be above or below the limits of survival for evolved life traced in Figure 6.3.

The axis of our planet precesses[3] slowly in time, without much immediate effect on the climate, and its moderate inclination permits, with the passing of the seasons (and the help of marine currents and the winds; Section 9.8), the distribution of heat over the entire surface of the planet. We have seen that the precession of the Earth's axis could be one of the causes of the periodic glaciation. Large variations in the inclination of the Earth's axis could have terrible effects on the climate. The presence of a large satellite that stabilizes the axis can therefore be one of the requirements of a habitable planet.

The satellites of a planet can also pose a risk if they have a retrograde motion — orbiting in a direction opposite to the rotation of the planet around its axis. In this case tidal effects will decelerate the satellite which will come closer and closer to the planet until it falls on it, causing an Armageddon. The tides are in fact caused by the force of gravity that ties a satellite to the planet. This force creates between the two bodies a frictional force. If the Earth and the Moon rotated in opposite directions, the Earth and the Moon would slow

---

3 *The direction of the rotation axis of the Earth changes slowly with time, making a complete rotation around an axis perpendicular to the plane of the orbit (dashed curve in Figure 10.1) with a period of 26,000 years. This motion is called the precession of the equinoxes because every year the equinoxes are displaced slightly and after 13,000 years (half the period) there will be an inversion of the seasons (summer will happen when the Earth is in the position where today there is winter). The movement is caused by the attraction of the Moon and the Sun. The same motion can be seen in the case of a top, which in addition to its rotation has a motion of the rotation axis around the vertical caused by gravity. Even if the motion is small it is well detectable. The first to note this motion was the Greek astronomer Hipparcus, who discovered it in 150 BC, looking at the measurements of the positions of the stars in the sky, which were registered in the annals by the astronomers who preceded him.*

down; the Moon would get progressively closer to Earth until it falls into it. In the Solar System there is only one such case — Neptune. One of Neptune's satellites, Triton, probably captured by the large planet after its formation, has a retrograde motion and gets closer to the planet every year; it is predicted to fall onto the planet within a hundred million years. Not all retrograde satellites will have the same fate; some small ones (probably former asteroids that are captured with time) are so far from the planet that they will last until the end of the Solar System.

This danger also exists if a satellite is inside the geostationary orbit (like many artificial satellites); they rotate around the planet more quickly than the motion of the planet about its axis.[4] It is as if the Moon would rotate around the Earth in less than 24 hours instead of the actual 28 days. Such satellites, because of the tidal effect, would drag the planet whose speed of rotation would increase; while the satellites would slow down until they fall onto the planet. An example is Phobos, a satellite of Mars (possibly an asteroid captured by the planet), which rotates at 9000 km from the surface of the planet and makes a complete circuit in 7 hours, while the Martian day lasts 24 hours, 37 minutes and 23 seconds. Its destiny is fixed.

The Moon, like most other satellites in the Solar System, orbits around the Earth in the same direction as the Earth rotates on its axis. In this case the Earth rotates more rapidly than the Moon (one period in 24 hours, compared to the 28 days of the Moon), dragging it along and accelerating it. So, after each year, our satellite moves a little farther from the Earth, and after about a billion years from now the influence of the Moon on the Earth's tides will be negligible. The movement of the Moon away from the Earth is done at the expense of the Earth's rotation (by the conservation of the angular momentum), which every year becomes a little slower so that the days grow longer.

There is no doubt that in a disordered system, which has planets and satellites with random directions of rotation and revolution, everything would be more unstable and more dangerous for evolved life-forms (bacteria survive cataclysms better than animals). The probability of such an occurrence in a planetary system is very low however, because the planets and their satellites are born from a disk (Chapter 3), which has a precise direction of rotation that is reflected in the rotations of all matter in its planetary system.

A few perturbations of this initial order can happen (like for Uranus and Triton), but they are exceptions and not the rule. Retrograde satellites can originate from two phenomena. The first is the capture of a wandering object by a planet, an event almost impossible for the Earth since the velocity of the asteroids and comets is too high (Section 10.3) for the weak gravitational field

---

4 *The period of rotation (the time to complete one orbit) diminishes when the diameter of the orbit is reduced (Kepler's Law).*

of the Earth to capture one like the Moon. The second is the birth of a satellite from the impact of a large asteroid (the most likely how the moon originated). As we will see in Section 10.3, such an impact is today highly unlikely and, in any case if it happens, it will be the end of every form of life on the Earth. Such events probably happened at the beginning of the life of the Solar System, but retrograde satellites born in this manner had fallen onto their planets after a few hundred million years, because of the tidal effects which we spoke of. There remains only those, like the Moon, that rotate in the same direction as their planet. For this reason, with the exception of Triton and of the satellites that are too far from their planets to feel the influence, there are no retrograde satellites in the Solar System.

10.2   The Importance of Jupiter

The contribution of Jupiter to make the Earth habitable is to sweep off the Solar System of all celestial bodies moving in a disordered fashion, with a high probability of striking the planets. Jupiter has eliminated these dangerous objects, throwing some out of the Solar System, others into the Oort cloud (Section 8.2) which was thus formed, and others into the Sun. Models show that in the absence of a large planet, the probability for the Earth of colliding with a body like the one that 65 million years ago led to extinction the dinosaurs is of the order of one impact every 20,000 to 100,000 years; enough to make impossible the development of evolved life.

Jupiter interacts with a body passing in its vicinity with a mechanism known as the *slingshot effect*. Because of the gravitational field of the planet, an object getting close to it will follow a hyperbolic orbit around the planet and then fly away with an energy that is higher or lower than the one before.

When the body is near Jupiter, the gravitational field of that planet drags it along, adding or subtracting energy from its motion. The closer to the planet the asteroid passes, the larger the impulse added or subtracted will be. An asteroid is accelerated when its motion is opposite to that of the planet. The acceleration occurs with a mechanism similar to what happens when a tennis ball (the asteroid) is struck by a tennis racket (the planet). The impulse given by the motion of the racket is added to what the ball had before the collision. The asteroid is instead slowed down when it is moving in the same direction as the planet (obviously at a greater velocity), like what happens when a ball strikes a racket moving in the same direction: in the collision the ball transfers energy to the racket and slows down. With these blows an asteroid can gain enough energy to be expelled from the Solar System, or slowed down to end up into the Sun.

This method is so efficient that it has been employed since the 1960s to send satellites to the outer zones of the Solar System (using other planets including the Earth and Venus), making possible missions that would otherwise be unthinkable because of the immense amount of fuel they would require.

The first satellites that used this effect were the *Pioneers*, which acquired from the external planets, particularly Jupiter, enough energy to leave the Solar System, starting a long journey toward the stars.

Jupiter, together with its helper, Saturn, performs this role of sentinel in a very discrete way, staying far from the orbit of the Earth. The proximity of Jupiter, as we have seen, is considered the cause of the missing planet in the asteroid belt (Section 8.3) and of the small mass of Mars (a tenth of the Earth). According to this theory, if Jupiter had played its role farther out in the Solar System, Mars would have become more massive, with more radioactive elements inside, forming a nucleus hot enough to cause continental drift and volcanoes (Section 9.1). Mars would thus have an atmosphere similar to that of the Earth, with a greenhouse effect and liquid water on its surface. The Solar System would thus have two habitable planets, even if Jupiter is farther out and its protection against the bombardment of meteorites would have been less effective and life could have been unable to evolve.

Will the necessity of having a massive planet to guard a habitable one reduce greatly the probability of finding habitable planets? It is difficult to answer such a question. We are talking about very speculative arguments, different from those in the previous chapters. One could suppose that massive planets, because of their ability to store angular momentum, are very common around stars (as the measurements of recent years have demonstrated, Section 3.9). In their absence, as the case is for planetary systems farther out in the galaxy (Section 8.1), the process by which bodies with disorganized motion are eliminated from a planetary system could be longer but not comparable to the life of a star, so that evolved life could develop, although perhaps later than on the Earth.

However, one thing is certain: if the habitable band of a star is occupied by a planet of the dimensions of Jupiter, a planet like the Earth cannot survive in the same area. Even if this happens, the planet will be in risk of being sent into the Oort cloud or into the Sun. There could, however, be satellites around one of these giant planets of dimensions similar to those of the Earth,[5] with an atmosphere and volcanism, and therefore with evolved life. Habitable satellites of large planets may exist, but it will not be possible to find them. The spatial resolution necessary (Appendix A.5) for separating their image, from that of the planet about which they rotate, would be 100–1000 times greater than that needed to separate an isolated planet like the Earth from its star — a goal that is out of reach today.

---

5 *In the Solar System, such satellites do not exist. The largest satellite of Jupiter has a mass four times smaller than that of Mars; Titan, satellite of Saturn, has a similar mass.*

Even if they are outside the habitable band, there is great interest in the massive satellites of the giant planets of the solar system, where one could find underground lakes heated by impact of comets, or by the heat of radioactive material beneath the soil, or by tidal effect induced by the giant planet about which they rotate. In these lakes there could be prebiotic molecules, if not bacteria. Among these satellites there is Europa, a moon of Jupiter, and Titan, a moon of Saturn. The latter is one of the most interesting objects because it has an atmosphere denser than the Earth's and a similar variation in its climate even if its temperature is 180°C below zero. To study this object, the European Space Agency has built the satellite *Huygens* which, after a voyage lasting many years (using the slingshot effect with Venus, the Earth and Jupiter), arrived together with the satellite *Cassini* near Saturn, and then, at the end of 2004, descended with a parachute onto the surface of Titan.

10.3 NEOs

NEOs is the acronym for Near Earth Objects; they are objects that orbit near the Earth and pose a threat to our planet. The threat is real, as witnessed by the extinction of dinosaurs 65 million years ago (Section 6.8 and Figure 6.12) and by the many craters found on the Earth, like the Manicouagan Crater in Canada (Figure 7.2), due to a relatively large asteroid fallen 200 million years ago, or the Barringer Crater in Arizona (Figure 10.2) caused by a small asteroid 40–50 m in diameter which fell 49,000 years ago. It reached the Earth's surface only because its iron core was not destroyed in the impact with the atmosphere.

Today, on the Earth, traces of 130 craters are found, with diameters ranging from 20 to 200 km (Figure 7.1), distributed across all continents, and are all due to impacts of large meteorites. Taking account of erosion, which has destroyed most of them, and of the impacts on the seas which are difficult to find, one can guess that the actual number is much higher.

An example of a recent impact is the asteroid that fell at Tunguska, Siberia in 1908. It was about 60 m in diameter, small enough to be disintegrated in the atmosphere before touching the Earth (it had not a metal nucleus). Nevertheless, it caused a shockwave capable of destroying 2000 km² of forest (Figure 10.3), an area 25 times larger than Manhattan island.

A spectacular impact, seen by all the astronomical observatories in the world, happened in July 1994, when the comet Shoemaker Levy 9 fell onto Jupiter, causing fiery splashes as big as the Earth.

The probability of an impact by a large object (as Eros, Figure 10.4) is very small, but the destruction it would bring is so great that it must be taken seriously into account. Starting with this consideration, some countries, among them the USA and UK, began in the late 1990s to evaluate the risk of such impacts and to see if the technology available today is capable of avoiding

Figure 10.2: The Barringer Crater in Arizona. The crater has a diameter of 1.2 km and a depth of 160 m. It was caused, 49,000 years ago, by the impact of an asteroid of diameter about 50 m. The energy released by the impact has been estimated about 10 megatons (roughly 700 times that of the atomic bomb that fell on Hiroshima). (© Meteor Crater, Northern Arizona, USA.)

Figure 10.3: Tunguska, Siberia. This is how the area appears 30 years after the impact of an asteroid. About 2000 km² of forest was destroyed by the shockwave. The asteroid of about 60 m diameter completely disintegrated in the atmosphere before reaching the Earth. (Photos from Kulik's aerial photographic survey (1938) of the Tunguska region; Bologna University.)

Figure 10.4: Eros: an asteroid 33 km long and 13 km across. The numerous craters on the surface show that impacts also occur among asteroids. The largest crater (in front) has a diameter of 5.3 km. If such an object struck the Earth the energy released would be 30–100 times greater than that released from the asteroid responsible of the extinction of the dinosaurs. (© NEAR satellite & NASA/JHUAPL.)

them. Some of the information contained in this chapter is taken from the report "*On Risks of NEOs*", presented to NASA in August, 2003[6] and from the report of the task force on the potential risk of NEOs presented in September 2000 to the British science minister. These commissions confirmed the nature and potency of the danger[7] and suggested plans to reduce the risk.

The research has already given some results. At the beginning of 2005, a 390 m asteroid was discovered; if it keeps to its orbit (it could be diverted by its passages close to Venus and the Earth which will take place in the coming years), it will strike the Earth in 2036, releasing an energy that is equivalent of 50,000 Hiroshima bombs. A computer simulation shows that the shockwave of the impact could destroy everything in a strip-several hundred kilometers large, crossing Europe and the Middle East or going from Canada to Mexico. Because of the great menace these objects represent, the researchers at JPL

---

6 *To go deeper we suggest visiting the website http://neo.jpl.nasa.gov/neo/*
7 *The first result of these studies was to furnish cinematographers in the 1990s with material for catastrophic films like* Armageddon *and* Deep Impact.

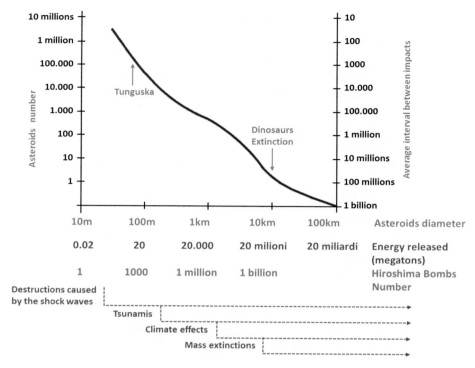

Figure 10.5: *The risk of NEOs and the damage caused by impacts. On the two vertical axes the number of NEOs that orbit around the Earth and the average interval between impacts are shown. On the horizontal axis the dimensions of the asteroid, the energy released by the impact in megatons of TNT, and the number of Hiroshima bombs necessary to inflict the same damage are illustrated. Below the horizontal axis, the diameter of the asteroid and the destructive potential are reported. (Adapted from the* "Report of the Task Force on potentially hazardous NEOs" *presented to the British government in September 2000.)*

have baptized it with the name *Apophis*, the Egyptian god of death and destruction.

The graph of (Figure 10.5) shows that near the Earth there are 10 million asteroids of diameter 10 meters and that every 5–10 years one of them fall into the atmosphere. Of asteroids like Tunguska (60–70 m) there are more than a million and, on average, an impact is expected every 500–1000 years. Asteroids like the one that caused the extinction of the dinosaurs (10–15 km in diameter) have an expectation of one impact every 100 million years. Asteroids with dimensions larger than 100 km have a very small probability of impact, comparable to the age of the Earth, so that it might never happen.

Under the horizontal axis, the corresponding released energy is reported in the scale of megatons. The scale is just an indication because the energy released in the impact depends on the mass of the asteroid and on the square of its velocity. Moreover two asteroids of similar diameter can have very different masses if one is made of ice and dust and the other of iron. For asteroids of the same type the destructive power increases with the cube of the diameter (i.e. the mass).

The figure shows that the destructive power reaches enormous values for relatively small asteroids. The cause is their large velocities. An asteroid that comes from a region relatively close to the Earth like the asteroid band (Section 8.3) can have an impact velocity ranging between a few km per second and 30 km per second; a comet that comes from a remote place like the Kuiper belt or the Oort cloud (Section 7.2) can exceed 70 km per second. For comparison, an airliner does not reach 0.3 km/s, the velocity of sound in air is a bit less than 0.4 km/s, the Pioneer and Voyager satellites, the fastest objects humans have ever created (in these years they are leaving the Solar System), travel at about 17 km/s.

In the figure, under the scale in megatons (a megaton is the energy released by an explosion of a million tons of TNT) is reported the number of bombs of power equivalent to that of Hiroshima necessary for producing the same effect (the Hiroshima bomb released 15 kilotons = 0.015 megatons).

Under the horizontal scale of Figure 10.5, and in greater detail in Table 10.1 the potential damage is reported.

Below 50 m, the effects are modest, the atmosphere is an excellent protection, and only a small part of the nucleus can, occasionally, reach the Earth (especially if it is metallic). These are the meteorites one sees in the natural history museums.

Above 50 m, the asteroids that fall on the Earth can produce great damage because of the shock wave produced by their velocity which can be 100 times the speed of sound. The *shock wave* produces very strong winds and causes damage worse than the worst of hurricanes. The 2000 km² of forest destroyed at Tunguska in Siberia is an example of this devastation; it was caused by an asteroid only 60 m in diameter, so small that it disintegrated before reaching the Earth. If that asteroid had fallen on a city, millions would have died. Two thirds of the Earth is covered with water, so that two thirds of asteroids fall in the sea. For asteroids of diameter less than 200 m the shock wave does not produce damage in the sea; the mass that reaches the water is too small to have much effect.

Above 200 m, the impact of asteroids on the sea causes seaquakes, which, with an increasing in diameter, produce greater and greater damage. With a diameter of the order of a kilometer, the effects are felt on a global scale with waves that move with a velocity greater than 700 km/h,[8] the speed of an airliner, reaching quickly large areas of the Earth. When these waves, generated by an asteroid impact, but also from earthquakes and volcanic explosions, arrive at a coast, because of the reduction in the depth of water their height is magnified 10–30

---

8 *The velocity of a tsunami depends on the depth of the sea. As a first approximation, the speed in m/s is the square root of 9.8 x (the depth of the sea in meters). At a depth of 4000 m the speed is 200 m/s, equal to 720 km/h.*

Table 10.1 Damage caused by the impact of an asteroid. (Adapted from the "*Report of the Task Force on potentially hazardous of NEOs*" presented to the British government in September 2000.)

Asteroid diameter	Megatons Produced (MT*)	Diameter crater (km)	Mean Interval between impacts (Years)	Consequences
<75 m	< 10			Explosion in upper atmosphere. Only meteorites with iron core (<3%) reach the earth surface.
75 m	From 10 to 100	1.5	1000	Destruction of an area the size of Washington, Moscow. (Tunguska, Meteor Crater arizona).
160 m	From 100 to 1000	3	4000	Destruction of an area the size of New York or Tokyo.
350 m	From 1000 to 10,000	6	16,000	Destruction of an area the size of Estonia. Seaquake.
700 m	From 10,000 to 100,000	12	63,000	Destruction of an area the size of Taiwan. Seaquake reaches world-wide scale.
1.7 km	From 100,000 to 1 million	30	250.000	Destruction of an area the size of California. Dust causes climate changes. Extinctions.
3 km	From 1 million to 10 million	60	1 million	Destruction of an area the size of India. Large fires caused by reentry in atmosphere of material sent by impact. Climate variations and mass extinction.
7 km	From 10 million to 100 million	125	10 million	Destruction of an area the size of a continent. Large effects on climate. Large mass extinction. (KT-event).
16 km	From 100 million to 1 billion	250	100 million	Dramatic effects on climate. Very large mass extinction.
> 100 km	> 1 billion			Evaporation of the oceans, destruction of every form of life.

times. For this reason, seaquakes are also called *Tsunami* (from Japanese) which means "port wave". The waves of a seaquake, as opposed to normal marine waves, displace a whole column of water (which can be thousands of meters tall). Therefore they carry lots of energy, even though in the open sea they are almost unnoticed. As soon as they reach a coast where the depth diminishes, the amplitude increases: a 1 m wave in deep water can become a 10–30 m tall wave on the shore (in Hawaii, a wave increased by 40 times has been recorded).

The most dangerous aspect of these waves is their wavelength, which can be hundreds of kilometers (compared to the hundreds of meters of an ordinary marine wave; Figure 10.6), and their period, which can be thousands of seconds or, in the case of large asteroids, hours. These waves manifest themselves on the coast as a rising of the sea like a long tidal wave which lasts the time necessary for the crest of the wave, hundreds of kilometers long (which travels at 700 km/h), to arrive completely at the coast. For this reason even seaquake waves of 3 or 4 m (a few cm in the high seas), much smaller than the waves of a normal storm, on their arrival at the coast, are transformed into violent rivers flowing deep into the land. In the case of a meteorite impact, the height of the seaquake wave can be a kilometer, and this rising of the sea is estimated to last for an hour so they will enter deeply into all continents.

At the end of 2004, some countries in Southeast Asia were hit by a seaquake that raised the sea for 10–20 min by a few meters (the maximum was 15 m

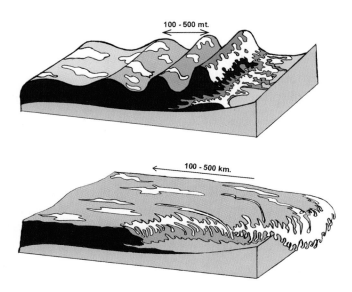

Figure 10.6: Top: a normal sea wave, characterized by a wavelength (the distance between two maxima) of a few hundred meters. Bottom: a tsunami wave, whose wavelength can be hundreds of km. Sea waves are superficial, even if they reach a height of tens of meters, they do not change the average sea level, so they can cause damage only to the coast. A tsunami causes an elevation of the average sea level similar to that of a tidal wave; if it encounters a zone where the coast is fairly flat, it can penetrate into the territory for kilometers.

at Sumatra, near the epicenter of the earthquake). A similar seaquake has hit Japan on March 2011, anyone who has seen images of the damage produced by these waves can imagine what damage a rising of the sea by 500 meters or a kilometer can cause. It is estimated that the asteroid that caused the extinction of the dinosaurs produced a wave of such a height. Traces left by that wave have been found after 60 million years in Haiti, Texas and Florida (Section 6.8).

Computer simulations of the fall of an asteroid (whose crater still exists today at the bottom of the sea), occurred 2.15 million years ago in the South Pacific, below Tierra del Fuego, show that after 5 h, the waves had traveled 2000 km (at a mean speed of 400 km/h) and hit the coast at a height of 70 m. Seaquake waves of similar height were caused by the implosion of the volcano of Santorini in 1500 BC which destroyed the island of Crete and the Minoan civilization.[9] The stories of inundations (like the universal flood), found in all cultures in the world are probably due to seaquakes.

Above 1.5 km, such impacts, whether at sea or on land, raise enough dust to have an effect on the climate and provoke mass extinctions. It is estimated that, on impacting the Earth, asteroids of diameter 1.5 km can destroy an area as big as France, asteroids of diameter 3 km an area as big as India, and asteroids of diameter 7 km an area of the dimensions of Europe, Australia or the United States.

Above 10 km, the climatic effects produced by the impact cause major mass extinctions. In Section 6.8 we have reconstructed the effects of the meteorite which caused the extinction of the dinosaurs. The material sent into space by the impact returned back to Earth, causing fires everywhere on the planet. The dust and water vapor emitted into the upper atmosphere, together with ash produced by fires, enveloped the planet in a dark blanket that caused a long winter through which only a few species survived.

Above 100 km, the effect would be the destruction of every form of life but, fortunately, once the planetary system has been formed, such an impact might never happen.

10.4 Can Impacts Be Avoided?

It is clear therefore that the problem is not the answer to the question "will the Earth will be hit by an asteroid?" because this will happen for sure, but the problem is to know when and possibly where. It is therefore natural to ask if it is possible, with the available technology, to avoid impacts, at least the most catastrophic. The answer seems to be yes, thanks to space technology.

---

9 *In 1500 BC in the Aegean, at the position which today is Santorini island, there was a circular volcanic island: Thera. After a long eruption the island collapsed hundreds of meters into the void left by the erupted magma. The entry of water into the crater produced a rebounding wave, which, when it reached the Island of Crete, had a height estimated at 70 m, destroying the Minoan civilization. According to geologists, the implosion of the island was not the first and will not be the last in the history of the Mediterranean. The emerged cone of the new volcano is already outside of the waters.*

The problem, as we will see, is to know the orbit of the asteroid or comet that could hit the Earth far enough in advance to organize the intervention.

Scientists are studying various techniques for pushing a dangerous object out of its orbit. Some foresee exploding a nuclear or chemical charge near its surface to give the object an impulse and displace it. Others foresee striking it with a high velocity projectile.

In January 2005, the *Deep Impact* mission of NASA was launched, with the principal objective of hitting the comet Tempel 1 with a mass of 371 kg of copper (Figure 10.7) to evaluate its consistency. The impact happened on July 4, 2005; the projectile struck the comet with a velocity of over 60,000 km/h, producing a crater bigger than a soccer field, showing that the comet was composed of matter less dense than expected.

ESA is studying a mission named *Don Quixote*, composed of two satellites, Sancho and Hidalgo which will act like the protagonists of the homonymic story: while Hidalgo strikes the comet furiously, Sancho will observe from a safe distance what happens, remaining for months in orbit about the comet, to evaluate with precision the effects of the impact. Deep Impact, instead, did not have enough fuel to stop near the comet and continued its way after having recorded the impact.

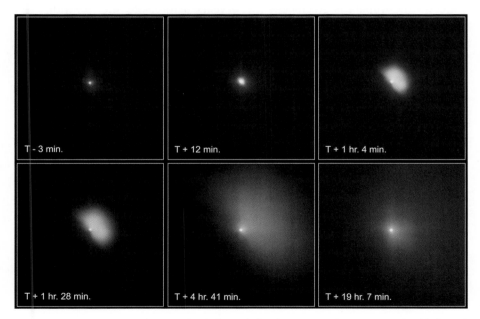

Figure 10.7: A HST sequence of images taken before and after the impact on the comet Tempel 1 of bullet with a mass of 371 kg of copper. The first image was taken 3 min before the impact; the others were taken after, and show the cloud of dust that surrounded the comet. In the image below, at the center, the cloud has reached a dimension of 3200 km. The comet is 13 km long and 4 km wide (similar in dimension to the object that caused the extinction of the dinosaurs) and was too far from HST to see its nucleus. (Photo by HST.) © NASA, ESA & P. Feldman and H. Weaver, Johns Hopkins University.

The problem with the "violent" techniques is the risk of fragmenting the comet or the asteroid into pieces that will be out of any possible control. Therefore, "soft" techniques are also developed. They foresee mounting, on the surface of the asteroid, or comet rockets[10] that would gently push the object into a safer orbit.

In May 2003, the mission *Hayabusa* was launched by a collaboration between the USA and Japan; it would land on an asteroid to study its composition and bring samples back to the Earth. The mission *Rosetta*,[11] launched by the ESA in March 2004, is expected to land on a comet, to study its composition and bring back a few samples. These missions, aside from collecting information of great scientific value, will put in place the technique of "soft landing" on these objects.

In years to come, these "nonviolent" techniques will be further refined. They require, however, much time to be realized. It is in fact necessary to decide what to do well in advance, in order to have the time necessary to prepare the mission and to reach the comet or the asteroid. Moreover, the object has to be reached when it is still far from the Earth, because only in that case it is possible to deviate its course with the weak impulse that our technology can produce. The impulse that is necessary for avoiding collision increases as the object approaches the Earth, because the deviation required becomes larger. If the object is too close, an impact becomes inevitable. It is estimated that if the notice of the impact is given only a few years before, the only solution is to evacuate the area that will be hit.

Five years in-advance is generally accepted to be the minimum time necessary for deviating the asteroid, with a good margin of safety, and using violent techniques (hitting it with a massive projectile or exploding in the vicinity a nuclear bomb).

Twenty years is the minimum necessary for reaching the asteroid and pushing it softly into a safer orbit. For example, for Apophis, an intervention before 2020 would be enough to alter its velocity by a ten-millionth and prevent a possible impact. A later intervention would require a change in velocity by a hundredth, beyond the possibility of our technology.

The real problem of defense against NEO is therefore to know its orbit in time. The only way to achieve that is to measure its orbital parameters in order to know in advance when it will pass by the Earth and at what distance.

The problem has been taken up with a campaign of observations, mostly by the USA to catalog the largest possible number of objects and to know their orbital parameters with sufficient accuracy to estimate their risk over the next

---

10   *Or a large sail that, reflecting the sunlight, will use the pressure of radiation as propeller.*

11   *The name Rosetta comes from the Rosetta Stone (conserved in the British Museum) which permitted Champollion to decipher the Egyptian language. By analogy, the mission Rosetta, by studying the composition of a comet that contains the matter from which the Solar System was constructed, will permit us to decipher its origin.*

hundred years, so that in case of danger we will have the time necessary to intervene. We are speaking of difficult measurements because the NEOs are very weak. The luminosity of an asteroid of 1 km diameter is on the average at magnitude 18, a million times weaker than the limits of the human eye. For this research, telescopes of large diameter are needed; they have to photograph, at a distance of a few hours or days, the same zone of the sky to see if one of the weakest points has moved relative to the fixed stars. That is the only way to distinguish an asteroid from a distant star.

Figure 10.8 presents the situation, published by NASA and updated through 2011. The vertical axis shows the number of NEOs for which we know the orbit, and the horizontal axis gives the year. In red the NEOs that have a diameter greater than 1 km are reported, in blue the total of all NEOs. From the figure one sees that the research activity is recent. The cataloged objects numbered 400 in 1998, 2600 in 2003, 7000 in 2009, and 8000 in 2011.

The red curve is nearly horizontal, meaning that the orbits of most objects larger than 1 km are known (fortunately none will pass near the Earth for the next 50 years). For smaller objects, the situation is definitely worse: we know only 40% of the objects of diameter 300 m and very little about those still smaller. The objective that the international community has posed itself for the coming years is to catalog all objects above 150 m. It is estimated that the number of these objects is about 14,000 ; Figure 10.8 shows that more than half of them have been detected. One hundred and fifty meters seems to be small for

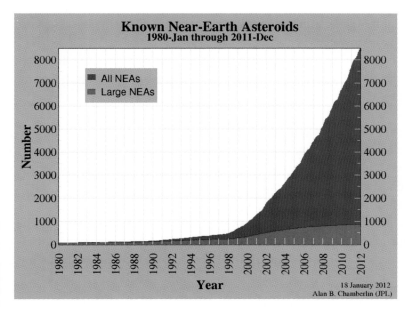

Figure 10.8: The number of NEOs for which we know the orbit. In blue is the total number and in red those larger than a kilometer. (© NASA & JPL.)

a diameter, but an object of this size falling into the sea would cause a tsunami worse than the one which devastated Japan in March 2011, and if it will fall on the surface it could wipe out a region the size of Haiti or Taiwan.

To reach this objective, an enormous amount of work is required with an estimated cost of US$400 million: it is necessary to give the orbits tens of thousands of very weak objects that only telescopes with a diameter of more than 3 m can see with difficulty.

We already know that a serious risk exists: it is the *Apophis*, the asteroid of diameter 390 m which we have spoken of in a preceding section. Its trajectory is known with reasonable accuracy, and if it will not be diverted by its coming passages through the orbits of Venus and the Earth,[12] it will be the first object that humans must target to displace from its orbit. If we fail, an area as large as Europe may cease to exist. Around 2014 we will have to decide whether and how to interfere with its orbit. By then we may have the results of the mission Don Quixote, and a possible intervention can be planned with great care. The impact is predicted to be in 2036, but even earlier, in 2029, *Apophis* will pass close enough to the Earth, and be visible to the naked eye.

The next step, the interval 140–150 m, is almost inaccessible to measurements from the Earth. They are certainly dangerous objects. However because they are more numerous, they have the greatest risk of an impact (Figure 10.5). They can cause a devastation, even if they do not risk the existence of humankind. Within 10 years, with the orbital stations functioning, a system of rapid response against these objects could be possible. Given their modest dimensions, it may be possible to hit and break them into smaller pieces that will disintegrate in the atmosphere. In this case it will not be necessary to detect them far in advance.

For comets the situation is different, because the risk is smaller; some estimates say that the risk of an impact by a comet is 100 times smaller than that posed by asteroids.

For comets of the Kuiper belt (Section 8.2), the periods are less than 200 years. So we are speaking of objects that have already been observed and whose orbits are all well known. With their luminous tails they are more easily observed than asteroids, so they have long been in the annals of astronomers. Our data show that none of them pose a threat to the Earth, at least for centuries.

The orbits of the comets of the Oort cloud are instead unknown. They have periods of millions of years (Section 8.2), and no one has observed their previous passages, and so no one on the Earth knows of their existence until they

---

12 *The probability of being hit varies with every close passage to a planet. For Apophis, at the end of 2004 the estimated risk of impact was 1/34, but at the end of 2005 it had diminished to 1/5000. It is predicted that after 2012, the destiny of Apophis, whether to hit us or move away from the Earth, will be settled.*

appear in the heavens. In 1996 two of them appeared in quick succession: Hyakutake, with a diameter estimated at 3 km, was noticed 5 months before its passage near the Earth; Hale Bopp (Figure 8.5), with a diameter estimated at 40 km (much bigger than the object that caused the extinction of the dinosaurs), was observed only 21 months before its passage across the orbit of the Earth. In both cases, the time was too short for a possible intervention. The comets that come from the Oort cloud generally do not move on the plane of the Earth's orbit (as do many NEOs and the comets from the Kuiper belt). In order to have an impact, they must cross the orbit of the Earth (Figure 8.5) at the instant in which the Earth is there; if it happens either a few hours after or before that, or a few thousand kilometers inside or outside of the Earth's orbit, the impact will not happen. The probability of a collision with one of these comets is therefore very small; this has permitted the Earth to survive until today, and probably will permit it to have a very long future.

It is difficult to evaluate how dangerous the impact of an asteroid or comet can be to the survival of a habitable planet. Based on our experience on the Earth, we can say that in the early stages of life on the planet, catastrophic impacts were more likely, but they have not kept life from developing because the first inhabitants of the planet were bacteria that are less affected by cataclysms than the evolved species. Once evolved and species like ours have been settled which are more vulnerable than bacteria, they may however be able to avoid the impact. In view of this, the ability of an asteroid or comet to prevent the development of evolved forms of life seems limited.

10.5 Fermi's Paradox

In the last section, we concluded our review of the elements that make life possible on a planet and those that could destroy it. Some, such as to be in the habitable band, of having an atmosphere, liquid water, and of being on an planet alive with volcanoes and earthquakes, are indispensable. Others, like the role of the Moon and Jupiter, are a bit more controversial.

Many researchers believe that the Earth is a happy combination of an incredibly high number of positive factors, and it is difficult for all to occur together; in such a case, the Earth would be a rare planet in the universe.[13] Among the pessimists on the possibility of finding evolved life in other planets there was Enrico Fermi. He said that if intelligent life in the universe were not rare, "the others" would be here (meaning by "the others" the inhabitants of another world, the extraterrestrials). Fermi did not know many of the things that are written in this book, but his objection remains valid and is based on simple considerations.

---

13 *These are theses held by P. Ward and D. Brownlee, in* The Rare Earth (*Springer Verlag, New York, 1999.*)

If our civilization continues to develop through the 5 billion years of the life of the Sun, it is reasonable to believe that, sooner or later, it will have the capacity, if not the necessity, to attempt the explorations of other planetary systems (Section 11.5). The nearest stars are at distances that, with the progress of technology, should be reachable in a time of the order of a century. Moreover, the measurements we will speak about in the final chapter should tell us where the habitable planets are. So the interstellar voyage will not be taken in pitch darkness.

If our civilization will continue to progress, it is probable that someone will try for such an adventure, even knowing that their grandchildren would be the ones that will arrive at the destination. Even assuming that all of our descendants will be wise enough to stay on the Earth, there will be a moment when the Sun will be close to become a red giant and at that time the only possibility for the survival of the populace will be to escape.

Around us there are numerous planetary nebulae that are the remnant of red giants; all are very young because the material of the nebula is dispersed in less than 10,000 years; they are the witnesses to the death of stars similar to the Sun. Before them, thousands of stars have died in our vicinity. If around those stars there were evolved civilizations, why did not they move away? There are also red giants near us, all easily visible to the naked eye, and their Greek and Arabic names are well known: Betelgeuse, Aldebaran, Arturo, Canopo, Capella and Pollux. They are all stars that became giants only a few thousand years ago. If around those stars there were inhabitable planets that produced evolved civilizations, their populations must have tried to emigrate. If these forced migrations happened for billions of years with the frequency with which we see red giants and planetary nebulae around us, it is reasonable to ask why "the others" are not here, and one answer could be that life is rare in the universe

However if one day we will receive radio messages from other stars, proving the existence of intelligent civilizations in the galaxy, the answer to the question "why are the others not here?" could be that life is common in the universe but the probability of success for voyages that would last centuries, is zero, and that "intelligent" species are condemned to die with their star.

10.6 Are We Alone in the Universe?

For some time astronomers have noted that certain zones of the sky appeared darker than others, as if they were less populated by celestial objects. Even with the best telescopes, it seemed that in those directions galaxies were rare. They asked themselves if those empty spaces are real, if the universe is less populated in those directions, or if those empty spaces are due to local variations in density that would vanish on a larger scale. If they would go deeper to where the instruments used up to now could not penetrate, would the density of galaxies return to what is seen in every other part of the sky? Are these empty spaces real, violating the cosmological principle (Section 2.4)?

To answer that question, the director of the Hubble Space Telescope decided to use the observation time he had at his disposal as the director to do something that would have otherwise been difficult to be accepted by evaluation committees of the observational proposals. He chose one of the darkest zones of the sky where no instrument had found a galaxy, a zone that was only 1/30 the diameter of the Moon, left the Hubble telescope pointed in that direction for 10 days (from 18 to 28 December 1995), and took three pictures in three different colors with an instrument (from which digital cameras were derived) capable of capturing every photon that arrived over this time.

The result of the measurement is shown in Figure 10.9. It is impressive: we see an incredible number of galaxies that no one had ever seen before: elliptical, spiral, irregular, some larger because they are closer, and others smaller, almost like points, because they are farther away. Each of these has, like the Milky Way, hundreds of billions of stars.

Figure 10.9: Hubble deep field; HST image (1996) of a piece of the sky equivalent to 1/30 the diameter of the Moon. It is the deepest image ever obtained. The least luminous object detected is 4 billion times weaker than the limit of the human eye. Almost every point you see is a galaxy like ours, containing billions of stars. Credit: Robert William, the Hubble Deep Field Team and NASA.

Not all of the galaxies in this square are visible. The most distant escape the sensitivity of the HST, because they emit most of their energy in the infrared and the submillimeter (Section 2.10) which the HST instrumentation does not see. The dimensions of the figure correspond to a square of 1 min of arc and, as we have said, one-thirtieth of the diameter of the full moon — so small that we see only one star in the image, at the center with the characteristic cross due to diffraction.[14] To cover the entire celestial sphere a hundred million pieces of such a dimension would be required!

If we count the points we see in the figure, which are all galaxies, whose existence was not known the day before the picture was taken and, multiplying by 100 million (the number of such squares necessary take for covering the celestial sphere), we arrive at 100 billion new galaxies, each with 100 billion stars. Few images, like this one, give an idea of the immensity of the universe, an immensity that we know to be finite.

Of all the billions of stars that populate these billions of galaxies, which are those around which life can emerge? The answer is simple: all those that lived long enough for life to evolve. We have seen that life emerged on the Earth 3.5 billion years ago, and that is the time it took to produce biologically evolved life (us). To be sure, we will impose a limit of 5 billion years (the Sun will live a total of 10 billion years). About 95% of the stars we see in the sky exceed this limit (Chapter 4); only the most massive have a shorter life, and they are very few in number. Therefore, if we return to the starry night of the first page of this book and look at the sky, we can say that practically all the stars we see can have life around them, and that this is true for all the stars in all the galaxies of Figure 10.8 and the millions of squares that cover the celestial sphere.

Thus, we know that there are many stars about which life could form, but how many of those have planets? Here the answer becomes more uncertain — all the stars have disks, but how many of them evolve into planetary systems? One is tempted to generalize from the case of the Earth and say that all the stars have planets, but without measurements this remains speculative.

Even if all the stars do have planets, there remains the following question: how many of those are habitable? We have seen that many conditions must be satisfied in order to render a planet habitable: to be in the right place in the galaxy (we are so, and stars around us also are; Section 8.1), at the correct distance from the star (Section 8.7), have enough water to let life develop, have a process like continental drift that restores eroded land, and volcanoes that again put into circulation atoms necessary for life that are trapped in sediments (Chapter 9). It is also useful for a planet to be able to avoid impacts with large

---

14 *The cross we see is due to the diffraction of light (Appendix A.4) coming from a brilliant point source (like a star) on the structure that holds the secondary mirror of the HST. It is an effect which is not seen for extended weak sources like galaxies.*

asteroids and comets which are capable of sterilizing it, and perhaps possess a satellite like the Moon capable of stabilizing its axis of rotation. An answer to these questions can only be given by a measurement: one that will measure enough stars to see if they have habitable planets, to give a statistically significant answer. We will speak of this measurement in the final chapter.

There are, however, many reasons for being optimistic, since we are in the optimal zone of the galaxy, so that if life survives on the Earth, it is probable that it also survived on stars nearby. The Sun is one of the most common stars in the galaxy, so it is reasonable to suppose, as we have often said, that what happens here also happens on the nearby stars. They are weak arguments, but in the absence of a measurement it is the best we can do.

On the question "Are we alone in the universe?" one can, however, respond that it is practically impossible that we are alone. If life is a consequence of the physical conditions found on our planet (Section 5.2), Figure 10.9 makes it impossible to guess that there are no other "Earths", knowing that each of those points represents 100 billion stars like the Sun. Even if the probability of having the conditions necessary for life would be low, there should certainly be other regions in the universe where these conditions exist. But probably this is not the answer we are looking for. We would probably want to know if there are habitable planets around us, that we can contact and perhaps, one day, even visit. The answer will come from the measurements we will make in the coming years.

Part III

In Search of Another Gaia

The Earth is the cradle of humanity, but you cannot always live in a cradle.

Konstantin Tsiolkovsky[1] Kaluga (1911)

---

1 *Konstantin Tsiolkovsky (1857–1935). Pioneer of Soviet astronautics, he first theorized many aspects of human presence in space.*

In Search of ETI

Chapter 11

11.1 Messages from the Stars

In 1959, two physicists from Cornell University, G. Cocconi and P. Morrison, published an article in the prestigious journal *Nature*. In the article they observed that if some planets around nearby stars was inhabited by intelligent civilizations (ETI, or ExtraTerrestrial Intelligence) with a level of technology similar or superior to ours, they would probably have sent signals into the cosmos to get in touch with evolved forms of life on other planets.

The band of frequency in which a possible ETI would have transmitted messages would probably have been among the centimeter waves (Figure A.1), the same one used by television transmissions, because they are least disturbed by the radiation emitted by celestial bodies. At longer wavelengths, those used for radio transmission, the noise (Section 2.5) increases with synchrotron radiation[2] emitted by ionized particles that move along the galactic magnetic fields. At shorter wavelengths — submillimeter, infrared and optical wavelengths — the noise increases because of emissions from giant molecular clouds (Section 3.4) and from stars. Centimeter waves constitute therefore the part of the radio spectrum with the least cosmic noise; for this reason they would have been chosen by a possible ETI for its transmissions.

---

2 *The name is derived from the synchrotron particle accelerator, where high energy particles spiral around the lines of force of the magnetic field, emitting the radiation.*

The idea was fascinating and was promised for great results with a relatively modest investment: it seemed sufficient to search inside the noise coming from space for a significant signal, to decipher it and then find a way of responding.

But how can one communicate with a completely different civilization using a simple radio message? The idea is to write messages based on mathematical concepts. In mathematics one recognizes a universal value; it is therefore natural to think that a civilization that has developed the technology necessary for attempting interstellar communications has acquired a notable mathematical capacity. An example of a possible message is the one transmitted by the great radio telescope of Arecibo in 1974, and for this reason is known as the Arecibo message. It was transmitted in centimeter waves in the direction of a globular cluster located in the constellation of Heracles and is still on its way. Even if traveling at the speed of light, it will take 10,000 years to reach its destination.

The message of Arecibo is written in binary code and consists of a sequence of 1679 characters (a sequence of 1's and 0's). The number 1679 is the key with which to start the interpretation of the entire message. It is the product of two prime numbers, 23×73; this should suggest a distant intelligence, who perhaps, some day will analyze the message to dispose the 1's and 0's in 73 lines of 23 characters each, as on the left side of Figure 11.1.

The first line of the message reports in binary the numbers from 1 to 10. Immediately under that, to confirm the numerical character of the message, the atomic numbers (the number of electrons that orbit around the nucleus) of the most abundant atoms in the universe (hydrogen, carbon, oxygen…) are written. This first part of the information has the purpose of confirming the correctness of the key used to read the message. Afterward the more complex part of the message, in which information is transmitted, begins. The message is about who we are, where we live, using images other than numbers. If in Figure 11.1 we color in black all the 1's (as is done in some children's books), we obtain the image on the right. In this manner under the atomic numbers the following information is reported graphically: the structure of certain chemical compounds, including the molecule of DNA with its double helix form, our physical aspect, the Solar System, the number of inhabitants on the Earth and the characteristics of the Arecibo antenna. The height of a man and the dimensions of the antenna are also expressed in binary numbers, as multiples of the wavelength which is used for the transmission (12.6 cm), to confirm the interpretation of the message.

The message may appear complex and will require time to be deciphered, but it will not pass unobserved. Whoever will analyze that piece of signal will notice that there is something ordered in the noise. Once the attention is attracted, it will not be difficult to decipher the message. A possible message from an ETI is expected to be similar to that of Arecibo.

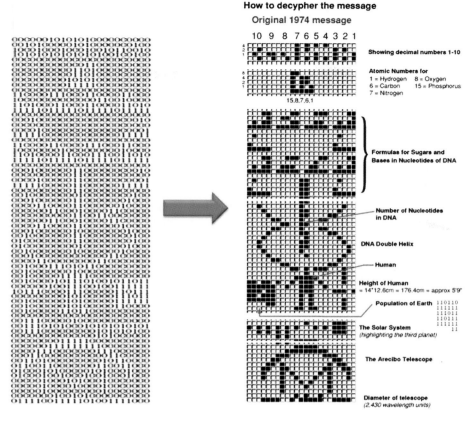

Figure 11.1: The message transmitted by the Arecibo radio telescope. Left: The 1679 characters are disposed in 73 lines of 23 characters each. Right: The interpretation of the message. (© Arecibo Observatory.)

The Cornell physicists, therefore, suggested searching in the radio wavelength the noise received from the cosmos (Section 2.5) for a sequence that repeats itself or a series of characters that do not dispose themselves in a random fashion (as one would expect for a signal that is only noise).

Since the publication of Cocconi and Morrison's paper, many researchers have analyzed the radio signals received from the stars to find out if something intelligent was there, in a sequence that could be interpreted in some way. They designed sophisticated computer programs that examined every signal and every noise coming from the cosmos, and tried to interpret these. The activity is very attractive, promising great results for a very modest investment. All that is required is an existing radio telescope, to point it at a star, to decipher the signal and to try to respond....

Some countries devote a lot of energy into this research. One of them was the Soviet Union, which in the 1960s set out to systematically observe large regions of the sky with antennae that were basically omnidirectional. The idea was to make a map of the whole sky so that no intelligent signal could escape.

The result was a total failure; despite the considerable resources invested in the program, the researchers found nothing. The cause was attributed to the fact that they tried to look at a piece of sky that was too large, with too many signals, both terrestrial and celestial. In fact, an antenna that looks at a large section of the sky is not sensitive to a signal coming from a point source like a distant planet. The antenna, seeing a large section of the sky, will record the noise from all the sources in that region, which grows with the dimension of the region seen by the detector (the field of view of the detector). The lack of success was therefore explained as being due to signals that were too weak in comparison with the noise of the antenna used. It was concluded that, in order to obtain a result, it was necessary to reduce the noise by reducing the angle of reception of the antenna as much as possible.

In the 1970s, the United States took up the problem by adopting this criterion. The researchers used antennae that were observing very limited zones of the sky to receive signals more distinguishable from the noise, and therefore easier to decipher. They decided, in practice, to examine one star at a time, starting with those that seemed to be the best candidates. The program was sponsored by NASA and had the name SETI (the Search for ETI). Not even in that search, using the most powerful radio telescopes of the world, and with a very careful selection of targets, was any significant signal revealed.

Shortly after SETI, a further system dedicated to the search for extraterrestrial signals was designed. Its name was Cyclops for its enormous dimensions; in fact, it should have been made up of 1500 parabolic radio antennae (similar to those used to receive television signals via satellite) with diameters of 100 m (7950 m²). They would have functioned like a single antenna, occupying an area of 60 km², with a collecting area of more than a square kilometer. It would have been the most sensitive receiver in the world — hundreds of times more sensitive than any existing radio telescope. In theory, it would have been capable of detecting every intelligent signal transmitted within a radius of hundreds of light years. Conceptually it was an instrument similar to ALMA (Section 4.10) except that it was much bigger and was conceived without the technology that exists today (mainly that of computers). Unlike ALMA, Cyclops was limited to working in centimeter waves.

These first projects involved some of the most prestigious American universities and the world's best research institutes, and contributed to a thaw these years in the USA–USSR relations. In 1971, in the middle of the Cold War, the United States and the Soviet Union jointly sponsored a conference with the promising title "*Communication with Extraterrestrial Intelligence*". It was the beginning of many exchanges of information between Russian and American scientists.

In 1984, the SETI Institute was founded in California. It is still active today, with the aim of supporting projects relevant to the search for life in the universe. It was the apogee of these activities in the world. As years pass, the lack of results brought many scientists to question the sense of the project, and their support from government funding was reduced.

The first victim of these failures was the project Cyclops, which was canceled. Then, at the end of 1993, after 30 years of failure, the United States congress decided to cut all funding for such research, including that for the SETI Institute. The decision was taken at the end of the celebration of the 500th anniversary of Columbus' arrival in America. The supporters of SETI underlined the irony of an America that was cutting funding and had lost confidence in a courageous project, while celebrating the explorer who had discovered America, thanks to the trust that the queen of Spain had in him and in a project that was more courageous than the search for ETI.

The funding cut was a mortal blow to SETI. The research continues today albeit in a lower tone with much struggle and with funds obtained from private parties. In those obstinate supporters of the project, there is the conviction that they are not to be discouraged by the current lack of success. Many great discoveries of the past were accompanied by incredulity and diffidence, and many great results were obtained through the constant support and the faith of a few in an idea.

11.2 The Research Continues

The heritage of the SETI project was gathered into another project called Phoenix, completely funded by private sources. This wishful name was given in the hope of making the search for extraterrestrial intelligence re-emerge after the difficulties met in the 1990s. Researchers who worked on this project knew that it was not easy to find signals of ETI as one had hoped 30 years earlier. They knew that they were looking for weak signals obscured by radio noise from celestial and terrestrial sources. But they also knew that they could not wait long, because every year the amount of radio signals produced on the Earth increases and soon it will be impossible to pick up extraterrestrial signals.

Project Phoenix followed the path of the SETI project, studying single objects. The researchers decided to concentrate on nearby stars similar to our Sun which have lived long enough for life to develop. In the 1990s they chose the sample to be studied, about 1000 stars similar to the Sun, all within 200 light-years from the Earth.

Among those radio telescopes that participated in the project were some of the largest in the world. The first measurements began in 1995 with the 210 ft (64 m) diameter telescope in New South Wales, Australia. Then other telescopes were added, like the Green Bank telescope in West Virginia (USA), and finally the largest radio telescope in the world, the one at Arecibo, Puerto Rico (Figure 11.2), which is the same telescope used to transmit the first human signal toward the stars. The Arecibo mirror has a diameter of 305 m and was constructed inside an inactive volcanic crater, by

Figure 11.2: The radio telescope at Arecibo (Puerto Rico). It is constructed inside a volcanic crater. With its 305 m diameter, it is the largest telescope in the world. (© Arecibo Observatory.)

dressing its sides with 38,778 aluminum panels, and by building an immense parabola that, being attached to the walls of the crater, cannot be moved. The telescope can point only at areas in the sky that lie within 15° of the vertical axis, by correcting its pointing with small movements of the secondary mirror (Figure 11.2), which is suspended by three cables supported by the pillions erected at the sides of the crater.

Project Phoenix is now completed. Nearly all the objects in the sample have been observed but, notwithstanding the sophisticated system of calculations used, nothing significant has been detected. The analysis of the captured data continues still today, thanks to the dedication of a few enthusiasts with the conviction that in the measurements taken, messages from extraterrestrials are hidden, messages which we do not yet know how to read. The problem, they say, is how to extract these messages out from the noise. Many are convinced that when we will find how to decipher the first signal, we will have the key to decipher all the others. Much of the work consists in analyzing the signals with mathematical algorithms and trying to reduce the noise. Given the enormous quantity of data to examine, great computational power is needed. Part of the work is done with the help of private parties who make their computers available when these are not in use, especially at night. Sometimes the computers are simple PCs that the owners leave online, at the disposal of the project. The range of data to examine is so great that every contribution is gratefully accepted. The continual failures seem, however, to indicate that if the ETI's signals exist, the system is not capable of extracting them from the noise.[3]

In the years to come, the fortunes of SETI may be reborn, thanks to an ambitious project in which numerous countries are participating. Its name is SKA (Square Kilometer Array). It is a new Cyclops that will be built 50 years later, with the experience of ALMA and the enormous power of the computers in the future. The frequencies of SKA are the centimeter waves, the same as Cyclops where there will be the minimum of cosmic noise (Section 10.1). If approved, the project should start off in 2020 with a structure similar to that of ALMA: a series of parabolic telescopes that will function like a single telescope (Figure 4.14). The difference is that they will be distributed over a much larger area, and a much higher spatial resolution can be reached. Half the collecting area ($500,000 \, m^2$) will be contained inside an area of $5 \, km^2$ and the rest in an area of $3000 \, km^2$. This will permit observations of a very small region of the sky, less than a millisecond of arc, thus eliminating the noise that comes from regions other than the source (but not what is coming from the Earth). The telescope will be placed in an area of very low radio disturbance, probably in the Australian desert. The reasons for constructing SKA, which will justify the enormous investment, are scientific. The instrument will be used to study the primordial universe, the evolution of galaxies, dark matter and interstellar magnetic fields, but part of its time is reserved for SETI, and this reignites much of the enthusiasm.

SKA will be the limit to what can be achieved from the Earth. After that, the search for ETI, if it will continue, it will have to be transferred to space, far from the radio noise of the Earth. In a few years, when space technology will become less expensive, it will be possible to build a spatial "Cyclops" far from the radio noise of our planet. It could be on the Moon rather than being built through free flyer satellites (like those of the Darwin mission; Figure 12.4). They will make possible an increase by a factor of 100–1000 in sensitivity. In that case, if ETI exists, it will not escape....

11.3 Drake's Equation

To evaluate quantitatively the probability of coming into contact with an ETI, the American researcher Frank Drake, founder and ex-director of the SETI Institute, put forward in 1961 a formula that bears his name. The formula expresses the number of civilizations, N, with which we can make contact through a series of parameters:

$$N = R \times F_p \times N_e \times F_l \times F_i \times F_c \times L$$

---

3 *With a similar declaration, the Australian radio telescopes in New South Wales concluded their participation in the project.*

The parameters on which N depend are:

R   *Number of stars born every year in the galaxy*
F$_p$ *Fraction of stars with planets*
N$_e$ *Number of habitable planets per star*
F$_l$ *Fraction of planets with life*
F$_i$ *Fraction with an intelligent form of life*
F$_c$ *Fraction capable of communicating by interstellar radio emission*
L   *Average lifespan of a civilization capable of communicating*

Of all these parameters, only the first two are known, so that Drake's equation, more than merely a tool for calculating the probability of finding an ETI, has in these years been used as a mechanism for reasoning about the problem. Many of the ideas suggested by these parameters have been discussed in the previous chapters. We will consider them briefly:

R: *Number of stars born every year in the galaxy.* We know that, on average, one star is born in the galaxy every year; this is evaluated using observations. With a reasonable degree of accuracy, we can say that $R = 1$.

F$_p$: *Fraction of stars with planets.* For this parameter, we can give a realistic estimate. In Chapter 4, we saw that the number of stars with planets is around 95%. In Section 8.1, we saw that this number varies from one point in our galaxy to another; the Sun is in the right zone, so we can say that at least in the vicinity of the Earth we have $F_p = 0.95$.

The evaluation of the next two parameters is difficult and we do not have anything concrete to measure, at least for now.

N$_e$: *Number of habitable planets per star;*

F$_l$: *Fraction of planets with life.* We know what is necessary to construct a habitable planet, but we do not know how probable that is (Section 10.6). These two parameters could however be estimated via measurements in less than 20 years, when the missions which we will speak of in the next chapter start. For now, we have to be contented with the arguments we presented earlier: if all the stars have planets, why should they be different from ours? If life on the Earth were born as soon as it was possible, why should it be different for stars similar to the Sun?

There remain the last three parameters which are the most difficult to evaluate: the questions they pose can be answered only by coming into contact with another culture.

F$_i$: *Fraction with intelligent life.* Is it possible that no form of life intelligent enough for interstellar communication could arrive on a planet? For example, if the dinosaurs had not gone extinct on the Earth, would a species that was intelligent enough for interstellar communication arise? Have the dinosaurs prevented the development of mammals? What level of intelligence and technology would the dinosaurs gain if they had not been extinguished? The answer to this last question is that the level would not be very high, because during the 135 million years the dinosaurs have dominated the planet, they did not reach, even remotely, the level that mammals reached in one third of the time. But that raises another question: Does the emergence of intelligent life depend on the impact of a meteorite?

F_c: *Fraction capable of communicating by interstellar radio emission.* Until we come into contact with other civilizations, the value of this parameter will remain at the same high level of uncertainty of F_i.

L: *Average lifespan of a civilization capable of communicating.* This parameter is also difficult to evaluate. There are many threats to an evolved civilization: the explosion of a nearby supernova, the impact of a large meteorite or comet (Section 10.3), self-destruction by war or an environmental cataclysm caused by ETI. For the Earth, the risk of destruction by a supernova is not high (Section 8.1), while that of an impact with a large asteroid or comet will diminish with the life of a planet (the risk is higher in the early times) and once an "intelligent" species is settled with the advancement of its technology (Section 10.4); with regard to self-destruction, it is hoped that ETIs, above all those of our planet, will be truly intelligent and wise enough to curb themselves in time… in any case, we are in an undefined field.

The uncertainty in the values to be given to the parameters of Drake's equation explains why some people doubt the significance of this research: habitable planets may be very rare, or there could be millions of them but with none having developed species equipped with technology necessary for communication with other planets. Narrowing the search to advanced forms of intelligent life, the possible cases may become very few, making the search very difficult.

To find out whether there are other intelligent beings in the universe, we are more likely to succeed if we proceed differently:

(1) Employ the techniques which we will speak of in the next chapter to search for habitable planets, regardless of whether there are forms of intelligent life or not.
(2) If the answer is positive, we will know which stars we should point our antennae at and try to contact a possible ETI. At that point, if even SKA does not see any signal, there is nothing left to do except to build large, dedicated spatial radio telescopes far from the Earth and its radio noise.

It is therefore reasonable to suppose that before the end of this century it will be possible to answer the two questions that we have asked: Do habitable planets like the Earth exist? Are these planets inhabited by ETIs?

During this time, we will also have developed lasers and masers of great power to launch signals into the cosmos with impulses, although for a short-time (less than a millisecond), could reach an intensity hundreds of millions of watts to approach many stars after a voyage of several years. We could thus start a new hunt for ETIs, knowing where to search, and with much more powerful weapons.

11.4 A Message to the Infinite

In 1972, 15 years after the launch of *Sputnik*, the United States launched *Pioneer 10*, the first human artifact destined to leave the Solar System. *Pioneer 10* was launched on 2 March 1972 and set numerous records. It was the first human satellite to pass undamaged through the asteroid belt and arrive at Jupiter (on 3 December 1973), passing at only 120,000 km from its surface. Moreover, it was the first satellite to experiment with the slingshot effect (Section 10.2) and obtain from the giant planet the impulse necessary for leaving the Solar System. A year later, on 4 December 1974, *Pioneer 11* was launched and followed the path of its predecessor, visiting Saturn after Jupiter. In 1977. *Voyager 1* and *2* were also launched, for a long voyage toward the stars. *Voyager 1* is today the man-made object farthest from the Earth; it is moving away from the Sun at 17 km/s[4] towards the constellation of Ophiuchus. The distance record (brought to date in 2009) of these satellites puts *Voyager 1* at 16,637 billion km from the Earth; *Pioneer 10* is at 14,570 billion km, traveling at a speed of 12.1 km/s toward Aldebaran in the constellation of the Bull; *Voyager 2* is at 13,493 billion km, going toward Sirius at 15.5 km/s; and, finally, *Pioneer 11* at 11,764 billion km, traveling at 11.5 km/s toward the constellation of the Eagle. The distances from the Earth do not follow the chronological order of their launches, but are influenced by the different routes they have taken inside the Solar System. These satellites are four messengers sent to different regions of the galaxy to witness our existence.

On the *Pioneer* satellites are loaded two inscribed plates carrying a message from the Earth to any inhabitant of our galaxy that may find them. The plates are identical on both satellites, in gold-coated aluminum and are placed in sheltered positions to protect them from impacts by dust and meteorites that might erode their surfaces.

On the plates (Figure 11.3), in the top-left hand corner, an atom of hydrogen is represented. Among the mechanisms that excite this atom, there is one that regards the spin of the electron. In a collision it can invert its direction with respect to the direction of the central nucleus spin. De-exciting it (i.e. realigning the two spin axes), the atom emits a photon at a wavelength of 21 cm, which corresponds to a frequency of 1420 MHz. This transition of hydrogen,

---

4 A very slow speed with respect to stellar distances. At that speed it would take 17,600 years to travel a light-year.

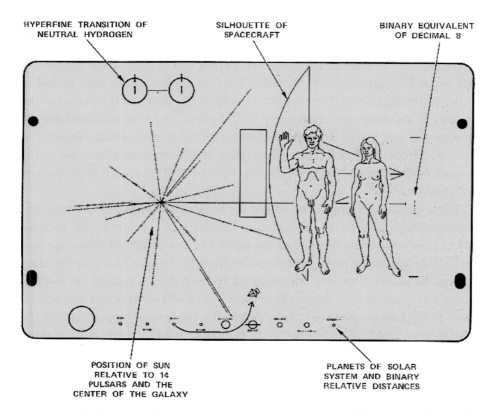

HYPERFINE TRANSITION OF
NEUTRAL HYDROGEN

SILHOUETTE OF
SPACECRAFT

BINARY EQUIVALENT
OF DECIMAL 8

POSITION OF SUN
RELATIVE TO 14
PULSARS AND THE
CENTER OF THE GALAXY

PLANETS OF SOLAR
SYSTEM AND BINARY
RELATIVE DISTANCES

Figure 11.3: The plate mounted on Pioneer *10 and 11, containing a message from the Earth, destined for whoever will find it, even if it will happen only a million years from now. (© NASA GRIN DataBase.)*

called "hyperfine", is used to measure the abundance of hydrogen, and would be known to any culture capable of recovering the *Pioneers*. The inscription of the hydrogen atom on the plate sets the unit of length (21 cm) and of frequency (1420 MHz), that will be used in the rest of the message for units of length and frequency.

On the right-hand side of the plate, a man and a woman are shown with averages features that should represent all the races of the Earth; their height is compared with that of the radio antenna and is also written beside them in binary with the number 8, which is the mean height of humans in units of the wavelength ($8 \times 21 = 168$ cm). The man has one hand raised as a sign of peace, a gesture that the authors of the design (Frank Drake, Carl Sagan and his wife Linda) wanted to attribute universal significance.

On the left of the design, the position of the Sun in our galaxy is indicated, making reference to 14 pulsars, easy to be recognized by an extraterrestrial civilization, because they send out radio signals with very precise frequencies defined by their rotation velocity (Section 4.5).

The frequency of the pulsars and theirs distance from the Sun are expressed in binary code, using as a unit of measure the wavelength and frequency of the hyperfine line of hydrogen. This should permit one to localize the position of the Sun with great precision. Moreover because pulsars reduce their period in time, according to a well known law, the difference between the period of the pulsars at the moment of recovery of the plates and that indicated on the design will give a very accurate evaluation of the year in which *Pioneer* was launched from the Earth. For a localization that is even more precise, at the bottom of the plate the design of the Solar System is shown, with the trajectory followed by *Pioneer*.

These plates were conceived to last for hundreds of millions of years, and perhaps billions. They could therefore survive our species. When tectonic plates have completely redesigned the surface of our planet, or when our species will be extinct or perhaps replaced by another, they could still be traveling across the galaxy. One day, they may be recovered by a culture far away in space and time. As Carl Sagan wrote then: "They will witness that in a precise year and a precise point in the universe the organisms represented in the design lived, and that these organisms wanted to share the news of their existence with other intelligent beings in the universe."

On the *Voyagers*, instead, it has been chosen to leave the message of our existence on an optical disk (similar to a CD) made with a technique that should permit it to survive for millions of years. In that case it was possible to register much more information. Aside from the information about who we are and where we are in the universe (written in a code similar to the message of Arecibo), a message of goodwill in all the known languages of the Earth is registered. The most beautiful music composed by humans are also inscribed, going from Mozart to the Beatles. Also, the music should have a universal value: the harmonies are based on precise ratios between different frequencies. On the surface of the discs the same message of the Pioneers plates is engraved.

The messages from the Earth will travel for thousands of years, perhaps millions. In the future more and more satellites will be launched: satellites that will arrive at the periphery of the Solar System, and then continue their trip toward the infinite, perhaps with velocities much greater than that of *Voyager*. If our civilization will last for a long time, millions of years (in theory it could last another five billion years, till the death of the Sun), the number of satellites of this type to be launched toward stars could be enormous. If the Earth will be inhabited for the next five billion years by intelligent life and if, on the average, a minimum of two satellites will be sent into space in every century, there will be more than 100 million satellites sent by the Earth during its existence. If in our galaxy there will be many habitable planets with technologically advanced civilizations similar or superior to ours, the number of satellites of this type crossing the galaxy shall be enormous, and some shall arrive at the Solar System. Will it be possible to find them?

If one of them is still active and emitting a weak radio signal, with the advancement of space technology we should be able to locate them easily. However, one asks "Can radio systems capable of operating after millions of years exist?" For terrestrial technology this is unthinkable. The problem is not the source of energy, since it can be supplied by radioactive material which can last a very long time (*Pioneer* and *Voyager* used plutonium, which decays rapidly, even so 40 years after the launch they still send signals to the Earth). The problem is to have an instrument functioning that long, which is inconceivable today. Maybe one could consider a passive radio reflector, a series of reflecting surfaces capable of sending the radio signal it receives back to the same direction they came from.[5] But it is difficult to think of a reflecting surface that remains efficient for millions of years.

It is not ruled out, however, that in the course of the Solar System's exploration, we will meet one of these messengers, sent perhaps by a civilization that lived around a star which died millions years ago and, as we have done, they wanted to tell us about their existence and let us know that we are not alone in the universe.

11.5 Toward the Conquest of Stars

The direct exploration of stars is an objective that will remain unreachable for a long time. Just consider the nearest star, Proxima Centauri, it is at a distance of 4.22 light-years from Earth (its light takes 4.22 years to reach us). Since a spacecraft constructed today does not reach a thousandth of the speed of light, the time required to reach this star and bring back the information would therefore be of the order of tens of thousands years; for other stars the time is even longer. To make interstellar exploration possible, we need therefore to reach velocities much higher than we do today — a goal far in time. Even in that case the trip would take more than a century and will be completed by a generation different from the one that started the trip. For this century, and probably for the followings, the exploration will therefore be limited to the Solar System.

The conquest of space by humans is not only difficult, but it will be impossible if we do not solve the problems caused by the absence of gravity and by the radiation in space which is four times higher than the level we have on the Earth.

The experience of the last years has shown that living in the absence of gravity over a long time causes cardiovascular problems, muscular atrophy, and a

---

5 *These reflectors have many uses, for example, returning a radar signal to the sender. In this manner, a small boat can easily be seen from the big ship and avoid a collision.*

decline of the immune system. The worst effect is the loss of calcium in the bones at about 0.5% per month (as a consequence there is an increase in kidney stone). Such a loss can reach up to 1.5% in the articulation of the hip. This problem, if it is not solved, will keep humans from going farther than the Moon.

Waiting for the solution of these problems, the first step toward stars should come around 2020. The "Return to the Moon", a program that foresees the construction of stable bases on our satellite, largely independent of replenishment from the Earth. They should be capable of exploiting the environment to produce everything they will need for living. These bases have to be constructed in places where water is found. On the Moon, water arrived when it was formed (Section 4.10), and survived till today in areas shaded from sunlight. Water is fundamental for an autonomous base; it is essential for the physiological needs of its inhabitants, for the cultivation of food and it will be the main source for the production of oxygen.

If humans will be able to build these bases on the Moon, the following step can be considered: the conquest of Mars. An adventure that may start around 2050 and will require efforts much greater than those necessary for the Moon. Such a mission will last for years, so it will not be possible to carry the provisions necessary for all the duration. It will even be impossible to carry the fuel needed for the return trip. All these should be produced in place. For these reasons, the travel to Mars will be only possible when the lunar bases have demonstrated the possibility of a long stay without any supply from the Earth. It also requires a solution to the physiological problems caused by staying long in space in the absence of gravity.

This mission has a minimum duration of 900 days, because of the Martian year, which is 1.88 Earth years. The length of the Martian year is such that, approximately every 2 years, the two planets are in *conjunction* (the position in which they are closer to each other). Only near this position the time to travel from one planet to another is minimized and, as a consequence, the physiological problems and exposure to radiation are minimized. Particularly when the Earth and Mars are on opposite sides of the Sun, the voyage is impossible).

In these optimal conditions the minimum duration of the voyage is approximately 150 days, to which 600 days have to be added, in order to wait for Mars to get close to the Earth again, and other 150 days for the return trip. The crew must therefore be prepared to handle every unforeseen event because, in a difficult situation, any help from the Earth will take years to arrive.

To minimize the risk, two years before the mission (when Mars and the Earth are in *conjunction*) one or more automatic modules will be launched containing the system to produce the fuel necessary for the return trip (oxygen and methane taken from the environment), the tanks to contain the fuel, a vehicle for the return to Earth and, finally, a module to live in on the surface of Mars. The astronauts will therefore leave our planet only when all of these equipment are working properly, so that they will find at their arrival a place to live and other things necessary for the return trip.

If permanent bases will be built on the Moon and on Mars in the second half of the century, the exploration will probably be extended to the asteroid belt and to some of the moons of the giant planets (like Europa, the moon of Jupiter). With the exploration of the asteroids, the commercial exploitation of space may begin. The materials that, for the meantime, will become scarce on our planet may be found on the asteroids and, because of their light gravity, can be easily sent to the Earth.

With time passing, the bases on the Moon, on Mars and the orbiting space stations will grow and increase in number and with these, will increase the experience of how to live in space, and we may also know how to accelerate a spaceship to a speed much higher than what is achievable today. At this point humans could be ready for the conquest of stars — a step that, as we said at the beginning, certainly will not happen in this century. With the experience gained on space stations, enormous spaceships could be constructed. They should be capable of transporting 400–500 people, a number big enough to ensure the genetic diversity of their descendants. The spacecraft should also be large enough to contain everything they will need to live inside for centuries. They will be little worlds where generations of peoples will be born and die without landing on a planet. These great structures (which cannot enter into the atmosphere of a planet) will rotate continuously to provide the gravity necessary for the stay in space.

The measures we will speak of in the next chapter will show the route to follow in order to find habitable planets to deposit the seeds of our species. In this way it might begin the conquest of stars, an event that would render even more emblematic the question posed by the Fermi's paradox (Section 10.5): if in the past there have been intelligent beings who lived on planets similar to ours, around stars that have disappeared, why did the inhabitants of these worlds not start a voyage before dying? Why are the others not here?

An answer to this question could come from the measurements we will be speaking about in the next chapter. We may discover that there are no habitable planets near us, that ours is the only planet in this part of the universe where biological forms can live. At that point, the great spaceships will not know where to go: unless they intend to begin of a never ending voyage. If habitable planets are rare, the search for them by traveling randomly in the galaxy is hopeless. Should this be the case, we will know that we will remain forever on our planet and we will probably better understand better that we have to take much more care of it, being the only place in the universe where we can live.

The Measurement

Chapter 12

12.1 In Search of Another Gaia

In the spring of 2004, the European Space Agency (ESA) performed an inquiry among the research institutes and universities of Europe to find out the research areas in astrophysics they considered to be fundamental for the years 2015–2025, the research in which the agency should invest its resources. This initiative was called "Cosmic Visions" and the results were discussed in September 2004, in Paris, at the UNESCO head quarter, a prestigious venue.

The responses were numerous and all of notable interest. At the end, the proposals were divided into three themes those, over the others, which would have an impact on Cosmic Visions humans would conceive of the universe. They were:

— *The search for other worlds and for life in the universe:* this great theme includes the study of the birth of stars and planets, the discovery and the study of exoplanets, and the search for extraterrestrial life.
— *The birth and evolution of the universe:* the birth of the universe, how the first structures were formed, how they built the first stars and the first galaxies, and the evolution of the universe up to our times.
— *The violent universe:* the study of matter in extreme conditions, black holes and their role in the evolution of galaxies, neutron stars, supernovae and the cycle of matter in the universe.

The first theme reflected the view of the scientific community that the search for life outside the Solar System had ended the pioneering age and that after 30 years of studies and projects, the technology was mature for this type of measurement. It was clear to all that the road ahead was still long but it was sensible for ESA (like NASA with the Origins program; Section 1.1) to consider this adventure possible, and invest in this research.

From the study of the history of the Earth (Chapter 6), we have seen that the presence of life has drastically modified the atmosphere of our planet (Figure 6.5), making abundant the gaseous oxygen that would otherwise be absent. It was therefore clear what to search for. Less clear it was how to carry out the measurements, how to identify the few photons indicating the presence of oxygen in a planetary atmosphere against those of the nearby star, and then at which wavelengths to perform these measurements.[1]

To this theme is this chapter devoted. It will necessarily be more technical than the others and will require more effort from the reader who is not accustomed to technical questions. Whoever does not feel up to this final effort can go directly to the Conclusion, knowing that science believes these measurements are possible and that they are foreseen in the programs of ESA and NASA.

For the more courageous who wants to know on what ideas these measurements are based and will read this chapter, the road will start with the examination of the samples of stars to be observed to give statistical significance to the answers (positive or negative). Then we will discuss on how to face the problem of diffuse emissions (which in the Solar System are called "zodiacal light"), caused by the remnants of the disk from which the planetary system are born. Finally, we will speak of the instrumentation, of how to separate the light of the star from that of the planet, and of the techniques necessary for setting the position of the satellites, of the space interferometer with the precision required for these observations.

The road to follow is not easy; there are many things to be understood and many measurements to be done to define the project. One fact, however, is certain: the search for another Gaia has begun. In the classical sense of the word, the search for an ecosystem where life, as we conceive, can exist.

12.2 The Stars of the Sample

Stars are far away; the closest is Proxima Centauri, which is approximately 4.22 light years from the Earth and is not suitable to be inhabited because it is a double star system (two stars that rotate around each other). Possible planets have, most probably, orbits so perturbed that they have a high risk of falling

---

1 *Oxygen is excited at many frequencies, from the ultraviolet to the submillimeter.*

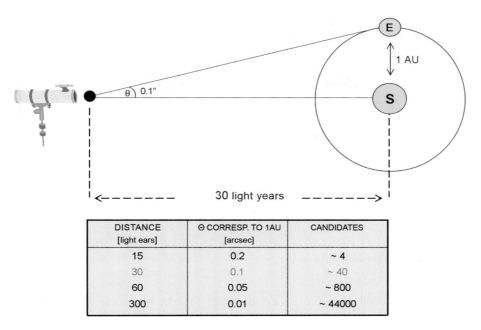

DISTANCE	Θ CORRESP. TO 1AU	CANDIDATES
[light ears]	[arcsec]	
15	0.2	~ 4
30	0.1	~ 40
60	0.05	~ 800
300	0.01	~ 44000

Figure 12.1: The angle between the Earth and the Sun as it would appear to an observer 30 light-years away. In the table the same angle (central column) is reported for various distances from the Earth, together with the number of candidates found within that distance.

into the star or of being let into space (Section 10.2). For this reason, double stars will probably be excluded from the first samples to be studied.

If we go even farther than Proxima Centauri, more than doubling the distance, and consider a sphere of 15 light-years, there are four candidates, all single stars that have lived long enough for life to possibly develop on one of the planets. Doubling the distance again, pushing out to 30 light-years, there are 40 candidates. Doubling the distance yet again, out to 60 light-years, there are 800 candidates. Finally, at 300 light-years, there are 44,000 candidates,[2] a number probably larger than can be considered in a single mission. To be certain of having a significant sample of stars, we should push ahead with a measure of at least 30 light-years, although 60 might be better since many of the candidates may be ruled out from our observations in upcoming years (Section 12.7).

The stars to be observed are close compared to the size of the galaxy, but are at very large distances as compared to the separation between a planet and its star. Imagine observing a solar system at a distance of 30 light-years: the orbit of the Earth would be like a euro coin seen at a distance of 20 km. This shows the difficulty of this measurement. In Figure 12.1, the angle at which the orbit

---

2 The number of candidates increases with the volume considered, that is, with the cube of the distance from the Earth.

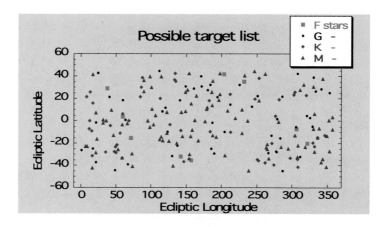

Figure 12.2: Distribution of the possible candidates in the sky, found within 60 light-years of the Earth. (From the homepage of the Darwin mission.)

of the Earth would be seen from a distance of 30 light-years is represented. The angle that separates the Earth from the Sun would be only 0.1 arcsec, which is 36,000 times smaller than 1°. In the table under the figure, the angle at which the orbit of the Earth would be seen is reported for various distances. In the last column, the number of possible stars similar to the Sun contained in a sphere of that radius is also reported.

The angles are very small, that makes us understand the difficulty of the measurement, because not only stars are close to the planet, but they are also billions of times more luminous. As we will see in the next section, this makes it impossible to perform the measurement in visible light; otherwise the resolution of the HST (0.08 arcsec) would have been sufficient to search for habitable planets around a hundred objects.

Reported in Figure 12.2 is the spatial distribution of the 800 candidates found within 60 light-years of the Earth. They are distributed uniformly in the sky (because they are so close to the Earth that they do not reflect the spiral structure of the galaxy).

12.3 The Problem with the Light of Stars

The light emitted by the Sun and by a few planets of the Solar System is represented in Figure 12.3. On the ordinate (vertical) axis the intensity of radiation emitted is reported in powers of 10 (1 corresponds to 10, 2 to 100, 3 to 1000, and so on); on the abscissa (horizontal) the wavelength which identifies the color of the radiation (Figure A.2). Also indicated on the axis are the bands of the visible and of the infrared.

The Sun, at a temperature close to 6000 K, emits most of its energy in the visible, while the planets, which are cooler, emit in the infrared. The figure

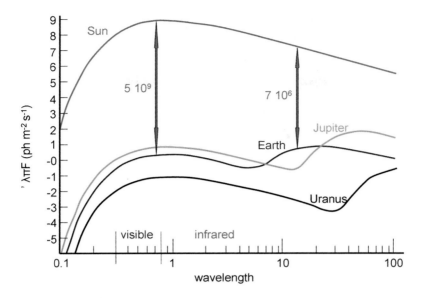

Figure 12.3: Emission from the Sun, the Earth, Jupiter and Uranus.

shows that while the Sun has one peak, the planets have two. The first, in the visible, at the same wavelength for all planets, is due to the sunlight reflected from their surfaces and reproduces, much attenuated, the solar spectrum. The second, in the infrared, is due to the intrinsic emissions of planets because of the temperature they have. The Earth, at about 290 K (17°C), has its peak around 10 microns; Jupiter, colder (Table 8.1), has a peak at 50 microns; Uranus, still colder, around 100 microns.

In the graph, the ratio between the intensity of light emitted by the Sun and that by the Earth in the visible and infrared bands is indicated with red arrows. In the visible, where HST works, the star is five billion times more luminous than the planet. It is therefore sufficient for a small imperfection in the optical system to have one photon from every billion coming from the star to be mixed with those coming from the planet, so that the star be five times brighter than the planet. No technique existing today can hide the star at such a level. For this reason it is not possible to carry the measurement in the visible and, therefore, HST cannot be used.

The situation is better in the infrared, where the planet is only seven million times weaker than the star (the difference is about 1000 times lower than in the visible) and technology can handle the situation.

In the infrared, at 10 microns there is the spectrum of Figure 8.16, which allows the detection of the presence of three molecules indicating the existence of life on a planet: CO_2, which proves the existence of an atmosphere, the easiest to be measured; H_2O indicating the existence of water, a necessary condition for life; O_3, an indicator of the existence of life (Section 8.6).

The infrared is therefore the ideal band for these measurements. At shorter wavelengths, we are dazzled by the light of the star; it is not worthwhile, to go to longer wavelengths because for the same angular resolution the diameter of the telescope (then the total dimension of the instrument) increases with the wavelength[3] (Appendix A.5).

The main difficulty of these measurements is the need for high spatial resolution (to separate the star from the planet which appear close to one another) together with the need to work at long wavelengths (to attenuate the light of the star). The result is that we have to use instruments of large diameter (hundreds of meters to kilometers). The problem would be resolvable if the measurement could be made from the Earth (with an instrument like ALMA Section 4.9). But this is not possible because, if measurements have to be carried in the infrared, it will be necessary to go outside the atmosphere.

12.4 The Zodiacal Dust

To measure light coming from a planet, it is necessary not to mix it with the light coming from the star during its long journey to the Earth. The problem here is the diffusion of light (Appendix A.5), which will occur in all the regions of dust, in which the beam will cross before reaching the observer. Whoever sees a car moving inside the fog during the night will notice that, if the fog is thick, few meters are enough to mix the light from the two headlamps and turn them into a single light. This happens because the particles of fog scatter the light, deviating it from the initial direction, mixing the two light. In our case, even a few grains of dust make the measurement impossible.

We have seen that the atmosphere is a good diffuser of light (Figure A.4). For this reason instruments on the Earth cannot be utilized to search for planets around stars and the measurement has to be done from space, because the diffusion of light by the atmosphere is too high.

The atmosphere is not the only diffuser of light the star encounters on its long trip. In the planetary system, there are remnants of the dust belonging to the disk from which planetary system are formed, that also diffuse the radiation we want to observe. In the Solar System, this residue is called zodiacal dust, because it follows, in first approximation, the distribution of the signs of the zodiac in the sky.[4]

In the infrared the diffusion of light by dust is less than it is in the visible (Figure 12.3), but this diffusion can still be sufficient to intermix the photons

---

3 *If in the infrared, at a wavelength of 10 microns (that of Figure 8.16), one looks for the same spatial resolution that HST has in the green color (~0.5 microns), the diameter of the telescope has to increase by a factor of 20 (=10/0.5), going from 2.4 m of the HST to 48 m. At 100 microns the diameter goes to 480 m, at 1 mm, to 4.8 km.*
4 *Zodiacal dust can be seen by the naked eye on clear nights with no Moon. It looks like a weak glow at the horizon, as if dawn were approaching when instead there are still many hours before the Sun rises. That glimmer is the zodiacal light, due to the solar radiation diffused (Appendix A.4) by the zodiacal dust.*

of the star with those of the planet. If this is the case, for the zodiacal dust the only solution is to move the instrument toward the edge of the Solar System. It is estimated that going beyond the asteroid belt, near the orbit of Jupiter, the zodiacal dust should not be a problem. The mission, however, becomes much more complex, expensive and farther ahead in the future.

If instead the light is diffused by the dust present in the planetary system we are studying, nothing can be done. The light from the star and the planet is mixed at the outset and nothing can be done to separate the two contributions. The only thing to do is to try to eliminate from the sample planetary system that have too much dust inside (Section 12.7).

In the infrared band, aside from the diffusion of light, there is a second problem: the emission of dust due to its temperature (Figure A.2). At 10 microns, where one wants to measure the molecules of Figure 8.16, the total emission of zodiacal dust is a ten-thousandth that of the Sun, but still 1000 times that of the Earth and 100 times that of Jupiter. This can be a problem for the detection of exoplanets, because their emissions could be a small fraction of the dust emissions.[5]

There is, however, an advantage: while the emissions of the exoplanet come from a precise point in its orbit, that of the exodust is diffused in the whole plane of the orbit. This allows to reduce the dust effect. The technique is to rotate the instrument during the measurement. The disk being uniform, will give a signal constant in time, while the planet will produce a peak every time the detector, in its rotation, passes in front of it. The signal of the planet will therefore oscillate at the rotation frequency of the satellite and will be easily detectable.

12.5 The Spatial Interferometer

To measure the 10 micron lines of Figure 8.16 coming from a planet and separate their light from that of the star, an instrument with a diameter between hundreds of meters and a few kilometers is required, depending on the distance to the star (Appendix A.5). Such an instrument cannot be built with a single mirror, even on the Earth. A solution like that of ALMA (Section 4.9) has to be implemented: many telescopes mounted in a relatively large area acting like a single telescope. In this case, the telescopes must be mounted on satellites whose mutual distances permit them to attain the spatial resolution necessary to separate the light of the planet from that of the star (Appendix A.5).

In Figure 12.4, the result of a study done by the French company Alcatel on behalf of the ESA is shown; the system consists of five satellites, each with a telescope onboard that sends the light to a sixth central satellite, where

---

5 *If the planetary system to be measured would be similar to the Solar System, the contribution of its dust would be negligible.*

Figure 12.4: *An artistic inpression of the space interferometer designed to detect exoplanets. The six satellites send their beams to a sixth central satellite, where they are recombined; the result is then sent to the satellite (left in the picture) and transmitted to the Earth. (From the homepage of the Darwin mission.)*

the beams are recombined for the measurement. For every new observation, the six satellites have to move together to point in the direction of the new star. The results of the measurement are then transmitted to a seventh satellite, which, using a large parabolic antenna, retransmits them to the Earth.

To make such an instrument work, the distances among satellites must be known to a thousandth of a millimeter (a fraction of a wavelength) — a precision that is clearly impossible for free flying satellites at distances that can be of kilometers. The idea under study is to control the distance among satellites, with a precision of centimeters (a very ambitious objective) and measure the relative variations in distances between the satellites with the precision of thousandths of a millimeter — a precision possible with technologies well known today.

In the central satellite there are small mobile mirrors, commanded by the instruments that measure the distance between the satellites. By micrometer displacements, these little mirrors lengthen or shorten the optical distance caused by possible displacements of the satellites. In this way the system behaves as if the relative positions of the satellites were stable to within a thousandth of a millimeter.

In Figure 12.5, details of the satellites of Figure 12.4 are shown.

12.6   The Null Interferometer

Once built, an instrument capable of separating the light of the star from that of the planet (the group of satellites which we have just spoken of) another

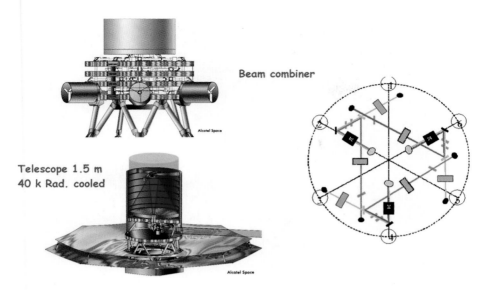

Beam combiner

Telescope 1.5 m
40 k Rad. cooled

Figure 12.5: Details of the interferometer. On the upper left is the central satellite, which contains the small mobile mirrors that compensate for the motion of the satellites and the null interferometer. To the right is its internal scheme. Below one of the five satellites is a 1.5 m telescope. (From the homepage of the Darwin mission.)

instrument has to be built to reduce the light from the star by 10 million times without reducing the light from the planet. To understand the problem, imagine that we are looking at an object near a lighthouse. When the lighthouse is off we see the object distinctly, but when it is on we are blinded and we do not see it anymore. To see the object we need to reduce the brightness of light coming from the lighthouse with a device, which efficiency has to increases with the lighthouse luminosity and the closeness of the object to be seen with the lighthouse.

To measure a weak object near a brilliant star, astronomers are using an instrument called a coronagraph. Its name derives from its first use: the measurement of the solar corona, the part of the solar atmosphere that is observed during an eclipse. And it is exactly from an eclipse of the Sun that the coronagraph is inspired: the instrument poses at a certain distance from the telescope an opaque disk that obscures the light coming from the star, making it possible to measure what is around the star. The instrument creates an artificial eclipse and it is still used today to study the solar corona without waiting for an eclipse.

Figure 12.6 shows an example of the use of a coronagraph in astronomy; it is a photo of the disk around β Pictoris, obtained from the HST (Section 3.6). This is a particular case because this disk is very large (for comparison, the Sun and the orbit of the Earth are shown on the left side). The star is not visible, but the disk illuminated by the light of the star is easily distinguished.

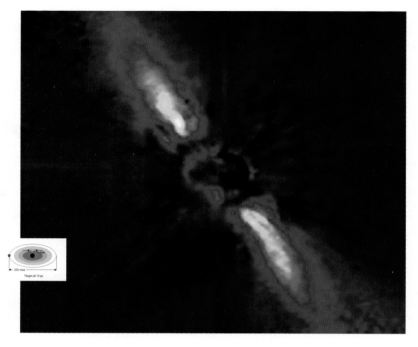

Figure 12.6: Image of the disk around the star β Pictoris obtained by a coronagraph. It is a very large circumstellar disk. For comparison the solar system is drawn to scale at the side. [© ESO & J. L. Beuzit (Grenoble Observatory & Southern European Observatory).]

It seems very difficult to reduce the light from a star by seven million times using a coronagraph. The problem comes from the light diffused by the edges of the disk covering the star. In Appendix A.4, we will see that every obstacle diffuses light; in our case it is sufficient for the border of the disk to spread out one photon in every 700,000 to have the photons of the star 10 times greater than those of the planet. In any case, a high-quality coronagraph that operates in space could give good results, at least for nearby objects. Given the simplicity of the instrument, NASA has recently approved a mission that, 10 years from now, will use such an instrument to search for habitable planets around nearby stars.

The optimal solution for eliminating the light of the star makes use of the wave nature of the light. In fact, if two groups of waves meet, and the maxima of the first correspond to the maxima of the second (they are in phase), the amplitude of the resulting wave will be equal to the sum of the two amplitudes (Figure 12.7 a). It is a fact well known to sailors that, in the middle of a modest tempest, sometimes they find "anomalous peaks" that can be 15 m high (like a four-story building) which disappear after a while. These peaks arise because at that point, waves coming from different directions arrive in phase and add their amplitudes up. After a bit, the anomalous wave disappears because successive waves are no longer in phase. There are zones in the sea where these

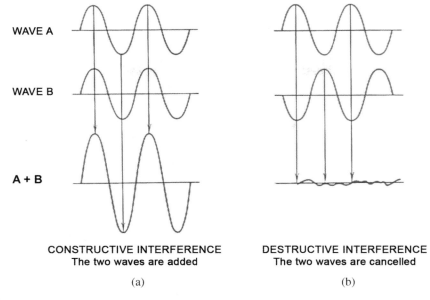

WAVE A

WAVE B

A + B

CONSTRUCTIVE INTERFERENCE
The two waves are added

DESTRUCTIVE INTERFERENCE
The two waves are cancelled

(a)

(b)

Figure 12.7: Examples of interference. (a) Constructive: the waves are in phase and they add up. (b) Destructive: the waves are in opposite phases and they cancel each other out.

waves are more probable than elsewhere: they are the points where marine basins meet, where waves formed by distant tempests interfere.

The opposite phenomenon also exists: if two waves meet and the maxima of one correspond to the minima of the other (they are in opposite phases), their intensities are subtracted (Figure 12.7 b). When this happens in the sea, there is a point of calm.[6]

We return to light. As with the waves in the sea, if two waves are in phase (their maxima meet) at that point the luminosity is the sum of the two maxima. If two waves are in counterphase we have a minimum: the difference between their amplitudes. If the amplitudes are identical, the light can be eliminated entirely. This phenomenon is called interference, and on this principle is based the null interferometer studied, to reduce by a million times the light from the star. In Figure 12.8 is shown the scheme of the instrument and a brief description of how it works.

---

6 In the sea there exists something more dangerous than an anomalous peak: an anomalous trough. When the troughs of different waves come together, they can create an anomalous trough, a hole tens of meters deep that, unlike an anomalous peak, is not visible to navigators onboard until they are close to fall-in. In the diary of the commander of an English destroyer crossing the English Channel during the second world war, such an encounter was described: "on board nothing was noticed until the ship inclined 30°, sending its prow into the void at a breathtaking speed. Arriving at the bottom the ship followed its course, going underwater for over half its length, then it rose again to the surface." A different ship, less robust and without the sealed compartments of a warship would probably not survive to this experience. These phenomena are today watched by satellites. They have been discovered to be more common than expected, to the point of pushing designers to modify the designs of new ships. Many mysteries of the sea may have their origins in such encounters.

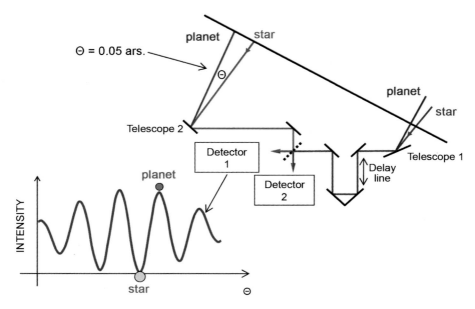

Figure 12.8: The scheme of the null inteferometer. Imagine to measure the same star with two telescopes. The two beams (colored blue and red) arrive at a semitransparent mirror (dotted line) that reflects half of the light and allow the other half to pass. From this mirror two beams emerge containing a mixture of light coming from both telescopes and ending on detectors 1 and 2. The blue beam, before recombining with the red, passes through a mirror in the shape of angle (the delay line) with back-and-forth micrometer movements. The path of the blue beam can thus be lengthened or shortened in order to have on its arrival at the detector a phase-displacement of half a wavelength (5 microns if the measurement is at 10 microns) with respect to the red beam. On detector 1 the situation of Figure 12.7(b) is reproduced and the light from the star is canceled. The planet (green ray) is in a different position from the star (at an angle of 0.05 arcsec). When its light arrives at the detector, it will not be in the opposite phase because it follows a different path, and will not be canceled. On detector 2 the situation is inverted; the light of the star has a maximum and can be applied to verify the pointing of the instrument.

In order to reduce the diffuse emissions (Section 12.4), the instrument will rotate around an imaginary axis that passes through the center of the star. In Figure 12.9 a computer simulation shows how our Solar System would appear in such a measurement. The three white points are Venus, Earth and Mars.

The null interferometer will be placed inside the central satellite of the fleet (Figure 12.5). It is a very ambitious instrument, difficult, complex and costly, but not impossible to construct. The necessity of carrying preparatory observations, before the mission is therefore understandable. About this preparatory work we will speak in the next section.

12.7 Preparatory Work

The measures detecting the presence of oxygen and water on a planet require a lot of time because signals are weak (to extract the signal from the noise days of observation can be necessary, for each star). The number of systems that one can study in such a mission is therefore limited. So it is very important to do

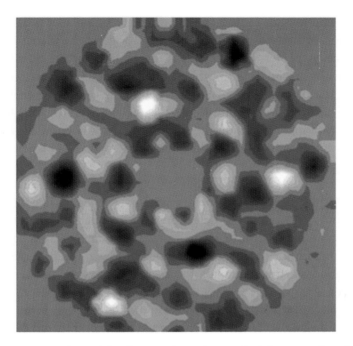

Figure 12.9: Computer simulation of the solar system from a distance of 30 light years, performed with the fleet of satellites (Figure 12.4) and a null spectrometer. The points correspond to the planets Venus, Earth and Mars. At the center we see the star perfectly hidden. (From the homepage of the Darwin mission.)

careful preparatory work to search, among the candidates (Section 12.2), those with the largest chance of success.

The ideal candidate for such a measurement is a planetary system similar to ours, with a star like the Sun, a large planet like Jupiter at the right distance, acting as a sentinel (Section 10.2), a planet like the Earth in the habitable zone (Section 8.7). To know in advance where such a combination exists is not impossible. Stars like the Sun have already been selected (Section 12.2), and we can know the existence and positions of massive planets (Section 3.10). Finally, in a few particular cases, it may be possible to see if there is a planet like the Earth at the right distance, using the shadow it produces on the star.

The existence of massive planets around stars is found with the technique we have spoken of in Section 3.9. In the years to come, the sensitivity of the measurements will increase, so that it will be possible to have a good information on the existence of massive planets in the systems intended for study. All planetary systems with massive planets in the habitable zone (as happens in many of the systems detected), will be eliminated, because in such a zone an Earth-like planet cannot survive (Section 10.2).

For a few of these systems, it would also be possible to determine whether there are planets like the Earth in the habitable zone. It will be possible if a planetary system is seen from the Earth when it crosses the line of sight between us and its star. In this case the existence of a planet is deduced from the reduction in luminosity caused by the passage of the planet in front the star's surface. A planet like Jupiter passing in front of the Sun will reduce the star's flux by one hundredth, while a small planet like the Earth reduces the light of the star by less than one ten-thousandth — a precision that should be reached by the NASA's satellite Kepler,[7] launched in 2009 with the goal of measuring with great precision the star's luminosity. A good measure of the variation of luminosity will allow to estimate the mass of a planet and its distance from the star.[8]

Kepler is expected to observe about 100,000 stars in about 4 years. It should generate a sample of 50–500 planetary systems with Earth-like planets in the habitable zone. The number of detections will be much less than the number of possible candidates because for many stars the plane where the planets stay will not be on our line of sight. The number, however, will be large enough to furnish a good starting point for the search for oxygen in their planetary atmospheres.

Another piece of work that will be done in the coming years is the study of the emissions of the zodiacal dust (at 10 microns) to select the candidates with the lowest emission (Section 12.4). The measurement will be done using the new generation of large ground-based telescopes that have diameters of the order of 8 m. An important contribution to this program will be made by the Mount Graham (Arizona) telescope, built in collaboration by Italy, USA and Germany. In Figure 12.10, the mechanical structure of the telescope is shown: a binocular telescope, with two 8.4 m mirrors mounted close to each other, which will operate as a single telescope; the arrow in the figure indicates that the instrument will have, in that direction, the angular resolution of a 22 m mirror. In the near infrared the resolution will equal that of HST in the optical.

A similar system works in the Atacama desert, not far from the site of ALMA (Section 4.9), where the European Southern Observatory (ESO) has constructed four telescopes of 8 m in diameter (Figure 12.11) that can operate independently, but also together by combining the beams. In this case, the beams coming from the telescopes are sent through the subterranean channels shown in the figure, and then they will reach the null interferometer. These two instruments, in the next few years, will offer an important contribution to the selection of the best candidates for probing nulling interferometry technology.

Another contribution to the preparatory work will be given in the coming years by some space missions, like LISA, an ESA mission that foresees the

---

7 *ESA also has a similar mission in progress, the satellite Eddington, but its launch is still very uncertain.*
8 *From the variation of luminosity it is possible to have the size of the planet and then the mass. Repeating the measurement over time, it is possible to measure the period of the orbit and via Kepler's law (in whose honor the satellite is called) to find the distance of the planet from the star, and then to know if it is in the habitable zone.*

Figure 12.10: The imposing structure of the LBT telescope in Italy. (Photo from the LBT team.)

Figure 12.11: The VLT telescopes that ESO has built in Paranal, in one of the most arid zones of the Atacama desert in Chile. (Photo provided by VLT, ESO.)

launch of three satellites to study gravitational waves.[9] These waves will be detected by the variation in the relative distances of the satellites, a measurement that requires knowing the positions of the satellites better than that it is necessary for the fleet of satellites of Figure 12.4. The success of LISA will prove that the technology is mature for building great spatial interferometers.

12.8 The Foreseen Missions

There are actually two missions foreseen to search for extrasolar habitable planets. Both will use the lines of Figure 8.16 to reveal the presence of CO_2, H_2O and O_3, and both will use a null interferometer to eliminate the light from the star. One mission is studied by NASA and is called TPF (acronym for Terrestrial Planet Finder). The other, by ESA, is called Darwin. Both missions have serious funding problems and heavy delays because of their high costs and some technical problems that are not yet solved.

ESA was foreseeing the launch of Darwin by 2015. In recent years, however, it has decided not to finance the initial phase of the project, but to postpone the launch to a date yet to be determined which will not be before 2025.

NASA has divided the TPF program into two missions. The first is simpler and has been approved for launching in 2015. It will use a coronagraph (Section 12.6) with an elevated capacity of canceling the light of the star. Less costly than the interferometer, it could obtain some results for the nearby stars and will certainly contribute to a better knowledge of the zodiacal emissions around the candidates to be observed. For the fleet of satellites with a null interferometers on board, NASA and ESA have delayed the launches to a date that surely will not be before 2025. Both agencies will therefore not be searching for habitable planets very soon.

Space missions require relatively long times to be realized. From the moment a mission is decided to the actual launch, 10–15 years in general will pass. For these missions the times can even be longer because of technical difficulties and the very high costs. So, if these missions will not enter in the design phase before 2015, the launch date will be further delayed.

It is also probable (and desirable) that because of the high costs and the technical difficulties, NASA and ESA will put together their resources. They could also be joined by other space agencies, such as the Japanese, Chinese, Russian and Indian ones. In this case, we would have the beautiful result of an international space union trying to know how rare our planet is in the universe. These years of waiting will not be lost, they will be devoted to the

---

9 *Gravitational waves have been predicted theoretically for a long time, but they have not been detected, because the instruments do not have sufficient sensitivity. They are variations in the gravity field caused by the motion of large masses, emitted by events that range from explosions of supernovae to the perturbation of the gravitational field induced by two stars rotating about each other.*

preparatory work (of which we have already spoken), so that, when the launch occurs, a well-selected sample for candidates of habitable planets will be available.

The observational strategy: once the best possible sample is defined, the first phase of the mission should first look for CO_2, the easiest molecule to be detected (Figure 8.16) around the stars of the samples. This measure will not only reveal the existence of a planet, but will also give the distance of the planet from the star. In few cases this information is already available from the measurements of satellites like Kepler. Knowing the distance of planets from their stars, it will be possible to identify those that are in the habitable zone, on which to carry the measurements of H_2O and O_3 which are more difficult and require more time. Such a mission will have a very long duration, more than 10 years if it is envisaged to go beyond the asteroid belt.

The number of sources to be observed should be large enough to have statistical significance. Positive or negative, the result will have an enormous impact on the vision which humans will have on the origin and evolution of life in the universe.

Astronomical Observations

Appendix A

A.1 Light

We have seen that the direct exploration of stars (Section 11.5) by humans is an objective that will not be achievable for a very long time if not forever. Even if it eventually becomes feasible, it will still be impossible to journey to a star and bring back the information on a timescale comparable with a human lifetime.

Excluding the solar system, therefore, which we can explore directly, all information about the universe is carried by radiation or particles that arrive on Earth after journeys lasting millions or even billions of years. Therefore, even if meteorites and cosmic rays furnish very valuable data, most of our information is provided by the radiation emitted by heavenly bodies, above all the part of this radiation seen by the eye and referred to as visible light.

Astronomy progress is therefore tied to the observation of this radiation and the development of instruments capable of detecting it. At the birth of mankind, the only instrument available for observing stars was the eye, a precise instrument that allowed many cultures — such as the Incan, the Mayan, the early Mesopotamian, the Indian and the Chinese — to calculate the length of the year precisely and enabled the Greeks to discover the precession of the equinoxes (Section 3.1). With the progress of technology, in the 1600s, Galileo constructed the first telescopes, with which he discovered the satellites of Jupiter; he named them Medicei in honor of his protector although they are now known as the Galilean. Then, as technology advanced still further, instruments became more refined resulting in today's great telescopes that, from Earth and in space, explore the cosmos in all wavelength bands, from radio to infrared, visible, X-rays and gamma rays; increasing their complexity; reaching in all bands spatial resolutions and sensitivities unthinkable a few decades ago.

To observe an astronomical object, be it a planet, a star, a huge structure like a giant molecular cloud, or a galaxy, it is therefore necessary to collect and analyze the light it sends us.

But what is light? Since ancient times, its nature has been subject of debates between philosophers and scientists. In the modern era there were two opposing views, the corpuscular model supported[1] by Newton (1642–1727) — for whom light rays were similar to particles, propagating in a straight line from the source to the observer — and the wave theory put forward by scientists like the Englishman Robert Hooke (1635–1703) and the Dutchman Christiaan Huygens (1629–1695); according to these, light is like sound, a mechanical vibration of the medium through which it is propagating. This theory therefore requires a medium that can sustain vibrations, a "cosmic ether" pervading the universe and allowing radiation to travel to us from heavenly bodies. It was an idea that survived until the nineteenth century and was adopted to explain the propagation of radio waves, too.

In the second half of the nineteenth century, the great Scottish scientist James Clerk Maxwell (1831–1879) succeeded in unifying electric phenomena — those resulting from static or moving electric charges. His famous equations showed they are different aspects of the same phenomenon, the "electromagnetic field". Maxwell proposed that light consisted of very rapid oscillations of this electromagnetic field as it would be produced by an oscillating charge, rather than a mechanical vibration of the ether. Because electric and magnetic fields can propagate in a vacuum, Maxwell's theory avoided the need for a — highly improbable — ether filling the universe. The theory was verified by many experiments and observations so that we now use the term "electromagnetic radiation" to refer to radiation of all wavelengths, from radio to gamma rays.

At the beginning of the twentieth century, the corpuscular theory made a comeback. To explain the emission and absorption of light by materials, it was necessary to accept that, in certain situations, light behaved like a particle; thus quantum optics arrived and photons were born. It was for his work on photons — and not

---

1 The first formulation of the corpuscular theory is due to Democritus (in the fourth century BC). He contradicted the ideas of other philosophers, like Pythagoras (582–500 BC) and Plato (427–347 BC), for whom the eye was a kind of lighthouse which sent out light to explore the surroundings. This theory, which now seems curious, illustrates the difficulty of explaining the phenomenon of light, a good understanding of which came only in the 1900s when wave theory was combined with quantum optics.

for his theories of relativity — that Albert Einstein (1879–1955) was awarded the Nobel Prize in 1921.

Undulatory phenomena are common in nature and they all follow the same laws. In the same way of electromagnetic waves, behave acoustic waves (which are oscillations of the density of a medium), sea waves and seismic waves (Section 9.2). Sea waves are, perhaps, the easiest to understand because we see them directly and because they propagate much more slowly than most of the other waves[2]; for this reason we often use them to give an intuitive idea of these phenomena.

Let us begin with the propagation of waves: observing waves on the surface of the sea, we immediately see that when they move they transfer energy without transferring matter; the water itself simply moves up and down whilst remaining at the same position; only when the waves arrive at the shore they dissipate their stored energy. Similarly, acoustic (pressure) waves propagate without any displacement of the material. If such displacement were to occur, the waves would be accompanied by winds.

Waves are characterized by three parameters, amplitude, wavelength and velocity of propagation. The *amplitude* is related to the energy transported by the wave, in other words to its intensity. The damage caused by a sea wave when it hits the shore is very different if its amplitude is 5 meters rather than 50 cm. It can be shown that *the energy carried by a wave is proportional to the square of its amplitude*. The amplitude of a wave, or its intensity, distinguishes a destructive sea wave from one that is not, a bright light or loud sound from weak ones and a good radio signal from one that is barely detectable.

The *wavelength* is the distance between two successive maxima. It is indicated by the symbol λ and is measured in meters, centimeters, etc. It can be a few meters in a lake, tens of meters in an enclosed sea like Mediterranean, hundreds of meters in the open ocean, and hundreds of kilometers in the case of the waves generated by a seaquake (tsunami, Section 10.3). For electromagnetic waves, the wavelength varies from kilometres for radio waves to less than 100 thousandths of a millimeter for gamma rays.

The *velocity of propagation* of a wave depends on the medium through which it propagates. In the sea, it depends on the depth, the density of the water and from the phenomenon that generated it. It varies between 700 km/h for seaquake waves and tens of km/h for normal waves. Sound propagates at around

---

2 *Experiments in "marine optics" have been carried out, in which the energy of waves covering a large area of the sea has been focused onto a point. The very high waves resulting at this focus could be used to generate electrical energy. Such optics have been constructed to model dams having profiles similar to the surface of a telescope mirror and which reflect waves to the focus. Optics having a shape similar to that of a lens have also been used to modify depth of the sea: waves arriving at the central part of the profile, where the depth is less, are slowed with respect to those outside, just as it is for the light hitting the central part of a converging lens.*

340 m/s in air, 1498 m/s in water and 5120 m/s in iron; the denser and more rigid the material, the faster the wave travels. Light travels at its maximum velocity in vacuum ($c = 299\,792.458$ km/s); in glass, the velocity decreases to about 110,000 km/s (depending on the type of glass) and it is still lower in water. Because space is more or less empty, light from the stars travels towards us at 299 792.458 km/s. The theory of relativity dictates that this velocity cannot be exceeded and that nobody having a finite mass can attain it. (This sets the limits on the time needed for interstellar travel, referred to at the beginning of this section.)

The wavelength and the velocity of propagation are related by another quantity, the *frequency*, denoted by the Greek symbol v, which determines the number of oscillations per second. Imagine, for example, that we are in an anchored boat and that we see a wave come toward us. We can count how many waves arrive every minute from the pitching of the boat: that number divided by 60 gives us the frequency of the wave. It is measured in units of Hertz — the number of oscillations per second. The larger the distance between successive maxima (the wavelength), the lower will be the frequency of oscillation of the boat. The frequency v, the wavelength λ and the velocity C are related by the simple equation $\lambda v = C$.

That which the eye perceives as color resides in different frequencies (wavelengths) of electromagnetic radiation. In fact, the eye is an instrument that measures the frequency of light and transforms it into a nervous signal that is sent to the brain, where the light is interpreted. In the visible band, the highest frequency corresponds to the color blue and the lowest to red.

The ear works in a similar way. It contains a membrane that oscillates in synchrony with the sound it receives. The brain distinguishes differing frequencies by the number of oscillations per second of this membrane. Higher-frequency sounds are those producing more vibrations per second, lower-frequency sounds producing fewer. Good music and beautifully-colored picture are nothing more than the brain's perception of an accord between various frequencies.

Sounds result from the oscillations in density of the medium through which the acoustic waves propagate. They cannot propagate in a vacuum. Light, on the other hand, is the oscillations of electric and magnetic fields and *can* propagate in vacuum.

At the top of Figure A.1 the names of the various "bands" of electromagnetic radiation are given. Underneath it the corresponding wavelengths in millimeters are shown. Finally, at the bottom, are familiar objects of sizes comparable with the corresponding wavelengths. From left to right, we move from radio waves, hundreds of meters long, like skyscrapers, to visible light, whose wavelength can be compared to protozoa, and finally to gamma rays, which have the dimensions of atomic nuclei. To obtain the corresponding frequencies, just divide the speed of light in vacuum ($c = 299\,792$ km/s) by the wavelength.

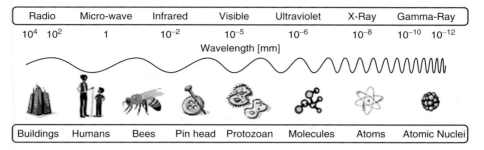

| Radio | Micro-wave | Infrared | Visible | Ultraviolet | X-Ray | Gamma-Ray |

10^4 10^2 1 10^{-2} 10^{-5} 10^{-6} 10^{-8} 10^{-10} 10^{-12}

Wavelength [mm]

| Buildings | Humans | Bees | Pin head | Protozoan | Molecules | Atoms | Atomic Nuclei |

Figure A.1: The electromagnetic spectrum. The wavelength is the distance between two successive maxima of the oscillation and it appears in the upper part of the figure in millimeters. In the lower part, each wavelength is compared with an object of similar dimensions. Imagine that the wave moves horizontally towards the left of this page at a constant velocity C: the frequency of the radiation (the number of maxima that pass the edge of the paper in one second) increases as we go from the radio towards the visible regions of the spectrum and continues to increase going towards the gamma-ray. This is because the distance between successive maxima, that is the wavelength, decreases in going from radio waves to gamma rays. (Adapted from NASA Beyond Einstein Roadmap Images Library.)

A.2   The Color of Light

The information we can obtain from electromagnetic radiation is, as we have seen, its frequency, that is its color, and its intensity, that is, a measure of the amount of energy coming from the source. These two quantities are the basis of most of the astronomical deductions. For example, we can deduce the mass of a star from the intensity of the radiation it emits (Section 4.3); intuitively, the bigger the star the more energy it emits.

The *color* or frequency of light emitted by a celestial object is of great importance in astronomy: it indicates the object's *temperature*.[3] It is an everyday experience that color and temperature are related. A blacksmith, for example, judges the temperature of the iron he is working from its color, as does a glass-blower with his glass. The blue of a flame indicates a higher temperature than does the yellow of a lamp, which in turn is hotter than the element of an electric heater, whose red light is seen with difficulty because the majority of its energy is emitted in the infrared, invisible to the human eye. The scale of temperatures follows the colors of the rainbow: the hottest objects emit blue light at the shortest wavelength and as the temperature decreases the color of light moves toward the red, which has the longest wavelength.

This phenomenon is enshrined in Wien's law, which states that the wavelength λ_{max} of a body's maximum emission is related to its temperature T by the equation $\lambda_{max} = 2.987/T$, where λ_{max} is measured in mm and T in Kelvin.

---

3 *Do not confuse* temperature *with* heat. *Heat is a measure of the energy emitted by a source, whilst the temperature is a characteristic of the source itself. A match is much hotter than a radiator, but it emits much less energy. We cannot warm a room by lighting a match, but we can burn our fingers if we touch it. We can touch a radiator because its temperature is much lower than that of a match, but it radiates much more heat and it can warm a room.*

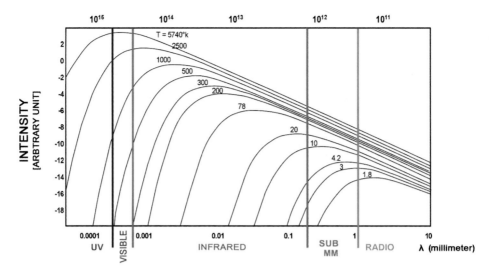

Figure A.2: *The figure shows the Planck function, which indicates how the radiation emitted by a body heated to various temperatures is distributed with wavelength. The Sun, at 5740 K, has its emission peak at the center of the optical band. All bodies at higher temperatures emit mainly in the ultraviolet. Bodies with temperatures between 20 K and 3000 K emit mostly in the infrared whilst bodies with temperatures less than 10 K, among them the fossil radiation from the Big Bang, emit in the submillimeter range (Figure 2.5). The peaks of the various curves follow Wien's law, $\lambda_{max} = 2.987/T$ (in mm).*

For example, an object at a temperature of about 6000 K, such as the Sun, emits its maximum energy at 0.0005 mm, a color close to yellow. The Earth, with a temperature of about 17°C (290 K) has its emission peak in the infrared. The law says that the temperature and wavelength are inversely proportional and that *cold bodies emit the majority of their energy at long wavelengths, and hot bodies at short wavelengths.*

The spectrum of energy emitted by a body has a characteristic form, called the Planck function, whose maximum moves, as we have said, as the temperature of the body varies. In Figure A.2 we can see this function for temperatures between 1.8 K (about –271°C) and 5740 K, the temperature of the Sun. An example of Planck radiation is the fossil remnant of the Big Bang (Figure 2.5).

Most stars have temperatures between 2000 and 10,000°C and therefore they emit most of their energy in the visible. The Sun, with its temperature of 5740 K, emits in the center of this band. The human eye is sensitive only to "visible light" — the wavelengths between the violet and red ends of the spectrum because these are the wavelengths at which the Sun emits most of its energy: the eye evolved so as to adapt itself to its surroundings. Beyond the edges of the visible band are the ultraviolet and the infrared, which the eye does not see.

The *ultraviolet*: objects with temperatures above 10,000 degrees are hotter than those that emit in the violet and they emit mainly in the ultraviolet. The human eye cannot perceive this radiation but it can be damaged by it. (The Sun emits a small amount of its energy in the ultraviolet but, fortunately for us, it is diffused (Figure A.4) and filtered by the atmosphere.) The tanning of the skin, when it is exposed to the Sun, is the way our bodies protect themselves from this radiation. Sun glasses are useful on high mountains or on snow, where the ultraviolet is more intense, to avoid damage to the eyes.

The *infrared*: objects at temperatures below 3000 degrees emit in the infrared (Figure A.1). Infrared is emitted for example, by an oven, by a radiator, by the human body and by a planet. Although the eye cannot perceive infrared radiation, our bodies sense the radiated energy. Infrared images are used to measure the loss of energy of a building or, in medicine, to detect any anomalous distribution of temperature on the human body.

Infrared and microwave radiation, that is the band adjacent to the infrared in Figure A.1, are particularly important for the topics treated in this book. Distant galaxies have their light received in the infrared because of the Doppler effect (Appendix A.3) and because they are wrapped around by dust that diffuse ultraviolet and visible light (Appendix A.4). Cold bodies, like the Earth, planets, asteroids and comets (with temperatures between −200 and 400°C) also emit in the infrared. The images of these objects, which we see in the visible light, are not created by their own emission but by sunlight reflected from their surfaces (Figure 12.3). It is for this reason that we said, in Chapter 12, that extra-solar planets cannot be detected in the visible: the light reflected from their surfaces is only a tiny fraction of the light of the star around which they orbit.

A.3 The Doppler Effect

Let us return to the example of waves in the sea and imagine we are in a boat moving in the opposite direction to that of the waves. The number of maxima we count in a minute (the frequency) increases with respect to those of an anchored boat. On the other hand, the number of maxima per minute will decrease if the boat moves in the same direction as the waves until, if we travel at the same velocity as the waves, it will become zero. We therefore deduce that the observed frequency of a water wave depends on our own motion with respect to it: observers moving at different speeds will measure different frequencies. The same applies to sound or light: observers moving at different speeds will hear sounds at different pitches or see light of different colors. This phenomenon, known as the *Doppler effect*, explains why the sound of a train or a car is higher in pitch when they are moving towards us than when they are moving away.

Johann Christian Doppler (1803–1853) professor of mathematics at the Realschule of Prague described this phenomenon theoretically in 1842, for both acoustic waves and light. A few years later, a Dutch meteorologist amusingly demonstrated this effect for acoustic waves. He embarked several buglers in the open carriage of a train running at high speed (for that time) in the countryside around Utrecht. Everyone along the route could observe that

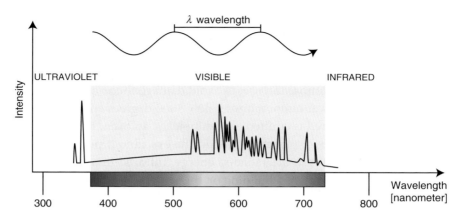

Figure A.3: *The spectrum of a neon lamp: the peaks we see are characteristic of this gas and are emitted at precise wavelengths. If the source moves toward the observer the frequency increases and the wavelength decreases so we observe the same graph but displaced toward left (the blue). The opposite happens if the source is moving away. By measuring this displacement we obtain the measurement of the relative velocity of the source. (Appendix A.1.)*

the pitch of the bugles was higher when the train was approaching and lower when it was receding.

The same thing applies to electromagnetic waves: if we move away from the source the frequency goes down and visible light moves toward the red and infrared region of the spectrum (Figure A.1); on the other hand, if the source moves towards us, its light is displaced toward the blue and ultraviolet regions.

Originally Doppler thought that the colors of stars could be explained by their motion relative to the Earth. The bluer ones were moving toward us whilst the redder were moving away. In reality, this is not observed because the emission of stars has no great variation in intensity across the visible region. Therefore in a star that is moving away from us, light that is shifted towards the red is replaced by light moving in from the blue; the overall color of the star does therefore not change appreciably.

Take the Sun, for example; to a first approximation, the curve corresponding to 5740 K in Figure A.2 represents the Sun's emission at various wavelengths. The Doppler effect produces a small horizontal displacement of the curve which does not produce an appreciable change of color.

As we have seen (Appendix A.2), the colors of the stars reflect their temperatures, the red ones being the coolest and the blue the hottest. The Doppler effect became important in astrophysics when, towards the end of the nineteenth century, astronomers started to apply it to what is called "spectrum". In Sections 3.3 and 4.9, we saw that, when an atom or a molecule becomes excited it emits light of characteristic frequencies called lines. The set of lines identifying a species (Figure A.3) are called its spectrum (Section 4.9). If the lines are emitted by a source moving towards us, its spectrum is displaced toward the blue; if it is moving away, it is displaced toward the red. Unlike the shift in a continuum spectrum, this shift is measurable. Using the Doppler effect, we can therefore obtain precise information about the motion of stars and of galaxies; we have seen, for example, that the discs of the galaxies rotate around their centers with a velocities up to 250 km/s, that the Andromeda galaxy is moving toward us at 300 km/s, and that Cappella is moving away at a velocity of 30 km/s. It was just such observations that allowed Carl Wirtz, in 1922, and Edwin Hubble, in 1929, to note that distant galaxies had all spectra displaced toward the red and therefore to conclude that the universe was expanding (Section 2.3).

A.4   The Diffusion of Light: Why the Sky is Blue

In Chapter 3, we saw that infrared, submillimeter and radio-frequency radiation can penetrate dense clouds like the placental one of Figure 3.1, whilst visible light cannot pass through these clouds, which therefore appear as dark patches on the sky. What prevents the light from transiting through the clouds?

This is another consequence of the undulatory nature of light: by increasing the wavelength, the capacity of electromagnetic waves to penetrate the regions of dust and gas also increases. This also explains why, in daytime, the sky is blue and luminous, rather than black as it is seen by astronauts in space or on the surface of the Moon. It also explains why the Sun is red at sunset and why radar (centimeter waves) can detect an object through the thickest fog.

When a wave strikes an object, it is scattered in all directions. One can observe this phenomenon by throwing a stone into calm water and seeing how the resultant waves behave when they encounter a rock poking up above the water. Wave spread out in all directions from the point at which the original wave was interrupted, spreading even to the region of water behind the rock. The phenomenon is more pronounced if the wavelength is comparable with the size of the object. If we have an island instead of a rock, much larger than the wavelength, the waves are not scattered very much, and there is a zone of calmness behind the island. If, on the other hand, the obstacle is much smaller than the wavelength, the waves pass by as if it were not there.

The same thing occurs with light waves. The atmosphere is composed of atoms that are little smaller than the wavelength of blue light, therefore, the blue component of visible radiation, closer to the dimension of atoms, is diffused much more than the yellow, which is diffused more than the red.

Figure A.4 shows what happens in the Earth's atmosphere. Sunlight, traveling from left to right in the diagram, is scattered out of the original direction of the beam. This scattering is greater for the blue component of the light than it is for the yellow and the red. Some of these waves, after many collisions with the atoms in the atmosphere, are actually scattered back out into space, giving our planet the blue color that one sees from space. Other rays, again after many collisions, reach our eyes, making the sky luminous and blue. Without this

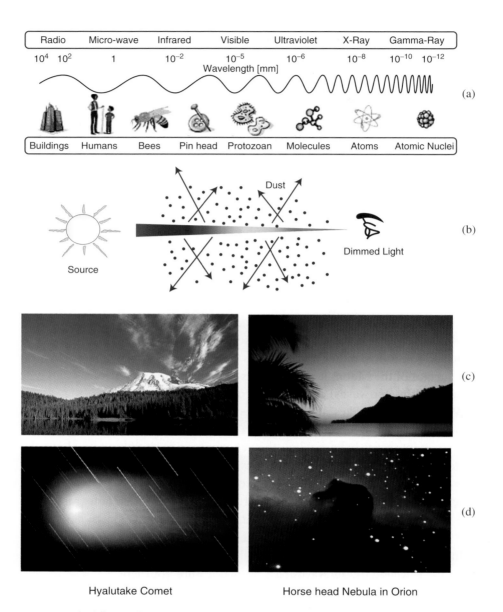

Figure A.4: The diffusion of light. (a) The electromagnetic spectrum. (b) Blue light is diffused more than red by the atoms in the atmosphere. (c) The blue light is diffused, making the sky appear blue, whilst the red light passes through the atmosphere and makes a sunset appear red. (d) Examples of celestial objects whose colors are the result of the diffusion of light: comets are blue because we see the sunlight diffused by the atoms of its tail; star-formation regions are red because only red light gets through the dust of the placental cloud. (Adapted by NASA Beyond Einstein Roadmap and Images Library, ESO.)

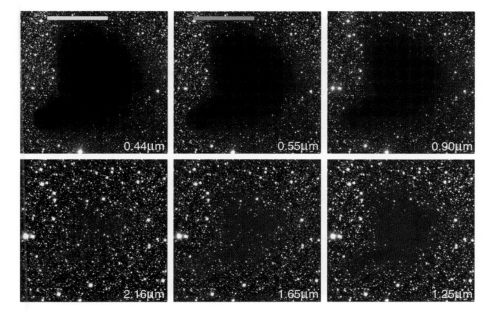

The Dark Cloud B68 at Different Wavelengths (NTT + SOFI)

ESO PR Photo 29b/99 (2 July 1999)

© European Southern Observatory

Figure A.5: The dense nucleus B68 (Section 3.1) photographed with a series of filters that pass only a very narrow band of wavelengths. The first two images are taken in yellow and green light, the others in the infrared. With increasing wavelength, the ability of the light to penetrate the cloud increases until, at λ = 2.16 µm, the cloud becomes invisible. (Photo by VLT-ESO, July, 1999.)

scattering, the sky would not be luminous and shadows would be much darker; this occurs on the Moon, where it is sufficient to go behind a wall to be in complete darkness because there is not the sky that diffuses the sunlight on the back of the wall. It is the Earth's luminous sky that enables a window to illuminate a room, even if the Sun is not shining directly on it.

At sunset, sunlight has to traverse a much longer path through the atmosphere than it does at midday. All colors are scattered but blue is scattered more than the longer wavelengths so that the light reaching us is red (Figures A.4 (b) and (c)).

This phenomenon also has a marked effect on the colours of astronomical objects. As we see in Figure A.4 (d), comets tend to be blue because what we see is sunlight scattered by their rough surfaces. Moreover, the regions where the stars are born, like that of Orion, tend to be red (Figures 3.3, 3.7 and 3.8) because the red light passes more easily through the dust and the gas of the placental clouds.

The ability of light of various wavelengths to penetrate regions of dust and gas is shown in Figure A.5, where the same dusty cloud of Figure 3.1 is photographed using filters of different colors. At the bottom right of every image is shown the wavelength in microns (thousandths of a millimeter); this is the wavelength passed by the filter. The first two images are taken in yellow and green light, in the visible band. The other images are taken with wavelength increasing as we go clockwise around the figure, reaching 2.16 microns in the last image. With increasing wavelength (decreasing frequency) the dark cloud appears smaller and smaller until it is practically invisible in the last image. This is because the longer the wavelengths the more it penetrates the cloud until, at 2.16 microns, the radiation crosses without obstruction the cloud that becomes invisible.

A.5 The Spatial Resolution

There are other quantities that are important in astronomical observations. One of these is spatial resolution, that is the detail that can be seen within an object. If we look at someone a few meters away, we can recognize him or her easily. If we move a kilometer away, we can still see the person but the details have been lost and we can no longer recognize whom we are seeing. If we use a telescope, however, the image becomes again clear and we see all the details.

If we move still farther away, we can try to observe the details of interest by using more powerful instruments, capable of greater and greater magnification, until finally the turbulence of the atmosphere prevents our seeing a clear image.

If you go into space where there is no atmosphere, the turbulence will be eliminated and the instrument's limit can be reached. This is the case of the Hubble Space Telescope, which we discussed in Section 1.1, and which provides the highest resolution so far achieved in the visible band. Many of the images of this book are examples of the extraordinary capability of the HST; another example is shown in Figure A.6.

Figure A.7 compares images of the same object obtained with the great 5-meter telescope of Mount Palomar (left) and with the 2.4-meter HST. The difference between them results from the absence of atmospheric turbulence in the HST image.

Above the atmosphere, we have said, a well-made instrument can achieve its intrinsic limit of resolution. This limit is imposed by the wave-nature of light and depends on the ratio between the diameter of the telescope and the wavelength one wishes to measure. This strict limit applies to all wavelengths and is most restrictive in the radio region, where wavelengths are considerable.

We have already seen, in Figure A.4, that electromagnetic waves are scattered by the objects they encounter. The same thing happens when light has to pass through a hole of diameter D, which, in this case, is the diameter of the telescope. It is shown that, beyond the hole, the light is diffused into a cone of diameter 2.4 λ/D, where λ is the wavelength — the distance between

Helix Nebula Detail
Hubble Space Telescope · WFPC2

PRC96-13b · April 15, 1996 · ST ScI OPO · C.R. O'Dell (Rice Univ.), NASA

Figure A.6: A mysterious image — astronomers have never seen anything like it! Only the extraordinary sensitivity and resolution of the HST has allowed to see these strange globules in the constellation of Acquarius, 450 light-years from the Earth. They are globules of gas and dust found near a planetary nebula. The globules are approximately twice the size of the Solar System in diameter but contain very little mass; if they should condense one day, they would form an object hardly larger than Jupiter. Their origin, their fate and how many similar objects exist in the galaxy, are questions that have still to be answered. (HST image, August1994. Credit: Robert O'Dell, Kerry P. Handron, Rice University, Houston, Texas and NASA.)

maxima. With increasing λ the diameter of the cone increases so that, to keep the same resolution in the image, the diameter of the telescope must increase in the same proportion. Figure A.8 shows how dramatic the effect can be: on the left is an image of Mount Cervino in Switzerland, obtained using a normal camera. On the right is the image obtained in the millimeter radio band, using the German KOSMA, a 3-meter diameter telescope, belonging to Cologne University, housed in the cupola that appears in the left photograph of Figure A.8.

It is impressive to see how much worse is the millimeter image, taken with a 3-meter diameter telescope, than that obtained in the optical with a lens a few centimetres in diameter. However, although millimetre observations have poor spatial resolution, they allows to measure the abundance of molecules in space (Section 4.10) and to probe the interior of a molecular cloud, where the visible light cannot penetrate. The modest spatial resolution is the price to pay.

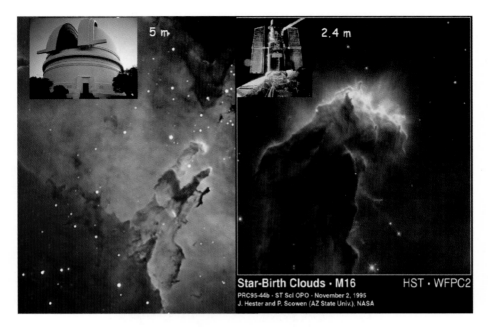

5 m

2.4 m

Star-Birth Clouds · M16　　　　　HST · WFPC2

PRC95-44b · ST ScI OPO · November 2, 1995
J. Hester and P. Scowen (AZ State Univ.), NASA

Figure A.7: The advantage of eliminating the turbulence of the Earth's atmosphere. On the left is an image of the Aquila nebula obtained by the 5-meter Palomar telescope. On the right is a detail of the same region observed from above the atmosphere by the HST, whose telescope is only 2.4-meters in diameter. The spatial resolution of HST is such that, if it were not for the atmosphere, it could be used to read a newspaper at a distance of 20 km. (HST Photo, November 1995. © NASA & J. Hester and P. Scowen, Arizona State University.)

The great spatial interferometers like ALMA, which we discussed in Section 4.9, provide the only opportunity of getting high spatial resolution at millimeter wavelengths. ALMA is equivalent to a single telescope, 14 kilometers in diameter, providing incredibly high resolution, even superior to that of the HST images we have seen in this book. We are talking of a few thousandths of an arcsecond, which is less than a billionth of a degree, a resolution sufficient to read a newspaper 200 km away or to resolve, at a distance of about 30 light-years, the image of a planet like the Earth from that of the star about which it orbits. Considering the immensity of astronomical distances, it is easy to understand the importance of high spatial resolution: it provides the capability of resolving the different components of an object that otherwise would appear as a single entity. The challenge of future years is to construct an instrument, with a resolution like that of HST, for observations in the far infrared and microwave regions, where the planets, the first stars and the first galaxies emit most of their energy. The most ambitious projects contemplate the construction of instruments working in the infrared with a resolution 10 times better than that of the HST.

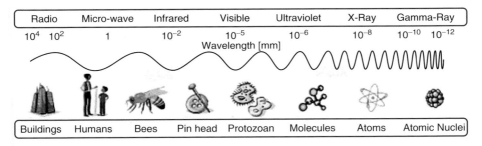

Radio	Micro-wave	Infrared	Visible	Ultraviolet	X-Ray	Gamma-Ray

10^4 10^2 1 10^{-2} 10^{-5} 10^{-6} 10^{-8} 10^{-10} 10^{-12}

Wavelength [mm]

Buildings	Humans	Bees	Pin head	Protozoan	Molecules	Atoms	Atomic Nuclei

D 3 cm in the visible (λ = 0,0005 mm)   D 300 cm (λ = 1 mm)

Figure A.8: The importance of large-diameter telescopes for radio observations. On the left is a photograph of Mount Cervino, taken with a normal camera. On the right is the image obtained in the millimeter radio band, using a 3 meter diameter telescope, the German KOSMA belonging to Cologne university, the dome of which is seen in the left photograph. (© Cologne University, Germany.)

This is the limit that mankind has to reach if we want to probe the edge of the universe (Chapter 2); to investigate what happens in protostellar discs (Chapter 3) and how they evolve; to understand how planets are born and; finally, to isolate the weak light reflected off a planet from the star about which it orbits (Chapter 12).

A Few Numbers

Physical Constants

2.99792458×10^{8} m/s : The speed of light in a vacuum

9.8 m/s^{2} : The acceleration of gravity on the Earth surface

Units of Measurement of distances

AU (Astronomical Unit) = 1.496×10^{11} m		Earth–Sun distance
ly (light year)	= 9.46×10^{15} m	Distance light travels in a year
pc (parsec)	= 3.086×10^{16} m	Distance at which an object of diameter of the Earth orbit is seen with an angle of 1 arsec
	= 3.26 ly	

Distances

Radius of the Observable Universe[1]:	10^{26} m
Radius of our Galaxy (the Milky way):	10^{21} m
Distance to Proxima Centuri:	4×10^{16} m = 2.7×10^{5} AU = 4.22 ly
Radius of the Oort cloud	1.5×10^{16} m = 10^{5} AU = 1.55 ly

---

1 *The diameter of the horizon (Section 2.7).*

Radius of the Kuiper Belt	1.5×10^{13} m = 100 AU
Semi major axis of Pluto's orbit	5.9×10^{12} m = 39,5 AU"
Semi major axis of Jupiter's orbit	7.78×10^{11} m = 5.2 AU
Semi major axis of Mars orbit	2.28×10^{11} m = 1.52 AU
Semi major axis of Earth's orbit	1.496×10^{11} m = 1 AU
Semi major axis of the Moon's orbit	3.84×10^8 m
Equatorial radius of the Sun	6.96×10^8 m
Equatorial radius of Jupiter	7.13×10^7 m
Equatorial radius of the Earth	6.38×10^6 m
Equatorial radius of Mars	3.39×10^6 m
Equatorial radius of the Moon	1.74×10^6 m
Length of Human DNA	1.5 m
Average diameter of one hair	$7 \ 10^{-5}$ m = 0.07 mm
Dimensions of bacteria	$0.1 - 600 \times 10^{-6}$ m = 0.1 - 600 micron
Dimensions of virus	$1 - 45 \times 10^{-8}$ m = 0.01 - 0,45 micron
Thickness of DNA	2×10^{-9} m
Dimension of hydrogen molecule (H2)	2×10^{-10} m
Radius proton	$\sim 10^{-15}$ m

Masses

Of the Universe:	10^{53} kg
Of the Milky Way:	10^{43} kg
Of the Sun:	1.99×10^{30} kg
Of Jupiter:	1.90×10^{27} kg
Of the Earth:	5.98×10^{24} kg
Of Mars:	6.42×10^{23} kg
Of the Moon:	7.35×10^{22} kg
Of the Earth's atmosphere:	5.13×10^{18} kg
Of a bacteria (average)	10^{-15} kg
Of a virus (average)	10^{-21} kg
Of a proton:	1.67×10^{-27} kg
Of an electron:	9.11×10^{-31} kg

Various Numbers

The number of different DNA molecules that can be constructed combining the bases of a single strand:	10^{100}
The number of atoms of the known Universe:	10^{80}
The number of different sequences that can be constructed combining the amino acids of a single protein:	10^{50}
The number of galaxies in the known Universe:	10^{11}
The number of stars in our galaxy:	1.6×10^{11}

Energy

Power radiated by the Sun:	3.85×10^{26} W
Energy incident on the Earth during one year (about 40% is reflected back to space):	1.52×10^{21} Wh
Energy generated in the interior of the Earth during one year (that feeds volcanoes, earthquakes and continental drift)[2]:	3.87×10^{17} Wh
Total energy consumed in the world during year 2009[3] of which	**1.41×10^{17} Wh**
Electric energy (1/7 of the total consumed energy) of which:	**2.01×10^{16} Wh**
Thermoelectric (67.1%):	1.35×10^{16} Wh
Hydroelectric (16.2%):	0.32×10^{16} Wh
Nuclear electric (13.4%):	0.27×10^{16} Wh
(Geothermal, Sun, Wind, Biofuel, Waste) (3.3%):	0.07×10^{16} Wh
Consumed energy inhabitant per year:	**2.0×10^{7} Wh**

---

2 *"Heat flow from the earth's interior" H. N. Pollack et al., Reviews of Geophysics 31, 267-280, August 1993.*
3 *The source for all the data on energy consumption are relative to year 2009 and come from the "Key world energy statistics 2011" of "Intenational Energy Agency" (IEA) on the web.*

Economic and Environmental Indicators

Listed in decreasing order of CO_2 emission per capita

Country	CO_2 Per Capita (tons/yr)	GPD Per Capita (US$/yr)	Population (million)	CO_2 Total (mtons/yr)	% of total
Qatar	40,09	28872	1,41	57	0,2
United Arab Emirates	31,97	25665	4,6	147	0,5
Kuwait	28,83	22725	2,8	81	0,3
USA	16,90	36936	307,48	5195	17,9
Australia	17,87	24219	22,1	395	1,4
Canada	15,43	25099	33,74	521	1,8
Saudi Arabia	12,23	9828	25,39	310	1,1
Russia	10,80	2802	141,9	1533	5,3
Finland	10,30	26434	5,34	55	0,2
Chinese-Taipei	10,89	17943	22,97	250	0,9
Korea	10,57	15443	48,75	515	1,8
OECD	9,83	24190	1225	12045	41,5
Singapore	8,98	28752	4,99	45	0,2
Israel	8,69	21797	7,44	65	0,2
Japan	8,58	38265	127,33	1093	3,8
United Kingdom	7,54	27142	61,79	466	1,6
Poland	7,52	6335	38,15	287	1,0
Iran	7,31	2169	72,9	533	1,8
Italy	6,47	18453	60,19	389	1,3
South Africa	7,49	3689	49,32	369	1,3
Ukraine	5,57	987	46,01	256	0,9
Malaysia	5,98	4992	27,47	164	0,6
France	5,49	22837	64,49	354	1,2
Venezuela	5,45	5638	28,38	155	0,5
Switzerland	5,44	36705	7,8	42	0,1
China	5,14	3328	1338	6832	23,7
World	4,29	5868	6761	28999	100,0
Argentina	4,14	9880	40,28	167	0,6
Uzbekistan	4,05	893	27,77	112	0,4
Romania	3,65	2607	21,48	78	0,3
Mexico	3,72	6742	107,44	400	1,4

(Continued)

Country	CO$_2$	GPD		CO$_2$	
	Per Capita			Total	%
	(tons/yr)	(US$/yr)	Population (million)	(mtons/yr)	of total
Iraq	3,41	795	28,955	99	0,3
Thailand	3,36	2567	67,76	228	0,8
DPR Korea	2,77	482	23,91	66	0,2
Cuba	2,40	4266	11,2	27	0,1
Latin America	2,16	4339	451	975	3,4
Egypt	2,11	1836	83	175	0,6
Brazil	1,74	4419	193,73	338	1,2
Indonesia	1,70	1170	220,97	376	1,3
ASIA (- China)	**1,43**	**1126**	**2208**	**3153**	**10,9**
India	**1,37**	**757**	**1155,35**	**1586**	**5,5**
Peru	1,32	2913	29,17	39	0,1
Vietnam	1,31	674	87,28	114	0,4
Morocco	1,29	1810	31,99	41	0,1
AFRICA	**0,92**	**888**	**1009**	**928**	**3,2**
Yemen	0,94	565	23,58	22	0,1
Pakistan	0,81	657	169,71	137	0,5
Philippines	0,77	1215	91,98	71	0,2
Sri Lanka	0,62	1233	20,3	13	0,0
Nigeria	0,27	506	154,73	41	0,1
Kenya	0,25	452	39,8	10	0,0
Sudan	0,31	537	42,27	13	0,0
Bangladesh	0,31	482	162,22	51	0,2
Tanzania	0,14	371	43,74	6	0,0
Nepal	0,12	261	29,33	3	0,0
Mozambique	0,10	371	22,89	2	0,0
Ethiopia	0,09	201	82,83	7	0,0
Republic Congo	0,04	97	66,02	3	0,0

The countries listed in the table correspond to the 90% of the world population.
In bold are the continents or the countries producing more than 1000 million of tons of CO$_2$ per year.
Source: IEA Key Word statistics 2011 (on the web).

Acronyms Used in the Text

AGU	American Geophysical Union, USA
ALMA	Atacame Large Millimeter Array, ESO, EU
CNR	National Research Council, Italy
DNA	Deoxyribonucleic acid
ESA	European Space Agency, EU
ESO	European Southern Observatory, EU
ETI	Extra Terrestrial Intelligence
GRIN	Great Images in NASA, USA
GSA	Geological Society of America, USA
IEA	International Energy Agency
IFSI	Institute of Physics of Interplanetary Space, INAF, Italy
ISO	Infrared Space Observatory ESA, EU
IRAM	Institute for Millimeter Radio Astronomy, France
JHUAPL	Johns Hopkins University Applied Physics Lab, NASA, USA
JPL	Jet Propulsion Laboratory, NASA, USA
JWST	James Webb Space Telescope, NASA, USA
HST	Hubble Space Telescope NASA, USA
KT	The Cretaceous-Tertiary Limit
LISA	Laser Interferometer Space Antenna ESA — NASA, EU–USA
LBT	Large Binocular Telescope, Italy, USA, Germany
LPI	Lunar and Planetary Institute, Houston, USA
INAF	National Institute for Astronomy and Astrophysics, Italy
LWS	Long Wavelength Spectrometer, ISO ESA, EU
NASA	National Aeronautics and Space Organization, USA
NEO	Near Earth Object
NEAR	Near Earth Asteroid Rendezvous, NASA, USA
NOAA	National Oceanic and Atmospheric Administration, USA
UN	United Nations
PNRA	National Program for Research in Antarctica, Italy
RNA	Ribonucleic Acid
SETI	Search for Extraterrestrial Intelligence
STScI	Space Telescope Institute, NASA, USA
TPF	Terrestrial Planet Finder, NASA, USA
UAI	Union of Italian Amateur Astronomers
UV	Ultraviolet
VLT	Very Large Space Telescope, ESO, EU

Conclusion

With the last chapter, more technical than the others, we have concluded this book, where we have introduced some of the extraordinary themes which science has faced up to the last century.

Several arguments have merely been mentioned because they are still not well understood by science, above all the origin of life, a process that we have said several times no scientist has been able to reproduce in the controlled conditions of a laboratory. The processes from which life originated are not known, not even is it understood how these processes could have begun in the hostile conditions of primordial Earth.

Other arguments are clearer, thanks to the progress made in the past decades. We know that the universe has a beginning, that it is still evolving (Chapter 2) and that perhaps it will exist forever. Nuclear processes that occurred in stars have consumed only a small fraction of the available fuel (Section 4.8), enriching the universe with the heavy elements of which planets and living beings are composed. We know that, through the biological processes, species evolve and modify the environment in which they live (Chapter 6); that stars and biological organisms are born, evolve and die, following cycles that make them always different over time. We know that for the Earth, everything will end in about five billion years, when the Solar System will be transformed into the dust from which new generations of stars and planetary systems will be born.

Will anything remain from the species that inhabited the Earth? Will there be seeds that will go elsewhere in space, or will everything disappear with the destruction of our planet? It is difficult to answer these questions, at least for the next 15 years. We have instead given an answer to the question: "Are we alone in the universe?" We say reasonably that we are not — an answer that arises from the first part of the book. We have seen that, excluding divine intervention, the beginning of life had to arise from a series of chemico-physical processes; therefore, given the *ineluctability* of the laws of nature, life must be born or will be born at every point of the universe in which it had existed given the present conditions on the primordial Earth (Section 5.2). Also the less probable hypothesis of panspermia (Section 5.10) leads to similar conclusion: if the seeds of life are everywhere in the universe, life must exist in all places where environmental conditions have permitted it to germinate.

"The History of the Earth" (Chapter 6) confirms this scenario. The fossil traces of the first bacteria that lived on the Earth show that this process, even if it is not clearly understood (it is not clear how the first protein, the first molecule of DNA, the first cell was born, or how the evolution of life on the Earth was possible), it was very efficient (life was born as soon as the Earth cooled; Section 6.3), in agreement with the word "ineluctability" which we have used above.

If the birth of life is the consequence of particular environmental circumstances, the answer to the question "How common is life in the universe?" is in the answer to the questions "*How rare in the universe is a planet like the Earth?*" and "*What makes the Earth habitable?*": these arguments are discussed in the second part of the book.

Unfortunately, today we can only study a single case, that of our own planet. We cannot consider the question of how frequent life in the universe is from a statistical point of view, because statistics limited to a single case makes no sense. The Earth is, however, close to one of the most common stars in the universe, which is born by accreting material from disks that become planetary systems (Chapter 3). One can therefore use statistics to say that, the Earth being the average case, the probabilities are biased toward representing our planet as the normal case and not an exception. This argument does not prove that inhabitable planets are frequent in the universe, but it provides a strong stimulus to search for them.

Through the study of our planet, we have seen that the existence of life on our planet depends on many factors. The first is the position of the Earth in the galaxy (Section 8.1) and in the Solar System (Section 8.6). If these conditions were different, life, even if it was born, would not have evolved toward animal forms. A second factor is that our planet has a hot core which keeps it geologically alive. The heat of the Earth induces continental drift which restores the land destroyed by erosion and volcanism, which preserve the atmosphere (essential for the survival of biological species and the temperate climate of the Earth; Chapter 9). The price to pay is the destruction wreaked by volcanic eruptions, earthquakes and seaquakes, but without them the Earth would be a dead planet.

Other important factors include perhaps the role of the Moon (Section 10.1) in stabilizing the axis of rotation of the Earth and therefore the climate; the presence of a large planet like Jupiter (Section 10.2) to defend a planet against large meteorite impacts. The mass of the Earth is probably essential for developing an environment favorable to life (Section 9.10), and perhaps is the reason why the Earth is geologically active while Mars is dead.

How rare is all this? Science is close to giving us many of the answers as outlined in Chapter 12, and in less than 20 years we will know whether or not there are biological forms of life on stars near us.

If we will find other forms of life on other planets, what will they be like? Certainly they will be biological forms based on organic chemistry. They could, however, be very different from us. The experience of the case of the Earth leaves behind many open questions. The history of our planet shows that evolution has been dominated by random factors, especially from the extinction of species (Chapter 7). What would have happened if the dinosaurs had not become extinct? Would mammals and humans have developed? The Cambrian explosion also poses questions that are not easy to answer. What caused it? Could it have not happened? Why did nothing similar happen in later eras? Maybe one day we will encounter a planet inhabited by intelligent dinosaurs whose evolution was not deterred by a meteorite....

A final reflection: in front of the immensity of the universe, we understand how little weight our species has on its history. We are only small castaways adrift on a grain of dust in the vastness of the cosmos. This should teach us not to challenge nature, but to use with intelligence the resources of the only planet we have. We should above all take care of our atmosphere, the thin layer of gas that protects us from a hostile universe. We must be aware that the development we are pursuing (Chapter 7) risks destroying this layer, provoking terrible climate changes that, once begun, will no longer be controllable. It must above all be very clear that if we should disappear because of our stupidity, nothing would change in the history of the universe which another, wiser civilization may someday write....

Index